Preface

Cooperative behavior is a hallmark of the primate order. Cooperation is therefore an area of intensive theoretical research in biology, anthropology, political sciences and economics, as well as a salient feature of the socially complex societies of humans and primates, where a large body of observational and experimental data has accumulated. This volume features a summary of recent work and progress in these related areas, integrating inter-related theoretical problems and their evolutionary and proximate solutions by humans and primates for the first time.

Cooperation refers to social interactions characterized by costs to an actor and benefits to other conspecifics. Because such behavior is, at first glance, difficult to reconcile with the selfish drive to maximize individual fitness, cooperation posed a problem for evolutionary biology until new theories in the 1960s invoked genetic relatedness (kin selection) and the logic of repeated interactions (reciprocal altruism). While these concepts have since been successfully applied to many cases and species, more recent reviews have emphasized the widespread nature of cooperation among unrelated individuals, for which humans provide many examples that cannot be explained by kin selection theory. Much recent research in a variety of disciplines has therefore focused on such alternative explanations for cooperative phenomena, ranging from prebiotic evolution to the evolution of human language (Hammerstein 2003a).

In this recent wave of inter-disciplinary research, biologists have adopted game-theoretical approaches from economics and analyzed the outcomes of evolutionary games in which frequency-dependent selection acts on genotypes. Anthropologists, economists and political scientists, on the other hand, have incorporated evolutionary logic into their models of learning and cultural transmission of cooperative behavior. However, there has been little direct contact between theoreticians and students of human behavior, and both groups have interacted very little with primatologists, even though non-human primates provide the best living models for many aspects of human cooperation. This volume provides a first attempt to initiate a more intensive dialogue among these three disciplines.

This volume has two immediate goals: (1) It documents and summarizes the range of cooperative behaviors among non-human primates and relates it to their diversity in social systems and genetic structure. Whereas some aspects of primate cooperation have been reviewed recently (Chapais & Berman 2004), many empirical and experimental data addressing other topics await to be synthesized. This volume, therefore, provides a comprehensive and up-to-date summary of the primate literature on social grooming, coalition formation, conflict management, cooperative hunting, alloparenting, food sharing and other relevant topics. (2) The range of behavioral mechanisms underlying cooperative behavior in primates and humans is documented and critically assessed to identify mechanisms of, and prerequisites for, cooperation that are uniquely human. Because primates exhibit such wide variation in social systems and cognitive abilities, they provide a natural link between humans and other animals to explore these questions productively. By clearly defining similarities and differences between human and non-human primates in such a salient aspect of social behavior, this volume will hopefully inform and focus future research in both disciplines.

These ambitious goals motivated us to organize a conference (Fourth Göttinger Freilandtage) at the German Primate Center in December 2003 to discuss these issues with more than 250 participants. Various aspects of cooperation in mammals as well as human and non-human primates were presented in more than 50 oral and poster papers, including 16 talks by invited speakers. Following the conference, 15 contributions were solicited in written form, and each one was subjected to rigorous peer review. They constitute a representative sample of the contributions to the conference, encompassing specific case studies, comprehensive reviews, theoretical analyses, as well as studies of non-primates that provide important comparative perspectives on general principles related to the issues raises above. We think that together they provide an up-to-date account of research on cooperation in primates and humans, as well as numerous stimulating suggestions for future research on these topics.

The conference, as well as the resulting volume, would not have been possible without the support of many people and organizations. The Fourth Göttinger Freilandtage were made possible by generous grants and support from the Deutsche Forschungsgemeinschaft (DFG), the Niedersächsisches Ministerium für Wissenschaft und Kultur, the German Primate Center (DPZ), the Universität Göttingen, the city of Göttingen and the Sparkasse Göttingen. Michael Lankeit crucially supported this conference from the first moment on in many ways. Claudia Fichtel did an amazing job of organizing every logistical detail before and during the meeting to everyone's satisfaction. The members of the Abteilung Soziobiologie at the DPZ, in particular Manfred Eberle, Eckhard Heymann, Ulrike Walbaum and Dietmar Zinner helped beyond the call of duty with the preparation of this conference.

The quality of the present volume is to a large extent due to the constructive comments of all contributors, who served as internal referees, as well as Rebecca Lewis, Craig Stanford and Roman Wittig, who provided additional comments on individual chapters. Christina Oberdieck double-checked every single reference. Julia Barthold prepared the index, and Claude Rosselet carefully checked it against the proofs. We thank all of them wholeheartedly. Finally, it is our pleasure to dedicate this volume to Claudia & Maria, Theresa & Anna and Jakob & Jaap, for their understanding, support and inspiration during the preparation of this volume.

Göttingen/Zürich, May 2005
Peter Kappeler and Carel van Schaik

Peter M. Kappeler

Carel P. van Schaik (Eds.)

Cooperation in Primates and Humans

Mechanisms and Evolution

With 61 Figures and 18 Tables

Professor Dr. Peter M. Kappeler
Department of Behavioral
Ecology & Sociobiology
German Primate Center (DPZ)
Kellnerweg 4
37077 Göttingen
Germany

Professor Dr. Carel P. van Schaik
Anthropological Institute
and Museum
University of Zürich
Winterthurerstr. 190
8057 Zürich
Switzerland

Cover:
Gray mouse lemur (*Microcebus murinus*) mother with infants, Forêt de Kirindy, Madagascar,
© Manfred Eberle, www.phocus.org

ISBN-10 3-540-28374-9 Springer-Verlag Berlin Heidelberg New York
ISBN-13 978-3-540-28374-4 Springer-Verlag Berlin Heidelberg New York

Libary of Congress Control Number: 2005930634

Springer is a part of Springer Science+Business Media
springeronline.com

Printed in Germany

Editor: Dr. Dieter Czeschlik, Heidelberg
Desk Editor: Anette Lindqvist, Heidelberg
Typesetting: Satz-Druck-Service, Leimen
Cover-Design: design & production, Heidelberg
Printed on acid-free paper 31/3151Re 5 4 3 2 1 0

Contents

Part IV
Mutualism

Part V
Biological Markets

Part VI
Cooperation in Humans

Contributors

AURELI, FILIPPO
Research Centre in Evolutionary Anthropology and Palaeoecology, School of Biological & Earth Sciences, Liverpool John Moores University, UK
f.aureli@livjm.ac.uk

BARRETT, LOUISE
School of Biological Sciences, University of Liverpool, UK
& Behavioural Ecology Research Group, University of Natal, South Africa
L.Barrett@liverpool.ac.uk

BOESCH, CHRISTOPHE
Max Planck Institute for Evolutionary Anthropology
Leipzig, Germany
boesch@eva.mpg.de

BOESCH, HEDWIGE
Max Planck Institute for Evolutionary Anthropology
Leipzig, Germany

BROSNAN, SARAH F.
Living Links, Yerkes National Primate Research Center
Emory University
Atlanta, GA 30329, USA

CHAPAIS, BERNARD
University of Montreal
Dept. Anthropology
Montreal, Canada
bernard.chapais@Umontreal.Ca

CLUTTON-BROCK, TIM
Department of Zoology
University of Cambridge
Cambridge, UK
thcb@cam.ac.uk

DE WAAL, FRANS B. M.
Living Links, Yerkes National Primate Research Center
Emory University
Atlanta, GA 30329, USA
dewaal@emory.edu

GÄCHTER, SIMON
University of Nottingham
CESifo & IZA
simon.gaechter@unisg.ch

HENZI, S. PETER
Behavioural Ecology Research Group, University of Natal, South Africa
Department of Psychology, University of Central Lancashire, UK

HERRMANN, BENEDIKT
Universität Göttingen
& Harvard University

Kappeler, Peter M.
Dept. Behavioral Ecology & Sociobiology
Deutsches Primatenzentrum
Göttingen, Germany
pkappel@gwdg.de

König, Barbara
Zoologisches Institut
Universität Zürich
Winterthurerstr. 190
8057 Zürich, Switzerland
bkoenig@zool.unizh.ch

Milinski, Manfred
MPI for Limnology
Plön, Germany
milinski@alpha1.mpil-ploen.mpg.de

Mitani, John C.
Department of Anthropology
University of Michigan
550 East University Avenue
Ann Arbor, MI 48109-1092, USA
mitani@umich.edu

Noë, Ronald
Ethologie des Primates –
CEPE (CNRS 9010)
University Louis-Pasteur
7, rue de l'Université
67000 Strasbourg, France
Ronald.Noe@wanadoo.fr

Pandit, Sagar A.
Department of Biological, Chemical and Physical Sciences
Illinois Institute of Technology,
Chicago, IL 60616, USA

Schaffner, Colleen
Department of Psychology
University College Chester, UK

Silk, Joan B.
Department of Anthropology
University of California
Los Angeles, CA 90095, USA
jsilk@anthro.ucla.edu

Trivers, Robert L.
Center for Human Evolutionary Studies
Rutgers University
131 George St
New Brunswick, NJ 08901-1414, USA
trivers@rci.rutgers.edu

van Schaik, Carel P.
Anthropologisches Institut & Museum
Universität Zürich
Winterthurerstr. 190
8057 Zürich, Switzerland
vschaik@aim.unizh.ch

Vigilant, Linda
Max Planck Institute for Evolutionary Anthropology
Leipzig, Germany

Vogel, Erin R.
Department of Ecology and Evolution,
SUNY at Stony Brook,
Stony Brook, NY, 11794, USA

Part I
Introduction

Chapter 1

Cooperation in primates and humans: closing the gap

Carel P. van Schaik, Peter M. Kappeler

1.1 Why does cooperation pose a challenge?

In common usage, we speak of cooperation if individuals actively assist or support others: the emphasis is on behavior. For evolutionary biologists, cooperation involves actions or traits that benefit other individuals. They stress the outcomes of these behaviors, in particular the consequences for the fitness of the interacting individuals. Cooperative acts that are beneficial for both actor and recipient are said to be mutualistic. A cooperative act that is costly to the actor is termed altruistic; if the recipient is a relative, the interaction is sometimes called nepotistic[1]. The behavioral definition and the outcome-based definition usually label the same phenomena cooperative.

Cooperation has been described at all levels of biological organization, from molecules, organelles and cells, to individuals or groups of the same species and even individuals of different species (Hammerstein 2003b). The contributions to this volume focus on cooperation in the form of behavioral interactions between individuals, largely within species. This kind of cooperation can be manifested through single behavioral acts, such as giving an alarm call or providing a conspecific with agonistic support, but also through long-term behavioral tactics or roles, such as helping relatives raise their offspring, or even through organismal adaptations, such as renouncing reproductive activity. Frequently encountered examples of cooperative behaviors in nature are coalition formation, the exchange of grooming or other forms of body care, alarm calling, predator inspection, protection against attacks by predators or conspecifics, supporting injured group members, helping in the reproduction of others (cooperative breeding), egg trading among hermaphrodites, nursing of other females' infants, communal defense of food sources or territory boundaries, interactions between neighboring territory owners, sharing of special skills or information, food sharing and cooperative hunting (see Dugatkin 1997, Clutton-Brock 2002).

[1] Note that we adhere to a broad definition of cooperation, in that both actor and recipient or only the recipient can benefit. The narrow definition requires the presence of altruistic acts, i.e. only the recipient benefits. We prefer the broad definition because it may be extremely difficult in practice to determine whether some action is altruistic; it includes mutualism, and it complies more closely with common usage.

As indicated by these examples, cooperative acts come in a myriad of forms. Nevertheless, they all share a central problem: the vulnerability of the cooperator to being exploited by selfish partners. Opportunities for exploitation come in two main forms, depending on the context of cooperation. First, they may arise due to the time delay inherent in reciprocity. When altruistic acts are exchanged reciprocally between members of a dyad, the partner who benefited from an earlier altruistic act can defect, either by reneging when his turn arises, or by returning less than he received. The second opportunity for exploitation is free riding, which arises when an individual does not (equally) contribute to the creation or maintenance of a shareable benefit or good (this can happen at the level of the dyad or at that of the group, in which case the benefit is called a public good). An additional threat to evolutionary stability of cooperation is risk-avoidance in mutualism. It arises when a mutualistic benefit can only be produced through some costly collective action by two or more partners, and one individual bows out at the moment of the dangerous collective action, thereby exposing the partner(s) to considerable risk of injury (see van Schaik, this volume). Agonistic coalitions or cooperative hunting of dangerous prey provide exemplary contexts for such risk. These three problems make cooperation less likely in nature. In some cases, such as high-risk altruistic support in agonistic conflicts or high-risk collective action, where opportunities for exploitation go hand in hand with risk avoidance, cooperation may be particularly unlikely or unstable.

However, cooperation is rife in nature, and an explanation for its origin and maintenance is therefore needed. Consequently, it has been the focus of much empirical and theoretical work for over a century. In the first section of this introductory chapter, we provide a brief overview of the history of the study of cooperation, from Darwin to the mid-1990s, for novices to the field. Although much progress has been made, this work has not led to a definitive solution of the cooperation problem. Nonetheless, much contemporary research on cooperation is building on three pillars of earlier efforts, namely nepotism, reciprocity and mutualism. We revisit these three pillars in the next section, which also serves as an overview of the contributions to this volume. However, it should not be forgotten that these explanatory models focus on selected acts of cooperation, and that animals in nature may be involved in multiple forms of cooperation with the same partners simultaneously.

A major rationale for this book is that an explosion of recent work on humans has done much to highlight the contrasts in cooperative behavior between humans and other animals, in particular great apes. In the next section of this introduction, we therefore explore the major differences and preview the chapters that focus on humans. We also address the important question as to why human cooperation became so fundamentally different from that among all other primates and non-eusocial animals. We close this chapter by drawing attention to some unresolved questions, in particular with respect to work on non-human primates.

1.2 Cooperation: a brief history of the main ideas

The struggle for life and the survival of the fittest are concepts that emerged from Darwin's (1859) reasoning that led him to identify natural selection as the agent responsible for adaptations. Accordingly, individuals who out-compete their conspecifics in the struggle for access to resources and mates enjoy greater reproductive success and, hence, pass on more copies of their genes to the next generation. Thus, competition naturally emerged as the main concept in explaining many aspects of organismal adaptation in evolutionary biology. Against this background, it is particularly difficult to explain the existence of behaviors that benefit others at the expense of the ego. Darwin was well aware that such cooperative acts do occur in nature at different levels, in different forms, and with different consequences for the actors involved, and he clearly recognized that altruistic behaviors presented 'a special difficulty', potentially fatal to his whole theory of natural selection. All subsequent work on the evolution of cooperation has focused on identifying the conditions under which altruistic acts can be evolutionarily stable against exploitation (see Dugatkin 1997).

Kropotkin (1902) re-affirmed the importance of cooperation in nature. He dealt with the defection problem, albeit implicitly, by relying on group selection or its even more improbable cousin, species selection, to explain all cooperative behavior in nature. Moreover, many of his examples would nowadays be ascribed to byproduct mutualism (see below).

Group selection continued to be invoked as an explanatory device for cooperation throughout the first half of the 20th century by influential scholars, such as Allee (1938, 1951), and later most explicitly Wynne-Edwards (1962). It was the rejection of group selection, inspired by Wynne-Edwards's book, more than any other development that pushed evolutionary and behavioral biologists who rejected group selection to systematically search for explanations for seemingly altruistic behaviors in nature (Hamilton 1963, 1964, Maynard Smith 1964, Williams 1966). By the early 1970s, these biologists had responded to this challenge by erecting two major explanatory frameworks to explain this kind of vulnerable cooperative behavior: kin selection and reciprocity (Hamilton 1964, Trivers 1971).

For ultimate explanations of altruism, the most fundamental distinction is that between interactions between either related or unrelated individuals. As first pointed out by Hamilton (1964), kin selection theory can provide a potent explanation for nepotistic behavior. Because a disposition to help close relatives will automatically enhance the propagation of genes in other individuals that are identical by descent from a common ancestor, the benefits of altruistic acts (B) towards relatives also accrue to the actor, discounted by the degree of relatedness, r, between the two, i.e. the probability that they share the same allele through descent from a common ancestor. This makes altruistic acts, with cost C, more likely to evolve between relatives, as expressed in Hamilton's now famous inequality $Br > C$.

The explanation of altruistic acts directed at unrelated individuals requires a different approach. Trivers (1971) offered the groundbreaking idea that reciprocal altruism, now generally called reciprocity, in which two individuals alternate between providing and obtaining benefits, can provide a simple, but sufficient evolutionary mechanism for many cases of cooperation between unrelated individuals. He suggested that reciprocity is especially common among long-lived animals, because they have more opportunities to exchange altruistic acts. Moreover, reciprocity should flourish in species that live in stable groups in which individuals recognize each other, as well as in species characterized by social tolerance, because dominants do not prevent others from reciprocating. In his contribution to this volume, Trivers reviews the evidence for reciprocal altruism that has accumulated over the last three decades.

Reciprocity differs from mutualism by the presence of a time delay between incurring the cost of the altruistic act and receiving the benefit when the partner reciprocates. As the duration of the time delay approaches zero, reciprocity grades into mutualism (e.g. Rothstein & Pierotti 1988). Thus, a discrete time delay is usually considered necessary before reciprocity needs to be invoked. However, as it gets longer, discounting of the benefits should make it harder for reciprocity to be stable (Stephens et al. 2002).

Reciprocity "may be the most perplexing and difficult category of cooperation to explain" (Dugatkin 1997). Accordingly, Trivers's idea has been explored in great detail (Trivers, this volume). Most tests have used the formal similarity of the problem to that modeled by the two-person Prisoner's Dilemma (PD) game developed in game theory (Axelrod & Hamilton 1981). The ESS (evolutionarily stable strategy: Maynard Smith 1982) solution to the one-shot PD game is to defect, but examination of the situation in which players interact again in the future suggested that cooperation could be robust (Axelrod & Hamilton 1981). In particular, a strategy called 'Tit-for-tat', which starts out as a cooperator and then simply repeats the move of the other player in the previous round, provided a robust solution in that it was never exploited by other strategies and produced high payoffs when paired with other cooperative strategies. Dissatisfaction with the lack of biological reality of this approach has spawned the development of the biological markets framework, in which the choice of partners and communication receive special attention (Noë et al. 1991, Noë & Hammerstein 1994, see below).

Kin selection and reciprocity remain the most important explanations for altruistic acts by animals, and for cooperation in general, to this day. However, more recently, a new and improved form of group selection, called trait-group, intrademic or multi-level selection, has been added to our explanatory arsenal (Wilson 1983, Sober & Wilson 1998). A trait group comprises all individuals that affect each other's fitness. Natural selection operates both within and between such trait groups. If groups with more cooperators out-produce other groups, cooperation can be favored by between-group selection, but only if this effect is greater than the result of within-group selection, which acts against cooperators. This approach did not acquire a great following, however, although it can be argued that selective association of cooperating dyads within a larger group (as in many primate groups) is equivalent to the formation of trait-groups.

A separate strand of thought drew attention to the possibility that we may misinterpret much animal behavior and see altruistic acts where none exist.

Thus, some of what is labeled as reciprocity may in fact represent byproduct mutualism (Dugatkin 1997). In such cases, one animal benefits from what a second animal is doing but would also be doing in the absence of the first animal. One good example is the phenomenon of group augmentation, where animals directly benefit from being in a group, and are therefore expected to coordinate their behavior (Kokko et al. 2001). The behavioral definition of cooperation excludes such byproduct mutualism from cooperation, because we cannot observe any special cooperative acts, even if the animals coordinate or synchronize their activities (cf. Clutton-Brock 2002). Usually, byproduct mutualism is easily distinguished based on this definition, but there are some cases that look deceptively like true cooperation. In several species of fish, piercing the skin, for example due to predator attack, causes the release of a compound ('Schreckstoff') that elicits alarm in other fishes. However, the compound has its own immediate function in protecting the fish against fungal infection, and its production is therefore not altruistic (Magurran et al. 1996).

A variation on this theme is that seemingly altruistic acts, such as grooming another individual or giving an alarm call, are not altruistic at all because they impose no costs on the actor or may even carry an immediate benefit (e.g. Dunbar & Sharman 1984). Thus, such interactions are in effect mutualistic. However, even if they are, this does not mean that there is nothing left to study; even in mutualistic interactions, there may be plenty of opportunities for conflict or asymmetric distribution of benefits. Moreover, the presence of undeniable examples of truly altruistic acts (e.g. risky alarm calls: Sherman 1977; blood donation: Wilkinson 1984; predator mobbing: e.g. van Schaik et al. 1983) suggests that this alternative cannot explain all forms and examples of cooperation.

Finally, individuals may be coerced into cooperative behavior. For instance, breeders may force younger relatives into helping them raise more young (Emlen & Wrege 1992), dominants may force subordinates into providing services (Tebbich et al. 1996) or group members may harass owners of food into food sharing (Stephens & Gilby 2004). However, the conditions under which such coercion leads to stable cooperation are probably quite restrictive (Kokko et al. 2001), so that cooperation for these reasons is probably rare.

1.3 The pillars of cooperation

1.3.1 Kin selection

Hamilton's (1964) fundamental insight was that altruistic behaviors could be explained evolutionarily if we focus on the gene rather than the individual as the unit of selection. Theoreticians have repeatedly re-evaluated Hamilton's rule by making the genetic assumptions increasingly explicit and realistic. Perhaps surprisingly, this very simple rule was found to hold up fairly well under such close scrutiny (Michod 1982). Empirically, as reviewed by Silk (this volume), many of the cooperative and altruistic acts performed by animals, including non-human and human primates, are directed towards relatives, and thus potentially best explained by kin selection (see also Griffin & West 2002). Silk also demonstrates for non-human primates that alternative explanations of behavior, or theoretical objections to preferential association by kin, do not obviate the need for kin selection.

Although many phenomena in animal behavior can be adequately explained by nepotism, this does not mean that all interactions between kin are nepotistic (West et al. 2002). Nor does it mean that all cooperation among kin is necessarily nepotism (unilateral altruism); kin also engage in mutualistic cooperation or in reciprocity (Clutton-Brock, this volume). The reason that this simple fact is often overlooked is that mutualism and reciprocity are often studied explicitly among non-kin in order to control for nepotism. Indeed, as stressed by both Silk (this volume) and Chapais (this volume), other forms of cooperation may also be more common among kin, because relatives tend to be available as partners, cooperation with relatives produces additional inclusive fitness benefits, and because kinship may act to stabilize mutualistic and reciprocal actions because it reduces the benefits of defection (cf. Wrangham 1982). Thus, reciprocity and risky mutualism may well have originated among kin and provided the lineage with the basic behavioral and emotional mechanisms, which were then in place to be applied to the same acts with non-kin. However, Chapais (this volume) warns that kin-biased cooperation may be less common than this argument suggests because only non-relatives may be competent partners for particular kinds of cooperation, for example agonistic coalitions.

Kin selection may also contribute to a deeper understanding of altruistic phenomena typically examined from other angles. For example, kin selection may be a critical component of reproductive skew theory, which, using different models, attempts to explain why reproduction is not equally distributed among the members of a social unit (Johnstone 2000). The concession model posits that moderate reproductive skew is the result of dominants granting some reproduction to subordinates. Genetic relatedness is a crucial variable when it comes to predicting which individuals should be granted which share of total reproduction. The most important prediction of the concession model is that high relatedness among the members of a social unit should produce high reproductive skew (Keller & Reeve 1994). Forfeiting individual reproduction in favor of a close relative could be interpreted as altruistic behavior. Such high reproductive skew is indeed found among related males in coalitions of lions or howler monkeys: the top-ranking male monopolizes all or most of the reproduction (Pope 1990, Packer & Pusey 1991, see also Cooney & Bennet 2000). However, viable alternative explanations for reproductive skew exist that do not involve concessions and do not make this prediction (Clutton-Brock 1998a, Johnstone 2000).

Kin may make the best collaborators, but at the same time they are the worst possible mates because incest carries a high risk of leading to deleterious effects (Keller & Waller 2002). Inbreeding avoidance is now known to be widespread and underlies sex differences in dispersal (Clutton-Brock 1989a, Lehmann & Perrin 2003). Sex-linked dispersal, in turn, may strongly affect the degree to which members of the dispersing sex remain spatially associated (e.g. Vigilant et al. 2001, Fredsted et al. 2004), the critical precondition for cooperation in all species but humans. The fact that mating with kin is to be avoided has imposed clear limitations on the reach of kin selection. Due to the modest fecundity of most individual birds and mammals, the number of relatives that can be clustered in space is rather small, especially if they subsequently mate with non-relatives and relatedness is diluted again. More obviously, inbreeding avoidance and sex-biased dispersal explain the rarity of strong intersexual kin-based cooperation (again with the exception of humans; cf. Rodseth et al. 2003). The exceptions to this rule among animals may be found where the stability provided to cooperative interactions by kinship is extremely important (Clutton-Brock, this volume).

1.3.2
Reciprocity

The debate on reciprocity over the past quarter century has been dominated by the two-player PD model, in both its one-shot and iterated versions (see above). This model assumes that defection in a one-shot game is the ESS, and efforts focus on overcoming this tendency to defect. Increasingly sophisticated mathematical models have been developed in increasingly fine and arcane detail to explore the conditions and consequences of reciprocity in this model (reviewed by Dugatkin 1997). However, Noë (1990, 1992) and Hammerstein (2003b), among others, have questioned the extent to which the PD adequately describes the situation in mobile organisms from fishes to primates (but see Trivers, this volume). In the words of Hammerstein (2003b), "some theoretical ideas appear to be so compelling that the lack of supporting evidence is indulged by major parts of the scientific community".

The main reason for this criticism is that animals in nature only rarely seem to engage in repeated PD games. The PD model focuses only on one component, partner control (decisions for future interactions based on outcomes of previous interactions), whereas there are additional important components of cooperative relationships among animals: partner selection and communication about willingness to undertake a cooperative interaction or about payoff distribution. Partner choice, for example in the form of switching to another partner when the current partner defected, allows for selective association of trustworthy players. The notion of partner choice naturally leads to consideration of the role of other potential partners available to the players, and hence to the idea of cooperation markets, where partners select the most profitable partners and the value of commodities or services depends on their relative demand and supply. Biological market theory (Noë et al. 1991, Noë & Hammerstein 1994, see Noë, this volume) therefore contributes to developing a broader alternative in general, and it provides a powerful explanatory tool for the understanding of primate social behavior, in particular (Barrett & Henzi, this volume).

Likewise, communication about the intentions of each player before the interactions and negotiation with them about payoffs is likely to make reciprocity much more stable than under the conditions of PD games. Thus, communication before engaging in risky cooperation is frequently observed in primates (Smuts & Watanabe 1990, Noë 1992). Subtle communication may also take place about the price of a service. For instance, in the grooming market of primates, discussed in detail by Barrett & Henzi (this volume), females must groom longer to get access to desirable infants of other females when there are fewer infants in the group, and the price is set by the refusal of mothers to provide access to the infants after shorter grooming bouts (R. Noë & T. Weingrill pers. com.).

Cooperation in nature offers a paradox. Lots of (unrelated) animals seem to engage in cooperation, yet only quite rarely do we see them engage in contingency-based reciprocity (Noë 1990, Hammerstein 2003b), even though experiments indicate that they are capable of it (Hemelrijk 1994). There may be two main reasons for this discrepancy. The first reason is still largely speculative. Animals in stable social units can use their previous experience with any of the group members to make decisions about whether to cooperate in the future, and thus engage in generalized reciprocity. This cognitively non-demanding behavioral rule is theoretically most likely in small groups (Pfeiffer et al., in press), and has been demonstrated experimentally (Rutte & Taborksy, in review), but it is not known how important this mechanism is in nature.

The second reason for the absence of contingency in cooperation that involves altruistic acts is well established. Pairs (dyads) of cooperating animals seem to be concerned with costs and benefits on a much longer time scale than that of the interaction; they form social relationships, such as bonds or friendships, within which a broad range of cooperative acts is usually exchanged. Thus, in addition to altruistic acts of the same kind, as envisaged by reciprocity, they also exchange altruistic acts of different kinds, for example grooming for support in agonistic conflicts (see Mitani, this volume) and various kinds of mutualism and perhaps byproduct mutualism. Individuals in a bond do not evaluate the immediate costs and benefits of their behavioral decisions, as demanded by the theory of reciprocal altruism, but rather evaluate the long-term balance of the benefits and costs of all the acts exchanged in the relationship (cf. Pusey & Packer 1997).

The presence of these bonds is well documented in primates (Cheney et al. 1986), and recent work has shown that bonds have a positive impact on fitness, even after controlling for rank effects (Silk et al. 2003). Similar observations are available for friendships in humans. Aureli & Schaffner (this volume) note that these bonds, because of the important benefits they provide to both partners (cf. van Schaik & Aureli 2000), must be protected against the negative impacts of conflicts. It is important to remember that animals in every cooperative relationship also encounter many opportunities for conflict, and thus face the challenge of maintaining their relationship, with the net benefits it brings, in the face of the potentially disruptive effects of these conflicts. This threat to the relationship explains the ubiquity of reconciliation in primates and other social animals (Aureli & Schaffner, this volume).

Because so many altruistic acts and commodities are exchanged in these relationships, it is difficult to imagine that the players can maintain careful score cards on these actions, let alone on the costs and benefits they entail. Animals and even humans usually seem to cooperate without carefully calculating the costs and benefits of each act. This perspective also reduces the concern about the cognitive demands of engaging in reciprocity (Dugatkin 2002a, Hammerstein 2003b, Stevens & Hauser 2004). As detailed by de Waal & Brosnan (this volume), most cooperating dyads in most species use emotion-based mechanisms involving attitudinal symmetries that are cognitively simple. Chimpanzees are capable of the 'calculated reciprocity' required by reciprocity models, as obviously are humans, but this mechanism may be rare among other species, if it occurs at all (see also Brosnan & de Waal 2002, Stevens & Hauser 2004).

The stress on social relationships should not be taken to mean that all reciprocity takes place in the framework of bonds. However, one would expect such cases to be associated with greater emphasis on strict reciprocity (see also Barrett & Henzi, this volume). Indeed, in humans strict reciprocity is seen only among 'casual acquaintances' (Silk 2003). Reciprocity in nature among animals that do not necessarily have bonds may likewise be rather strict (e.g. grooming among impala, which have unstable associations: Hart & Hart 1992; egg-trading among simultaneously hermaphrodite fishes: Fischer 1980). These cases may derive their stability from the fact that the altruistic services or commodities are parceled out in small packages, leading to frequent alternation taking place in rapid sequence.

1.3.3 Mutualism

Mutualism as an explanation for cooperative behavior is theoretically simple. Numerous examples exist, from living in groups, which dilutes predation risk, to coalitions, where all participants gain in rank or gain access to limiting resources (Clutton-Brock 2002). However, this simplicity is only apparent. Mutualism is vulnerable to free riding, where partners (in the case of dyadic mutualism) or other group members (if group-level, or public benefits are produced) can harvest benefits without providing corresponding benefits in return. In dyadic mutualism, the costs are often opportunity costs because partner switching might produce greater benefits. In the case of group-level benefits, the costs tend to be real because the acts themselves, while providing a clear net benefit to the actors, are costly. The free riders who do not join-in in producing the benefit, thus harvest a larger net benefit. This problem is known in the social sciences as the collective action problem, and it is also demonstrably present in primate groups (van Schaik 1996, Nunn 2000, Nunn & Deaner 2004). We should only expect to see mutualism where these threats are somehow dealt with.

Mutualism and byproduct mutualism can be seen within and between species, and our focus here is on intra-specific interactions. Byproduct mutualism (e.g. individual escape behavior against predators that serves to alert other group members) does not require the presence of bonds or even stable association. However, dyad-level mutualistic exchanges usually take place within an existing long-term relationship, in which both partners have an interest in keeping the beneficial cooperation going, and incentives to large-scale defection are therefore minimal. Hence, the distinction between reciprocity and mutualism becomes somewhat artificial and may be of no concern to the animals. Similarly, as discussed for the case of reciprocity, kinship may shore up the stability of these relationships.

At least among non-human primates, examples of dyadic cooperative relationships are far more numerous than mutualism that involves more players or even entire groups. And where particular cases of mutualism can involve two or more players, those involving only two tend to be more common. For instance, the agonistic coalitions among primate males described by van Schaik et al. (this volume) almost always contain only two members, especially the risky varieties where coalition members attack a higher-ranking male to take over his top-dominant position. Similarly, the communal nursing among female house mice described by König (this volume) most commonly involves only two females. The relationship perspective may explain why this is so. First, when animals cooperate in pairs, it is easier to exert control over the partner's behavior. In pairs, the costs of partner control, for example by punishment (Clutton-Brock & Parker 1995), can be recouped again when the partner subsequently behaves in a more cooperative manner. In group-level mutualism, this punishment is altruistic (Fehr & Gächter 2002), because all other group members benefit as well without incurring any costs. Second, in dyadic cooperation, it is also easier to exert partner choice. A dissatisfied individual can usually switch to another partner in the group, whereas in group-level mutualism it would require either expulsion of free-riding partners or dispersal to other groups with more cooperative partners, both of which are likely to carry considerable cost. The rarity of smooth collective action among animals other than eusocial species is perhaps the main distinction between humans and other animals in this context.

One of the few well-documented cases of multi-player mutualism in primates is the cooperative hunting described among chimpanzees in the Taï Forest by Boesch et al. (this volume). The very existence of this behavior shows that the individuals somehow deal with free riding, whereas among chimpanzees elsewhere, dominant males, who did not necessarily participate in the hunt itself, tend to end up with the prey and control its distribution. Multiple males also participate in other areas, but it is only in the Taï Forest that individuals take on complementary roles, resulting in the ability to subdue larger prey (Boesch et al., this volume). The authors note that the forest structure in Taï makes such close cooperation critical to achieving success. At other sites, group hunting is more like byproduct mutualism; males merely hunt simultaneously but still end up better off, despite attempts by dominants to monopolize the distribution of meat. The true cooperation in the Taï Forest is made possible by the 'fair' distribution of meat, but why this works there and not elsewhere is not clear. The answer is eminently important for the evolution of the strong tendency to mutualism we see in humans.

Other instances of mutualism near the group-level end of the spectrum also exist. For instance, helpers in cooperative breeders that are not related to the breeders may help because of the advantages of being in the social unit (group augmentation: Kokko et al. 2001, Clutton-Brock, this volume). Residents allow them to join and stay, not only due to benefits gained from the help, but also from reduced risk of predation or attacks by neighboring groups. Helpers gain these same benefits, but are expected to contribute to the semi-public goods through helping, such as providing sentinel service. Experimental evidence on helpers in a cooperatively-breeding cichlid fish suggests that helpers prevented from helping are attacked more and work harder upon return (Balshine-Earn et al. 1998, Bergmüller & Taborsky 2005).

In all successful cases of mutualism, free riding is kept in check. In the behavioral examples discussed above, this is done through behavioral control. Sometimes, however, mutualism works due to restraint by dominants. Thus, in groups, dominants may peripheralize the subordinates to gain greater safety, but the benefit of the selfish herd tends to be a sufficient incentive for the subordinates to stay (Hamilton 1971), if only because dominants refrain from stronger peripheralization because that would entice the subordinate to leave and join other groups.

In cases without obvious behavioral control, the presence of successful mutualism requires that the conditions restrict either the opportunities or the incentives for free-riding. A good example is provided by the distribution of communal nursing described by König (this volume). Here, females are unable to recognize their young; they are therefore unable to favor their young over those of others. Because this ability to recognize young emerges some time before weaning, however, it is probably no coincidence that most of the observed cases of communal nursing involve related females. A more subtle example is provided by the formation of fruiting bodies in normally solitary amoebas that form colonies to reproduce. The cells of *Dictyostelium discoideum* cooperatively form fruiting bodies that produce spores. These sit on top of stalks, which are therefore reproductive dead ends. Yet, all cell lines are represented equally in the production of stalks and fruiting bodies (Foster et al. 2004), probably because defection is prevented biochemically. The gene *DimA* is involved in the production of stalks. Hence, the absence of *DimA* would potentially allow the cell to forgo participation in stalk production. However, absence of the gene also pleiotropically results in exclusion from the stalk, thus keeping such a benefit to defection in check.

Perfectly stable mutualism should be found where defection is impossible, and hence no additional mechanisms of partner control are required. The cooperation among components within entities, such as the organelles within a cell, or by cells within a body, might be stable because the opportunities for defection by partner cells have largely been eliminated. The very long delay between the origin of simple unicellular organisms, and the eukaryotic cell and multicellular organisms, however, suggests that this transition may not be easy, and that active policing remains necessary (e.g. Michod 2003).

1.4 Cooperation among humans

Primates differ from many other animal lineages in that they show rather good evidence for cooperation, especially in long-term relationships (beyond simple protection of offspring by mothers), although it remains to be seen to what extent this picture is due to poor documentation for other lineages (Dugatkin 1997). One thing is clear, however; humans are dramatically different even from other primates. "Human cooperation represents a spectacular outlier in the animal world" (Fehr & Rockenbach 2004). We are a species in which there is far more cooperation than in any other non-eusocial species. In this section, we will try to document exactly how humans differ from other primates, then examine the proximate mechanisms (emotional, cognitive) that underlie these differences, and finally briefly address the possible selective agents that gave rise to these differences.

First, humans tend to engage much more commonly in group-level cooperation, whereas most cooperation in nature is at the level of dyads. Human groups can behave almost as superorganisms (allowing functionalism in sociology to treat social groups, rather than individuals, as the unit of analysis), setting communal goals and engaging in communal tasks. One expression of the strong organization at the group level is individual specialization and division of labor, often by sex.

Undertaking cooperation at the group level rather than that of the dyad poses more serious cheater detection problems. As we noted earlier, it is easier for an individual to control the behavior of a partner in a dyad than it is to control the behavior of a group of individuals; selective association or punishment are likely to be costlier, and the required coordination in the case of group-level action may be cognitively complex as well (see also Boyd & Richerson 1988). Humans must therefore possess cognitive and emotional mechanisms that act to detect even subtle ways of defection and control the behavior of group members. Gächter & Hermann (this volume) review an array of mechanisms that act to stabilize the intrinsically very fragile group-level cooperation.

Second, humans tend to engage in extremely high-risk cooperation, much more than other animals, even than chimpanzee males. Coalitionary killing by male chimpanzees is otherwise unique among primates, but tends to involve serious asymmetries in the collective strength of the opposing parties (Wrangham 1999, Wilson & Wrangham 2003). In the typical case, three or more males from one community attack and kill a single male from a neighboring community. As a result, risk of injury to the attackers is limited. Chimpanzee males also attack large and potentially dangerous prey (adult red colobus monkeys: see Boesch et al., this volume), but the literature contains no references to males getting injured. In both cases, the risk of injury is kept low because of the close coordination of the attacks.

Human war is similar to coalitionary killing of males in many respects, and probably predates the origin of states (Keeley 1996), although it is perhaps not homologous with that among chimpanzees. However, human coalitionary killing, at least among contemporary humans, differs from that among chimps in that it also occurs between parties with much more symmetric collective strengths. The more balanced power of human armies implies higher individual risks to fighters. The appalling loss of life in many historically-documented wars attests to this, yet in numerous cases soldiers are not forced into battle and fighting is largely voluntary.

The third difference is more gradual than the other ones, but still worth noting. Humans tend to cooperate with non-kin more than other primates. In nonhuman primates, "the most costly forms of cooperation are reserved for close kin" (Silk, this volume). There is some evidence that male baboons and Barbary macaques that form leveling coalitions are non-relatives (see van Schaik et al., this volume). Chimpanzee males represent the strongest exception to Silk's generalization. As we saw, they engage in risky collective combat, yet surprisingly, the collaborators need not be close (maternal) kin (Mitani, this volume). Humans, of course, are arguably even more extreme than chimpanzees in this respect. Human military history is littered with descriptions of acts of amazing bravery aimed at comrades who are not relatives, although descriptions often invoke kin-colored terminology, such as brothers-in-arms. There is no firm explanation for these anomalies as yet, although Chapais' (this volume) competence principle may play a major role; where the competence of the partner becomes an increasingly important factor in deciding the success of cooperative actions, it is increasingly less likely that a close relative is at hand that is sufficiently competent. Yet, there is probably far more to it than that.

Fourth, humans are willing to incur some cost to punish non-cooperators in the group-level kind of cooperation in which individuals contribute to common goals and free riders risk the breakdown of all cooperative effort. Thus, strong reciprocity (Gintis 2000) combines altruistic rewarding of cooperators with altruistic punishment of defectors (called moralistic aggression in Trivers 1971), both of which are costly to the actor.

So far, there is no evidence for altruistic punishment among animals in nature, as suggested by studies of species engaging in collective, high-risk defense of territories against neighboring social groups, in ring-tailed lemurs (Nunn & Deaner 2004), lions (Heinsohn & Packer 1995) or even chimpanzees (D. Watts pers. com.). However, de Waal & Brosnan (this volume) describe experimentally-induced costly refusal to cooperate, thus challenging the categorical uniqueness of altruistic punishment. However, even if confirmed in capuchins and/or chimpanzees, this does not mean that its presence in other primate species can be generalized, because these two genera are among the most socially tolerant and intensely cooperative among all primates. Moreover, it is possible that altruistic punishment in non-human primates is always directed at cheating partners, whereas humans often direct altruistic punishment at individuals they observed cheating in interactions with third parties. The difference critically depends on the presence of societal norms, for which there is no evidence so far in non-humans.

A fifth difference concerns the role of reputation in facilitating reciprocity. Reputation is almost certainly much more important in human than in non-human primates. The three basic preconditions for reputation are individual recognition, variation in personality traits, and curiosity about the outcome of interactions involving third parties. The first two of these are met in most non-human primates, but the third may require awareness of third-party relations, which involves cognitive abilities so far demonstrated in only a few species (Cheney & Seyfarth 2003), although it may be more widespread. There is good evidence that primates use information on their experience with others in the past to predict their behavior in the future (Silk 2002a), and it is almost inevitable that this information is also used to select partners in whom they invest in order to establish social bonds. Yet, there is no evidence that they use reputation based on third-party interactions. Obviously, this does not mean that none do, but it would take careful observations and experiments to demonstrate it.

Humans, in contrast, commonly engage in indirect reciprocity (Alexander 1979), in which an ego's tendency to cooperate with a partner depends on the latter's reputation, which is established not only based on the ego's direct experience with the individual but also on this individual's behavior toward others, which is either observed directly by ego or reported to ego by third parties. No doubt, this use of reputation is enhanced by language. The displacement quality of language allows one to learn about the behavior of others even if the acts were not observed and the actors are not present, although the reliability of this information is subject to manipulation due to the very same quality of displacement.

Reputation is vital for an individual's success in society, and individuals show great concern over their reputation. Milinski (this volume) shows that reputation is also an unexpectedly powerful mechanism for maintaining group-level mutualism (the creation of public goods), which is especially vulnerable to the free-riding problem. In experiments, players became more cooperative when such public goods games were alternated with indirect reciprocity games. In other words, the concern with maintaining a good reputation, with its obvious benefits in indirect reciprocity, spills over into the public-goods situation. Since humans are normally engaged in multiple cooperative relationships simultaneously, this finding spells hope for improvement of the management of common or public resources.

The final difference is that humans exchange goods and services using token-based ('mercantile') exchange; we trade. At least among members of the same society, this usually works, even if the participants are perfect strangers without too much risk of exploitation or worse, because of guarantees put in place by societies. This trade requires not only the ability to weigh the value of goods or services relative to those of other goods or services of different kinds, but also to manipulate symbolic representations of values, and subsequently to accept in themselves arbitrary tokens as intermediary payment that can later be exchanged for other goods or services (Ofek 2001). These abilities could not have evolved if a system of trust had not been put in place; our subsistence style would be all but impossible without it, since we critically depend on the products and services of others. Obviously, nothing among animals in nature compares to this system, although the generous food sharing and trading of these favors for subsequent services in chimpanzees (see Boesch, this volume; Mitani, this volume) is clearly the foundation upon which our trade is built.

These differences can be summed up as follows: humans are far more likely to cooperate, both at the dyadic and especially at the group level, and we do so with non-relatives and often in situations of extremely high risk, apparently even with strangers (but see Trivers, this volume). This tendency would seem to expose us to unacceptably great risks of defection, but we have evolved special mechanisms, including cheater detection, the use of reputation to gauge the quality of potential partners and, most spectacularly, altruistic punishment to keep the tendencies toward defection by partners in check. According to Fehr & Fischbacher (2003), all of this boils down to our unique capacity to establish and enforce social norms: rules of social conduct that are internalized and are upheld even if the individual is not directly affected.

Given this uniquely derived level of cooperation, we must expect the presence of derived psychological mechanisms that provide the proximate control mechanisms for these cooperative tendencies (cf. Trivers 1971). It is tempting to look to other uniquely derived human features, such as language, advanced intelligence or cultural transmission of social norms, as functional prerequisites for the evolution of these mechanisms. While they are undoubtedly involved, the evidence suggests a critical role for unique emotions as the main mechanisms. Extreme vigilance toward cheaters, a sense of gratitude upon receiving support, a sense of guilt upon being detected at free riding, willingness to engage in donation of help and a zeal to dole out altruistic punishment at free riders – all these are examples of mechanisms and the underlying emotions that are less developed in even our closest relatives. Moreover, many of the emotions reflect societal equivalence in norms, such as those of fairness and justice. Emotions can be seen as the mechanisms used by evolution to achieve the optimum ('rational') outcomes without explicit reasoning or calculation (cf. van Hooff 2001). This might explain the emotional flavor to virtually all human decision-making (as no doubt in animal decision making).

Some of these mechanisms are also present in animals (see Brosnan & de Waal 2004), but they are certainly much exaggerated in our species (see also McGuire 2003). Uniquely, humans dispense the costliest of tendencies, altruistic punishment, even toward perfect strangers whose free-riding did not affect them (Fehr & Gächter 2002). Evolutionary biologists have a serious challenge explaining these tendencies, and Gächter & Hermann (this volume) briefly review the lively debate that has ensued about altruistic punishment, although no doubt the last word on this has not yet been spoken (see also Fehr & Henrich 2003).

The critical difference as we now see it, is that in humans these emotions have become normative, i.e. we have these emotions also when we are not directly involved, whereas the bulk of the evidence for animals still supports the notion of self-centered emotions, although great apes may turn out to be an exception (Flack & de Waal 2000).

The question obviously arises as to how humans could have become such a spectacular outlier in just a few million years. This question has recently spawned an active research effort. The dramatic differences with even our closest relatives suggest that the regular processes of kin selection or relationship-based cooperation involving reciprocity and mutualism are inadequate. Moreover, because the greatest qualitative difference is in group-level cooperation, one could argue that coordinated group activities, such as cooperative hunting and gathering accompanied by a division of labor and especially warfare, may underlie the amazing willingness to invest in cooperative relationships among humans. This has led to suggestions involving language and culture as the key pacemakers of group-level cooperation.

The most detailed hypothesis is the cultural group selection model offered by Richerson et al. (2003). They argue that conformist transmission (Boyd & Richerson 1985), a regular adaptation, can create groups that differ systematically and persistently from other groups, even in the absence of genetic differentiation and in the presence of migration between groups. Suppose that the Pleistocene saw major, rapid changes in the environment, destroying local adaptations. Further suppose that a successful novel invention gets established in one group, and is subsequently maintained by conformist transmission, and that some of these inventions favored group-level cooperation and strong group coordination (Boyd & Richerson 2002), with its attendant massive fitness benefits in a hostile environment. Successfully cooperating groups could now out-compete and displace other groups that lacked this invention.

A simpler alternative has recently been proposed (Panchanathan & Boyd 2004). Our interdependence at the dyadic level, including extensive indirect reciprocity with people we hardly know, has led to a critical reliance on reputation. Our pursuit of a good reputation in all contexts has, as a byproduct, created prosocial behavior at the group level, including altruistic punishment (which enhances reputation: see Milinski, this volume). We should expect to see major advances in this area in the near future. Trivers (this volume) provides a set of constructive suggestions and criticisms of previous approaches that should help focus future work on human cooperation on realistic assumptions and predictions.

1.5 Summary and outlook

Let us now briefly review the important advances that have been made in research on cooperation among individuals in dyads and groups, as well as flag issues that still remain to be solved by future work. The frontier of cooperation research has recently moved to humans, and many of the theoretical problems surrounding cooperation in animals seem to have been settled. However, that does not mean that there is no need for further empirical work on animal cooperation. Here, as in the rest of this book, we focus largely on primates.

Although altruism directed at kin is theoretically straightforward, some questions nonetheless remain. Thus, it is still unknown whether animals have a cut-off relatedness for all kinds of altruistic acts (i.e. consider those above a particular r value family, regardless of the kinds of acts they direct at them), or whether they differentiate among relatives depending on the cost of the acts involved (e.g. protecting only closer kin against higher-ranking opponents or predators: see Chapais, this volume; see also chapters in Chapais & Berman 2003). Likewise, debate continues to rage as to whether non-human primates recognize and classify kin entirely on the basis of association history or whether they also use other clues, summarized under the header phenotype matching. These questions are obviously related; differentiation among degrees of kin requires mechanisms of kin recognition that permit finer estimates of relatedness than using a simple cut-off point. The recent development of non-invasive genetic techniques will no doubt help us to find answers to both these questions.

Kin discrimination can strongly improve the power of kin selection and thus the behavioral reach of nepotism. In particular, if kin recognition is extended to the paternal component of kinship, we should expect richer patterns of nepotism in group-living animals. New studies that estimate the paternal component of relatedness suggest that kin discrimination among non-human primates is based on more than familiarity through association (Widdig et al. 2001, Buchan et al. 2003), but because other studies suggest otherwise (see Paul & Kuester 2003), it is important to identify the causes of these discrepancies.

At a more practical level, the use of non-invasive genetic techniques brought to light some clear mismatches between clustering of kin due to differential dispersal and the importance of kinship in social behavior. Thus, several prosimians (and perhaps orangutans) show evidence for female philopatry (Kappeler et al. 2002, Wimmer et al. 2002), but at least in some (e.g. *Mirza coquereli*), females do not seem to engage in any social behaviors that might benefit from having kin as partners. On the other hand, male chimpanzees are both philopatric and form strong alliances, yet they do not seem to select close kin more than expected (Vigilant et al. 2001). The necessary re-evaluation of the importance of nepotistic cooperation in philopatry (e.g. Wrangham 1980, Waser 1988) suggested by these cases will be facilitated by more descriptions from the field.

Turning now to cooperation among unrelated animals, we saw that animals generally do not play a PD-kind of game. Some of the theoretical work spawned by the PD model has actually begun to address these concerns. Models suggest that the ability to select partners, and subsequent selective association with them, favors the evolution of cooperation (e.g. Peck 1993), as does, obviously, communication, for instance in the form of the ability to punish non-cooperators (Boyd & Richerson 1992).

Our discussion was organized according to the three classic explanatory models of cooperation: nepotism, reciprocity and mutualism. However, cooperation among animals observed in naturalistic conditions often contains a mix of these categories of cooperation. Variables affecting the presence of cooperative behaviors in nature are the number of players and the degree to which the cooperative acts are enacted within a stable local context. Thus, three categories of naturalistic cooperation can readily be recognized: (i) dyadic cooperation, but without social relationships between the partners, (ii) dyadic cooperation in the context of long-term relationships, and (iii) group-level cooperation in stable groups. Table 1.1 groups some of the examples mentioned in the text so far into these three categories, which are characterized by different threats to the stability of cooperation and hence by different behavioral solutions. Not included in the table are the cases where the solutions do not require behavioral action. These include, for dyadic, anonymous cooperation, byproduct mutualisms, such as a selfish herd effect, and for group-level cooperation, cases of mutualism that are stabilized by structural safeguards against free-riding, such as in the amoeba example.

The value of this table should be heuristic, in that it suggests new approaches. Noë (1990), Hammerstein (2003b), Silk (2002b) and others have pointed at the large gap between theory and empirical observations on animals that are mobile and form long-term relationships. Market models provide a promising avenue (Bshary & Noë 2003) to examine partner choice, especially for dyadic cooperation, but active attempts to model real cases might produce new questions for

Table 1.1. Cooperation among independent animals, such as non-human primates.

Category	Threat	Solutions	Examples, labels
Dyadic, without relationships	Absence of reciprocation	1. Partner control: parceling altruism in small packets; 2. Leave	1. Strict reciprocity; 2. Anonymous donations in humans
Dyadic, in long-term relationship	Asymmetric investment	1. Partner change; 2. Negotiation about payoffs	1. Unilateral altruism toward kin; 2. Friendships, alliances
Group-level	Free-riding; taking of larger shares than 'fair'	1. Switch groups; 2. Altruistically expel or punish free-riders; 3. Altruistically reward cooperators	1. Helping by paying for staying + group augmentation; 2. Group-level mutualisms (e.g. communal defense, nursing); 3. Generosity: establishment of reputation through alarm calling, mobbing, food sharing, etc.

other aspects of cooperation, for example for partner control through communication (see also Bowles & Hammerstein 2003). For primates, we need more work on the natural history of alarm calls and mobbing; it is quite conceivable that a rich interpretation is needed, in which animals undertake these acts in order to establish and maintain reputation (of the immediate, not the third-party variety), which would favor their being accepted as group members or as partners in dyadic bonds (cf. Maklakov 2002).

Of special interest to biological anthropology would be a more precise mapping of differences between humans and great apes. We noted that great apes seem to be more prone to collaboration with non-kin than other primates; this is not only true for male chimpanzees but also for female bonobos (Hohmann et al. 1999). Great apes also show remarkable tolerance toward curious imitators of their skills (van Schaik 2003). If these copiers are unrelated, and if their acquiring the skills improves their fitness, are those who allow themselves to have their skills copied not altruistic? Generous food sharing, as seen in chimpanzees (Boesch, this volume; Mitani, this volume) can be regarded as investments into a good reputation (as in humans: Bird et al. 2001), but if reputation is the key to understanding group-level cooperation, this might explain why chimpanzees show more of it than other non-human primates. The same holds true for the possibility of norms in chimpanzees. It is therefore conceivable that more detailed work on primates will show that cooperation provides yet another case where the actual gap between human and non-human primates is less wide than generally perceived.

Acknowledgments

We thank Tim Clutton-Brock, Barbara König, Ronald Noë and Claudia Rutte for discussion and comments.

Part II
Kinship

Chapter 2

Practicing Hamilton's rule: kin selection in primate groups

JOAN B. SILK

" 'Kin selection' ...has become a bandwagon, and when bandwagons start to roll attitudes sometimes polarise. The rush to jump on provokes a healthy reaction. So it is today that the sensitive ethologist with his ear to the ground detects a murmuring of skeptical growls, rising to an occasional smug baying when one of the early triumphs of the theory encounters new problems. Such polarisation is a pity..." (Dawkins 1979, p. 184).

2.1 Introduction

The kin selection bandwagon has been rolling for several decades in behavioral ecology, gaining speed and momentum. There is abundant evidence for kin biases in behavior (nepotism) in many animal species, and most behavioral ecologists have taken this as crude, but convincing evidence that kin selection (Hamilton 1964) is operating. However, Dawkins' 'sensitive ethologist' may have begun to detect a new chorus of 'skeptical growls' about the role of kin selection in shaping cooperative behavior in animal groups (Chapais 2001, Clutton-Brock 2002). These critiques focus on three issues. First, in cooperatively breeding vertebrates, behavior that has been commonly attributed to kin selection may actually enhance individual fitness directly (Clutton-Brock 2002). In a broad range of species, some types of behavior that have been categorized as altruistic may actually benefit the actor (Chapais 2001, Chapais & Bélisle 2004, Chapais, this volume). In these cases, mutualism or simple self-interest may operate, not kin selection. Second, competition between relatives may counteract the effects of kin selection, limiting the prospects for the evolution of altruism by this route (West et al. 2002). Third, kin biases in behavior may be the product of processes besides kin selection. If social interactions are enhanced by familiarity, then initial kin biases in association patterns may lead to kin biases in behavior, but the dynamics of interactions may be shaped by mutualism, reciprocity, or individual benefits rather than kin selection (Chapais 2001, Chapais & Bélisle 2004).

All three of these issues are potentially relevant to non-human primates. Cooperative breeding occurs among the callitrichids, and group members share the burden of carrying, protecting, and provisioning offspring (Goldizen 1987b). It is not clear whether kin selection fully explains the deployment of help in these species. Within-group competition is thought to be a major force influencing the evolution of social organization in primate groups (e.g. Sterck et al. 1997). In cooperatively-breeding species, competition among female kin is manifest in reproductive inhibition. In solitary prosimians, the balance between cooperation and competition among mothers and their daughters influences social organization (Kappeler et al. 2002), and may shape birth sex ratios (Clark 1978). Female philopatry, strong female bonds, and matrilineal dominance hierarchies are expected to occur when it is beneficial for females to form nepotistic alliances in within-group contests. Finally, primate females form strong and enduring ties with their offspring, and this may inadvertently generate high rates of association among siblings and other types of maternal kin (Chapais 2001, Chapais & Bélisle 2004). Kin biases in association may be reflected in kin biases in behavior, but interactions among kin may be regulated by processes other than kin selection.

Here, my goal is to review the evidence for nepotism and kin selection in primate groups from the perspective of these recent critiques. This exercise is complicated by the fact that it is very difficult to quantify the costs and benefits of behavior in long-lived animals like non-human primates. It is usually impossible to be certain whether particular categories of behavior or specific instances of particular kinds of behavior are altruistic, mutualistic, or selfish. Attempts to estimate the costs incurred and benefits derived from social interactions are further confounded by the fact that cost-benefit ratios may not be constant over time or across individuals. The first scrap of meat that one chimpanzee shares with another may be more valuable than the second or third bit. In addition, the benefits derived from particular types of service may depend on the recipient's age or reproductive status. Thus, agonistic support may have greater value to an adolescent female who is trying to establish her dominance position than to an aged female whose reproductive value has declined almost to zero. The phenotypic gambit (Grafen 1991), which has been one of the behavioral ecologist's most productive tools, falters when there is so much uncertainty about the fitness consequences of behavior.

Comparative analyses of the extent of kin biases in behavior are also complicated by difficulties in estimating relatedness. Until recently, measures of relatedness were based entirely on genealogical data. Kinship was traced only through maternal lines because information about paternal kinship was generally unavailable. Only recently, have molecular genetic techniques made it possible to generate true estimates of genetic relatedness (Morin & Goldberg 2004, Woodruff 2004). Further, although relatedness varies continuously from zero to one, investigators often consider kinship as a categorical variable. For example, 'close kin' may be compared to 'distant kin', or 'kin' may be compared with 'non-kin'. For the most part, these categories are based on arbitrary criteria about what constitutes close or distant kin ties or what distinguishes kin from non-kin, and do not necessarily represent categories that are meaningful to the animals themselves.

Nonetheless, I will argue that there is evidence for kin selection in primate groups, and that this evidence extends beyond the extensive list of examples of matrilineal kin biases in cooperative behavior. At the same time, as many of the other chapters in this volume attest, there is growing evidence that kin selection is not the only force shaping cooperation in primate groups.

2.2 Cooperative breeding in primates: one for all or all for kin?

Kin selection is usually invoked to explain the evolution of cooperative breeding in vertebrate species (Emlen 1997). Griffin & West (2003) recently completed a meta-analysis of helping behavior in vertebrate species. Their analysis demonstrates that individuals discriminate among their relatives, and the extent of discrimination is linked to the benefits derived from helping. However, Griffin & West's analysis was based mainly on data from cooperatively-breeding birds, and included few mammalian species. Clutton-Brock (2002) argues that much of the 'cooperative' behavior that occurs in communally-breeding mammals may actually enhance individual fitness. This conclusion is based partly on the observation that genetic relatedness does not influence the extent of helpers' contributions as predicted by Hamilton's rule. For example, in meerkats, *Suricata suricatta*, immigrants who are not related to other group members, take their turn at standing guard as often as natal group residents (Clutton-Brock et al. 1999b). Careful analyses of helper's activity budgets and weight gains indicate that the cost of helping is quite low. In addition, group size is positively associated with reproductive performance. Thus, Clutton-Brock argues that the benefits of living in large groups, and the relatively low costs of helping, may make it profitable for individuals to perform behaviors that benefit other group members.

Callitrichids typically live in small, territorial, bisexual groups of 4-15 individuals (French 1997, Tardiff 1997). Although there is considerable variation in the mating systems of these species, most groups of most species contain only one breeding female, one or more breeding males, and some number of non-breeding helpers (Dietz 2004, French 1997, Tardiff 1997). Callitrichids are unusual among anthropoid primates, because females typically give birth to twins and can produce two litters per year. Group members cooperate in infant care and defense of home ranges and important resources.

In females, high reproductive skew is maintained by pheromonal reproductive suppression, behavioral inhibition, or inbreeding avoidance (Garber 1997). For females, high reproductive skew is probably linked to competition for helpers. In free-ranging tamarins and marmosets, the reproductive success of females is influenced by the presence of helpers, particularly adult males (Garber 1997, Dietz 2004). In one population of common marmosets, the breeding efforts of subordinate females were only successful if their infants were born at a time that the dominant female did not have dependent infants (Digby 1995). In two cases, dominant females have killed infants produced by subordinate females (Digby 2001).

There is good evidence that helping is costly. Helpers suffer reductions in feeding time, vigilance and sociality (Achenbach & Snowdon 2002). In captive settings, males lose weight while they are caring for infants, sometimes losing as much as 10% of their bodyweight, and the extent of their weight loss is linked to the number of other available helpers (Achenbach & Snowdon 2002). This weight loss is striking because these males lived in undemanding environments with free access to food, no need to move from one feeding site to another, and no predators.

Both kinship and direct benefits may sustain cooperation in callitrichid groups (Garber 1996). Kinship seems to affect the extent of reproductive skew among females in one population of golden lion tamarins. Approximately 10% of females share reproduction with subordinate females for one or two years. Females are most likely to share breeding with their own daughters, less commonly with sisters, and rarely with unrelated females. Only mothers and daughters were both successful in rearing infants in the same season (Baker et al. 2002, cited in Dietz 2004). When sisters or unrelated females bred, only one female's infants survived. Overall, the number of surviving infants per female is lower in groups in which two females breed than in groups in which only one female reproduces (Dietz & Baker 1993). However, mothers gain inclusive fitness benefits when their daughters reproduce, particularly when their daughters are unrelated to group males, as was typically the case when both mothers and daughters mated. Demographic models suggest that the cost of allowing daughters to breed is relatively low when unrelated mates are available and daughters do not pose a threat to the mothers' social status within their groups.

Infant care is one of the most important forms of cooperation in callitrichid groups. If groups are composed of a breeding pair and their descendant offspring, then alloparental care toward all group offspring may be favored by kin selection. On the other hand, if the genetic composition of social groups is more complicated, then we might expect helpers to discriminate among infants. Genetic data are now available for a small number of wild callitrichid groups. In two groups of common marmosets, *Callithrix jacchus*, the dominant male fathered all of the infants and all group members were closely related (Nievergelt et al. 2000). Elsewhere, the pattern seems to be more variable. Six of 10 wild groups of common marmosets included adults that were unrelated to the breeding pair (Faulkes 2003). In a genetic survey of the composition of eight groups of wild tamarins, three included adults unrelated to the breeding pair and a fourth group included an adult male only distantly related to other group members (Huck et al. 2004).

The effect of kinship on helping behavior is variable. In one group of wild tamarins, infants were carried equally often by related and unrelated helpers (Savage 1990, cited by Dietz 2004). In seven groups of golden lion tamarins (*Leontopithecus rosalia*), kinship influenced the amount of help provided by males, but not by females (Baker 1991, cited in Dietz 2004). In captivity, helpers provide care for all group infants, and do not discriminate on the basis of relatedness (Cleveland & Snowdon 1984, Wamboldt et al. 1988, Achenbach & Snowdon 1998, reviewed by Dietz 2004) or even familiarity (Cleveland & Snowdon 1984). Dietz (2004) concludes that, "although there is considerable individual variation in degree of help provided, callitrichid alloparents generally care for all infants in the groups, regardless of kinship".

Clutton-Brock's analysis of cooperative breeding in vertebrates emphasizes the importance of documenting the genetic relationships among group members, the contributions of individuals to communal activities, and the opportunity costs associated with helping. It is not yet known how relatedness influences the amount of work that group members perform in callitrichid groups. However, the genetic data raise the possibility that helpers may derive individual benefits from cooperation. This is an important area for future work in cooperatively-breeding primates.

2.3 Competition among kin

Socioecological theory tells us that gregariousness will evolve when the benefits of associating with conspecifics, such as improved ability to defend access to food or reduced risk of predation, outweigh the costs which include greater competition over access to resources from group members, cuckoldry, contagion, infanticide, and cannibalism (Krebs & Davies 1993). The social systems that we see in primate groups reflect the balance of these forces. Even when primates live in social groups, competition among kin is not entirely muted.

The idea that competition among kin may offset inclusive fitness benefits derived from helping kin is not a new one. The effects of density-dependent mortality on the evolution of social behavior have been considered before (Hamilton 1970, Boyd 1982). The local resource competition model of sex ratio adjustment (Clark 1978, Silk 1984) is explicitly built on the assumption that competition over access to highly concentrated resources reduces the benefits derived from producing daughters, who compete locally, and favors sex biases in favor of males, who disperse over longer distance and compete globally.

In most cases, it is impossible to assess the balance between the benefits of associating with kin and the costs of competition among kin because we cannot measure the fitness outcomes associated with particular behavioral acts. But there are some cases in which the tradeoffs are dramatic. For example, in callitrichid groups, subordinates provide valuable help in rearing young, often their own siblings, but do not breed successfully themselves (references above). Helping may be the best response when prospects for dispersing successfully are limited, as in saturated environments, or when helpers are closely related to the resident breeders (Dietz 2004). However, increments to the breeding pairs' fitness are partly offset by decrements to the fitness of related helpers.

Strong nepotism may coexist with competition among kin. For example, Japanese macaque (*Macaca fuscata*) females typically rise in rank over their older sisters, and sisters are eventually ranked in inverse order of their ages. Young females 'target' their older sisters for rank reversals (Chapais et al. 1994), and older sisters sometimes resist vigorously. When females intervene in disputes involving their older sisters and subordinate non-kin, they are as likely to intervene against their sisters as they are to support them. In contrast, when females intervene in conflicts involving kin that are not targeted for rank reversals, females are much more likely to intervene on behalf of their relatives than their opponents (Chapais et al. 1994). Chapais (1995) notes that females "apparently solve the conflict of interest between egotism and nepotism by maximizing their own rank among their kin on the one hand, and by maximizing the rank of their kin in relation to non-kin on the other" (p. 129).

2.4
Kin biases in primate groups

2.4.1
Nepotism among females

Among primate females, kin biases in behavior seem to be common whenever females form lasting associations (Stewart & Harcourt 1987). Female nepotism emerges in the tightly structured matrilineal dominance hierarchies in macaque groups (Chapais 1992), alloparental care in vervet monkeys (*Cercopithecus aethiops*, Fairbanks 1990a), jockeying over recruitment of natal females in red howler groups (*Alouatta seniculus*, Crockett 1984), and the composition of local communities of solitary lemurs (Kappeler et al. 2002). Although dispersal patterns, social organization and reproductive skew structure the genetic composition of social groups and influence females' opportunities to interact with kin, kin biases in females' behavior emerge under a wide range of circumstances.

2.4.1.1 Solitary species

In some 'solitary' primates, kinship structures the social environment of females. Coquerel's dwarf lemurs (*Mirza coquereli*) in the Kirindy Forest of Western Madagascar forage alone during the night and make their own nests where they rest during the day. Females occupy home ranges that may overlap with the home ranges of one to eight other females. Genetic analyses of mitochondrial DNA reveal that closely related females cluster in neighboring home ranges (Kappeler et al. 2002). Females tend to settle near their mothers, creating a matrilineal community of females, which includes females of several generations. Males tend to disperse farther than females, sometimes over large distances. The matrilineal structure of this population emerges from genetic analyses, not from behavioral observations as females are not gregarious and do not participate in cooperative activities or share nests.

In grey mouse lemurs, *Microcebus murinus*, this pattern is elaborated slightly further. The home ranges of matrilineal female kin are clustered in space, as they are in dwarf lemurs, and females forage alone at night. However, during the day females often gather together to sleep. These sleeping groups are relatively stable. New genetic analyses indicate that sleeping groups are composed of matrilineal female kin (Radespiel et al. 2001, Wimmer et al. 2002, Eberle & Kappeler 2004). Similar patterns have been suggested in solitary galagos (Clark 1978).

2.4.1.2 Species with female philopatry

Female philopatry, close female bonds, cohesive matrilines, and high degrees of nepotism are well documented in several genera of Old World monkeys, particularly baboons, macaques and vervets. In these species, females remain in their natal groups throughout their lives and members of several generations may live in the same group at the same time. In baboon, macaque and vervet groups, rates of grooming, association, co-feeding, reconciliation, and coalitionary support are substantially elevated among close maternal kin, such as mothers and daughters (reviewed by Chapais 2001, Gouzoules & Gouzoules 1987, Silk 1987, 2001, 2002a, Chapais & Bélisle 2004, Kapsalis 2004).

These sorts of kin biases may be important to females for at least two reasons. First, sociality enhances reproductive success of female baboons (Silk et al. 2003), and these effects are independent of variation in female dominance rank or ecological conditions. Second, coalitionary support has a direct impact on the acquisition of maternal dominance rank and the formation of matrilineal dominance hierarchies (see Chapais 1992 for a detailed analysis of this process). Juveniles receive support from their mothers and other close kin, and this support enables them to acquire dominance ranks just below their mothers. Coalitionary support is generally thought to play some role in stabilizing dominance relationships among adult females (e.g. Silk 1987, Kapsalis 2004), although the existence of stable matrilineal hierarchies in baboon groups where very few coalitions occur among adult females suggests that coalitionary support may not be necessary for maintaining dominance ranks among adults (Henzi & Barrett 1999, Silk et al. 2004). High dominance rank is positively linked to female reproductive success in a number of cercopithecine groups (Harcourt 1987, Silk 1987, van Noordwijk & van Schaik 1999, Altmann & Alberts 2003, Setchell et al. 2003), although significant relationships between dominance rank and reproductive success are not found in all groups (Packer et al. 1995, Cheney et al. 1988, 2004).

We know much less about the effects of kinship on the behavior of females in other species with female philopatry, which include Hanuman langurs (*Presbytis entellus*), capuchins (*Cebus* spp.), blue monkeys (*Cercopithecus mitis*) and squirrel monkeys (*Saimiri sciureus*). These species have not been studied as extensively as the cercopithecine primates, and information about kinship relationships among adult females is available for relatively few individuals in a small number of groups of each of these species (Kapsalis 2004).

In Hanuman langur groups, females typically remain in their natal groups throughout their lives. Unlike cercopithecine primate females, female Hanuman langurs form age-graded dominance hierarchies. Young adult females hold high-ranking positions, while older females occupy lower-ranking positions (Hrdy & Hrdy 1976, Borries et al. 1991). In one long-term study, mothers and daughters groomed one another more often than they groomed all other females or other females of similar dominance rank (Borries et al. 1994). Coalitions among adult females are very rare, but when they do occur, they nearly always involve close kin (Borries et al. 1991, Borries 1993).

Female capuchins typically remain in their natal groups and are very sociable. Females groom, form coalitions with other females, and establish dominance hierarchies. In three groups of wedge-capped capuchins, *Cebus olivaceus*, offspring of high-ranking females inherit their mother's ranks, but offspring of low-ranking females do not. Moreover, mother-daughter dyads do not have the strongest social bonds, unlike what we see in cercopithecine primates. Instead, females seem to have the greatest affinities for unrelated females who hold adjacent ranks (O'Brien 1993, O'Brien & Robinson 1993, Kapsalis 2004). We know little about the nature of kin relationships among white-faced capuchins (*Cebus capucinus*), but at one site, two known mother-daughter pairs groomed often and were closely ranked, while one pair at another site held disparate ranks (Manson et al. 1999).

2.4.1.3 Species with female dispersal

When females disperse from their natal groups, opportunities for nepotism are much more variable. However, selective associations among female kin seem to be common, even when females do not remain in their natal groups. Most mountain gorilla (*Gorilla gorilla berengei*) females emigrate from their natal groups (Stewart & Harcourt 1987), but dispersing females sometimes join groups that contain females who formerly belonged to their natal group. In some cases, females who are likely to be full or half-sisters emigrate together (Stewart & Harcourt 1987). Thus, even though females are not philopatric, nearly 70% of females spend at least some of their reproductive years in the company of female kin (Watts 1996). When females live with related females, they tend to show strong nepotistic preferences. Adult female mountain gorillas spend more time resting and feeding near their relatives than non-relatives, rarely fight with kin, and are more likely to groom and support kin than non-kin (Harcourt & Stewart 1987, 1989, Watts 1991, 1994, 1996).

Female chimpanzees (*Pan troglodytes*) usually leave their natal groups when they reach sexual maturity while males are philopatric (Nishida 1996). However, at Gombe not all females emigrate permanently, and some females maintain close bonds with their adolescent and adult daughters (Goodall 1986, Williams et al. 2002). Recent analyses of the long-term records indicate that mother-daughter dyads spend much more time together than pairs of unrelated females do (Williams et al. 2002).

Adult orangutans (*Pongo* spp.) are largely solitary. There is no evidence that females form affiliative relationships with one another or establish differentiated social bonds (van Schaik & van Hooff 1996). However, at some sites, some females' ranges overlap extensively, while other females' ranges overlap very little. Researchers suspect that males disperse further than females, and that females who share much of their home ranges are close kin (Galdikas 1988, Watts & Pusey 1993, van Schaik & van Hooff 1996, Singleton & van Schaik 2002).

Female hamadryas baboons, *Papio cynocephalus hamadryas*, disperse from their natal groups. Females move from group to group several times during their lives, and often move to groups that include females that they lived with in the past (Sigg et al. 1982, Stammbach 1987). Hamadryas baboons are usually described as cross-sex bonded (Byrne et al. 1989) because females have strong relationships to their 'leader' males and weak links to other females (Abegglen 1984, Stammbach 1987). New data from the Awash National Park in Ethiopia, however, present a different picture. There, females spend as much time socializing with other adult females as they do with resident males, and sometimes socialize with females from other one-male groups (Swedell 2002). It is not known whether females interact at higher rates with kin than non-kin, but levels of genetic relatedness among females are high in other hamadryas groups in the same area, (Woolley-Barker 1998, 1999, cited in Swedell 2002). Moreover, evidence from captive groups and other wild populations suggests that contacts among females from different one-male groups may often involve kin (Swedell 2002).

In the Gambia, female red colobus monkeys, *Colobus badius temminckii*, disperse from their natal groups, often in the company of female age-mates. After dispersal, these sets of females continue to socialize together, and rarely behave aggressively to one another (Starin 1994). Females cooperate in intergroup encounters and collectively attack potentially infanticidal nonresident males. If there is high reproductive skew within red colobus groups in the Gambia, then kinship might underlie high rates of cooperation within these female peer groups.

In red howler (*Alouatta seniculus*) groups, opportunities for nepotism vary over the course of time. New groups are formed when solitary migrating females meet, form ties, attract males, establish territories, and begin to reproduce (Pope 2000a). Female group size is strictly limited, as small groups cannot defend their territories and large groups have difficulty maintaining access to sufficient resources to feed them all and are more vulnerable to male takeovers and subsequent infanticide (Pope 2000a). Recruitment of natal females depends on the number of adult females already present. When groups are smaller than the modal group size, natal females may remain in their natal groups. But when groups are larger than the modal size, all natal females are forced to leave (Crockett & Pope 1993, Pope 1998). Dispersal can be a risky strategy for young females, particularly when habitats are saturated and available territories are limited. Dispersing females do not produce their first infant until they are about 7-years-old (Crockett 1984, Crockett & Pope 1993), while females that breed in their natal groups produce their first infant at the age of 5 years (Crockett & Pope 1993). In addition, females in newly-established groups have fewer surviving infants per year than females in well-established groups do (Pope 2000a).

The high costs of dispersal are linked to intense competition among females over recruitment opportunities for their daughters. Adult females actively harass maturing females in an apparent effort to force them to emigrate. Females actively intervene on behalf of their daughters in these contests, and daughters manage to stay only if their mothers are present in the group (Crockett 1984, Crockett & Pope 1993). Nonrandom recruitment of breeding females influences the genetic relationship among resident females. In newly-established groups, females are unrelated to one another. In long-established groups, the average degree of relatedness approaches 0.5 (Pope 1998, 2000a, 2000b). This has adaptive consequences, as female reproductive success is correlated with the degree of relatedness within their groups (Pope 2000a).

In summary, female sociality is grounded in kinship. Kin biases seem to emerge reliably when females have an opportunity to interact with their relatives, and when there are benefits to be gained by cooperating with other females. For females, nepotism is an integral part of the structure of social life.

2.4.2 Nepotism among males

Evidence for nepotistic associations among male primates is more limited than among female primates. There may be several related reasons for this. If nepotism is linked to philopatry (e.g. van Hooff & van Schaik 1992, 1994, Boinski 1994, Hill 1994), then nepotism among males may be uncommon simply because male philopatry is rare. It is also possible that nepotism among males is uncommon because males derive fewer benefits from cooperating with their relatives than females do (van Hooff 2000, van Hooff & van Schaik 1992, 1994). Finally, it is possible that constraints on paternal kin recognition limit the potential for nepotism among males.

2.4.2.1 Species with male philopatry

Male philopatry and female dispersal characterize members of the genus *Pan* (Nishida & Hasegawa-Hiraiwa 1987), members of the subfamily Atelinae (Robinson & Janson 1987), red colobus monkeys (Struhsaker & Leland 1987), and Central American squirrel monkeys (*Saimiri oerstedii*; Boinski 1994). In these species, kinship among members of the philopatric sex is expected to facilitate the development of close bonds and cooperative behavior. However, there is a key difference between species with female and male philopatry. In species with male philopatry, kinship will accumulate along paternal kin lines. Because paternity is difficult to detect from observational data, this complicates efforts to evaluate the effects of kinship on behavior among males. Moreover, paternal kinship will only influence behavior if animals are able to discriminate between paternal kin and others (see below).

Male chimpanzees form strong and durable social relationships. Males spend much of their time with other males, and males groom, hunt, share meat, aid, and patrol the borders of their territories with one another (reviewed by Nishida & Hasegawa-Hiraiwa 1987, Nishida & Hosaka 1996, Watts 2002). At some sites, maternal brothers form particularly close relationships (Nishida 1979, Goodall 1986), but these dyads may be the exception rather than the rule. Analyses of mitochondrial DNA patterns among chimpanzees in East Africa indicate that males do not selectively associate with maternal kin (Goldberg & Wrangham 1999, Mitani et al. 2000, 2002a). Males do seem to prefer age-mates and males close in rank to themselves. If reproductive skew is high, then age-mates might be paternal kin (Mitani et al. 2002a). However, the average degree of relatedness among males in chimpanzee communities in the Taï Forest and in Budongo, is surprisingly low (Vigilant et al. 2001), potentially reducing the scope for maternal or paternal nepotism among adult males.

In bonobos, *Pan paniscus*, males are the philopatric sex, but bonds among male bonobos are weaker than bonds between males and females or between females (reviewed by Hohmann & Fruth 2002). Like chimpanzees, maternal brothers in bonobo groups do not consistently form close bonds (Furiuchi & Ihobe 1994, Hashimoto et al. 1996).

Male muriquis (*Brachyteles arachnoides*) are well-known for their peaceful temperaments and high degree of tolerance (Strier et al. 2000). Males spend much of their time in proximity to other males, embrace one another, and share sexual access to receptive females (Strier et al. 2002). Aggression among males of the same community is very uncommon. However, limited evidence from the only long-term study of muriquis suggests that maternal kinship is not linked to the quality of male social relationships; the only two known pairs of maternal brothers did not form particularly close bonds (Strier et al. 2002).

Among red colobus monkeys, females disperse from their natal groups. Males may leave their groups temporarily at adolescence, but are rarely able to join new groups; many eventually return to their natal groups. In the Kibale Forest, males groom one another relatively more often than they groom females and jointly fight against members of other groups (Struhsaker & Leland 1987, Struhsaker 2000). In contrast, in the Gambia, social bonds among males are very weak. Adult males rarely groom one another, do not rest together, and do not support one another in agonistic interactions (Starin 1994).

In Costa Rican squirrel monkeys, males remain in their natal groups. Males form close social bonds, and rarely behave aggressively to other males within their groups (Boinski 1994). Males cooperate in sexual inspections of females, intergroup encounters, and defense against predators. Males tend to form the closest bonds with age-mates. In these squirrel monkey groups, the dominant male monopolizes reproductive activity (Boinski 1987), so age-mates might be paternal kin. However, it is not known whether maternal or paternal kinship influences the quality of males' social relationships.

2.4.2.2 Species with male dispersal

In species with male dispersal, males sometimes emigrate in the company of other males. This seems to be particularly common in seasonally-breeding species with well-defined birth cohorts (Pusey & Packer 1987). South American squirrel monkeys (*Saimiri sciureus*) form migratory alliances of two to four males (Mitchell 1994). These sets of males tend to be age-mates and occupy adjacent positions in the dominance hierarchy. One or two males monopolize reproductive activity each year, so age-mates might be paternal relatives (Mitchell 1990, cited in Mitchell 1994). In red-fronted lemurs (*Eulemur fulvus rufus*), the dominant male fathers the majority of infants (Kappeler & Wimmer 2002) and natal males often migrate together (Ostner & Kappeler 2004). Reproductive skew is high in black-capped capuchins and males sometimes disperse with natal kin (Izawa 1994, cited in Strier 2000, Jack & Fedigan 2002, 2004). Peer migration is also observed in vervet monkeys (Cheney 1983a, Cheney & Seyfarth 1983), several species of macaques (van Noordwijk & van Schaik 1985), ring-tailed lemurs (*Lemur catta*), and sifaka (*Propithecus verreauxi*; Pusey & Packer 1987).

At some sites, Hanuman langurs form one-male groups. Males disperse from their natal groups as juveniles, and then join all-male bands. These all-male bands attempt to take over bisexual groups and oust resident males. Typically, the most dominant male in the band becomes the new resident male (Rajpurohit & Sommer 1993), and the others return to the all-male band. Resident males monopolize reproductive activity within their groups, and are thought to sire the majority of infants conceived during their tenure in the group (Rajpurohit & Sommer 1993, Launhardt et al. 2001). Paternal kinship influences dispersal patterns of young males; 85% of juvenile males disperse with paternal brothers, and about half live for some period in all-male groups with their fathers. Mortality is high in the months after emigration; males who leave with their fathers tend to be more likely to survive than other males (Rajpurohit & Sommer 1993).

In other langur species, the demographic composition of groups changes over time. Some one-male groups become two- or three-male groups. As the resident male ages, one or more natal males may remain, creating an age-graded group. Sterck & van Hooff (2000) argue that these cases "are not based on an inability of the dominant male to exclude others, but must be based on tolerance on his part and an interest in their staying" (p. 128). Resident males may tolerate young natal males because they cooperate in intergroup encounters and protect the group's infants from infanticide (Sterck & van Hooff 2000, Steenbeek et al. 2000). Interestingly, in Thomas' langurs, *Presbytis thomasi*, age-graded groups were not formed unless the original resident male's tenure was long enough for his male offspring to mature (Steenbeek et al. 2000). This suggests that nepotism underlies males' tolerance of younger males, and younger males' participation in group defense and protection of infants.

There is limited evidence of nepotistic social behavior among males in species with male dispersal. Juvenile male rhesus macaques (*Macaca mulatta*) associate, affiliate, and form alliances with their maternal brothers while they are in their natal groups (Colvin 1983). Male bonnet macaques (*Macaca radiata*) intervene frequently in conflicts involving other males (Simonds 1974, Rahaman & Parthasarathy 1978) and groom, greet, and huddle with other males at high rates (Sugiyama 1971, Simonds 1974). In one captive group, maternal kinship increased rates of support (Silk 1992a) and decreased rates of aggression (Silk 1994), but did not affect rates of affiliative behavior (Silk 1994). Dispersing macaque and vervet males often join groups that members of their natal group have previously moved to. The presence of familiar, or possibly related, males may enhance males' chances of gaining entry into the group or enhance their chance of achieving high rank after they join (Pusey & Packer 1987). In one case, it is known that maternal siblings tend to support one another in conflicts in non-natal groups (Meikle & Vessey 1981).

Examples of nepotistic behavior can also be found in species in which members of both sexes disperse. In red howler groups, males sometimes form cooperative alliances, and these partnerships often involve paternal kin. When local habitats are saturated and groups are large, single males have difficulty defending groups of females. In these situations, red howler males sometimes form coalitions. Males collectively defend females against incursions by foreign males and males jointly challenge residents for access to groups of females. Some coalitions are composed of related males, often fathers and sons (Sekulic 1983). Kinship enhances the stability of coalitions as pairs of related males stay together about 3.5 times as long as pairs of unrelated males (Pope 1990). In addition, rank relationships are more stable when allies are related than when allies are unrelated. This has important adaptive consequences because the dominant male monopolizes reproductive activity within the group (Pope 1990). Thus, males seem more likely to tolerate their subordinate status and reduced individual fitness when their allies are close kin than when their allies are unrelated.

Nepotistic associations are also suspected to occur in mountain gorillas and hamadryas baboons. Hamadryas baboons typically form one-male units, which are clustered together in clans, and clans are clustered in bands (Stammbach 1987). 'Leader' males are sometimes succeeded and subsequently tolerated by males who are likely to be their sons (Sigg et al. 1982). Although one-male units spend much of their time near one another, leaders of one-male units rarely attempt to take females from other males in their clans and are quite tolerant of the males in their clans. Based on phenotypic similarities, males in the same band are thought to be related (Stammbach 1987).

Gorillas generally form one-male groups, but multi-male groups sometimes occur (Stewart & Harcourt 1987, Watts 2000a). In mountain gorilla groups, multi-male groups can be formed when silverbacks have 'followers', who are younger and initially subordinate to them. In some instances, followers are recruited from the natal group, but they can also come from an all-male group the resident previously lived in. Followers cooperate with dominant males in aggression against males from outside the group (Stewart & Harcourt 1987, Watts 2000a). In the majority of cases, natal followers are the putative sons of the dominant male, or occasionally putative paternal siblings (Watts 2000a). When lowland gorilla (*Gorilla g. gorilla*) males leave their natal groups, they may settle nearby. New genetic analyses indicate that silverbacks in neighboring lowland gorilla groups are often closely related (Bradley et al. 2004). This pattern of relatedness may be linked to the relatively low levels of competition and aggression observed among male silverbacks from neighboring territories.

Callitrichid groups typically contain one breeding female and one or more adult males (Dietz 2004). In one well-studied population of golden lion tamarins (*Leontopithecus rosalia*) in Brazil, multiple adult males were present in groups about half of the time (Dietz 2004). Like red howlers, dominant male tamarins tend to monopolize reproductive activity. In the majority of cases, co-resident males were close kin, and related pairs tended to remain together longer than unrelated pairs (Baker et al. 2002, cited in Dietz 2004).

In summary, evidence for nepotism among males is much more limited than evidence for nepotism among females. This may be partly due to the fact that males often disperse from their natal groups and this limits opportunities for extended relationships with maternal kin. The problems inherent in paternal kin recognition may limit nepotism among males in species in which males are the philopatric sex. Nonetheless, it is striking that cooperation among males is often grounded in kinship, although chimpanzees seem to represent an exception to this general rule.

2.5 Processes underlying nepotism

Kin biases in grooming, coalition formation, food sharing, and association patterns are common in primate groups (references above). In general, primates interact at higher rates with kin than non-kin, and interact more often with close kin than with distant kin. Both these patterns are consistent with qualitative predictions derived from Hamilton's rule. When $r = 0$, the inequality cannot be satisfied; and the conditions for altruism become considerably more restrictive as r declines. When $r = 0.5$, the degree of relatedness between mothers and offspring or full siblings, the benefits must exceed just twice the costs in order to satisfy the inequality $br > c$. But for the offspring of half-siblings (half-cousins), who are related by $r = 0.0625$, the benefits to the recipient must exceed 16 times the cost to the actor. In nature, situations in which the benefits to the actor are so many times greater than the costs to the actor may be rare; so altruism may commonly be limited to relatively close kin.

Even though we cannot measure or even make reasonable estimates of the fitness consequences of these kinds of behavioral interactions on actors or recipients, and we cannot evaluate Hamilton's rule precisely, most primatologists are confident that these kin biases are the product of kin selection. However, there are at least two other processes that could generate high rates of interaction among kin. Kin biases could reflect an attraction to animals of similar rank or kin biases could be a byproduct of extended associations between mothers and their offspring. I consider each of these alternatives in more detail below.

2.5.1 Nepotism reflects an attraction to individuals of similar rank

In species with matrilineal dominance, hierarchies may reflect an attraction to animals of similar rank, not an attraction to kin *per se*. This argument was first proposed by Seyfarth (1977, 1983). He hypothesized that females might exchange grooming for support in agonistic conflicts. Because the most powerful females make the most attractive allies, females will direct their grooming efforts toward the highest-ranking females in their groups. However, time budgets constrain the amount of time available for grooming (Dunbar 1991), so females must compete for access to the highest-ranking females. High rank confers competitive advantages, so females are forced to settle for grooming partners of adjacent rank and trade grooming in kind. In species that form matrilineal dominance hierarchies, kin occupy adjacent ranks. Thus, kin biases in grooming reflect the outcome of competition for high-ranking partners, not the action of kin selection. However, Seyfarth (1983) recognized that both kin selection and reciprocity might contribute to high rates of interactions among females of adjacent ranks. Later, de Waal (1991b) and de Waal & Luttrell (1986) suggested that females will be attracted to those who most closely resemble them in rank, age and kinship because these individuals are the ones with whom exchange relationships are most likely to be profitable.

There has been considerable debate about Seyfarth's hypothesis. Monkeys trade grooming for support in some situations and they interact at high rates with those of similar rank. Vervets are more likely to respond to tape-recorded screams of animals that have previously groomed them than they are to respond to the tape-recorded screams of the same partners under other circumstances (Seyfarth & Cheney 1984). Similarly, when disputes among long-tailed macaques are artificially induced, females are more likely to support monkeys who have recently groomed them than they are to support others (Hemelrijk 1994). However, in nonexperimental settings, associations between grooming and support have been harder to find (reviewed by Schino 2001), producing some skepticism about the importance and regularity of this process in nature (Henzi & Barrett 1999).

However, there are several reasons to suspect that even if females do trade grooming for support, this is not sufficient to fully explain the patterns of grooming and affiliation that we see in nature. First, in Seyfarth & Cheney's experiment, the contingency between grooming received and subsequent responses to screams disappears when partners are close kin. That is, close kin respond whether or not they have been groomed recently. This suggests that reciprocity may promote high rates of interaction among unrelated females of adjacent rank, while kin selection may promote high rates of interaction among maternal kin.

Second, in baboons and macaques, females' attraction to maternal kin is stronger than their attraction to females of adjacent rank. For example, my colleagues and I compared the effects of rank distance and maternal kinship on rates of grooming among female baboons in the Moremi Reserve of Botswana. We found that females groomed close maternal kin at much higher rates than they groomed unrelated females who occupied adjacent ranks (Silk et al. 1999). Similarly, female bonnet macaques (Silk 1982) and rhesus macaques preferentially groom maternal kin, and these preferences persist when rank distance is controlled (Kapsalis & Berman 1996a, 1996b).

Third, dominance rank and maternal kinship are disassociated in Hanuman langur groups in Jodphur, but strong maternal rank effects persist. Adult females groom their own daughters at higher rates than they groom other females in their daughters' rank class, and adult females groom their mothers at higher rates than they groom other females in their mother's rank class (Borries et al. 1994). Nepotistic biases are also reported in gorillas, another species in which kin do not occupy adjacent ranks (Watts & Pusey 1993).

Taken together, these data demonstrate that matrilineal kin biases in grooming are not simply an artifact of females' attraction to animals of similar rank. At the same time, kin biases may be amplified in some cases by the correlated effect of attraction to animals of similar rank.

2.5.2 Nepotism is a byproduct of extended maternal ties

Chapais (2001, this volume) and Chapais & Bélisle (2004) have recently suggested that some forms of nepotism may be byproducts of mother-infant association patterns. They point out that mothers form close and enduring ties with their older offspring. As mothers wean one infant and produce another, they continue to associate with their juvenile offspring. Mother-daughter ties continue into adulthood. Maternal siblings are drawn together by their joint association with their mother. If familiarity promotes the formation of social bonds, by simply creating ample opportunities to interact or providing information about the reliability or motivational state of potential partners, then kin biases in interactions patterns among siblings, cousins, grandmothers and grandchildren, and other kinds of maternal kin may emerge. The dynamics of social interactions among maternal kin may be regulated by reciprocity, mutualism, or self-interest, not by kin selection.

This hypothesis is plausible. Extended maternal associations are thought to be the foundation of maternal kin recognition in primates and other mammals (Sherman et al. 1997). This means that it is difficult to untangle the evolutionary mechanisms that underlie kin biases in many contexts. We can neither eliminate nor confirm the possibility that nepotism is the product of kin selection. However, there are certain instances in which we can confidently eliminate other kinds of explanations for kin biases. These include: (i) mother daughter rank reversals in baboons, (ii) paternal kin biases in behavior among baboons and macaques, and (iii) examples of unilateral altruism.

2.5.2.1 Mother-daughter rank reversals

In cercopithecine primate species, females acquire ranks just below their mothers early in their adult lives. As young females mature, they usually rise in rank over their older sisters, creating matrilineal units in which mothers rank above all their daughters, and sisters are ranked in inverse order of their age (reviewed by Chapais 1992). The rank ordering among sisters may reflect their present reproductive value; reproductive value is high just after females mature, and then declines (Chapais & Shulman 1980).

Combes & Altmann (2001) applied a similar logic to explore the timing and pattern of rank changes between mothers and daughters in baboon groups. Female baboons usually maintain their ranks throughout their adult lives, but some females drop in rank well before they die. They generated two hypotheses to account for mother-daughter rank reversals. First, senescence might contribute to rank loss. As females get older, their physical powers may decline, making them less able to maintain their rank. If this is the case, all females should be equally vulnerable to rank challenges as they grow older. However, there is another possibility. Kin selection might favor 'consensual' rank reversals among maternal kin. As mothers grow older, their reproductive value declines; the reproductive value of their daughters follows the same trajectory but is offset in time. Thus, the senescence hypothesis predicts that rank decline will be independent of the presence of kin, while the kin selection model predicts that rank decline will depend on the presence and age of kin, particularly daughters.

Combes & Altmann used Hamilton's rule to derive a prediction for the timing of kin-selected rank reversals between mothers and daughters. They set b equal to the reproductive value of the female who wins the rank challenge (VW) and c to the reproductive value of the female who loses the rank challenge (VL). So, rank reversals would occur when $br > c$, or VW/VL > 2. Note that females are not expected to drop in rank below unrelated females, for whom $r = 0$.

Combes & Altmann were able to test their prediction using the extensive long-term demographic data from the Amboseli baboon population. They found that the threshold for mother-daughter rank reversals is reached at about 15 years of age for females who have daughters approximately 8-12 years younger than themselves, but is not reached until about 17 years for females who have daughters just six years younger than themselves. Females who have no adult daughters typically maintain their rank throughout their lives. However, for females who have adult daughters, the picture is very different. Females with mature daughters are much more likely to decline in rank. This pattern is particularly striking because we might expect the presence of mature female kin, who provide coalitionary support, to enhance female dominance rank and enable older females to maintain their ranks. However, females without mature daughters are more likely to maintain their rank than females with mature daughters.

This set of results cannot be explained as a byproduct of early kin biases in association between mothers and their offspring. Instead, the occurrence and temporal pattern of rank decline fits predictions derived from Hamilton's rule.

2.5.2.2 Paternal kin preferences

Conventional wisdom tells us that primates do not recognize paternal kin (Rendall 2004). According to the standard view, paternal kin recognition is precluded because females often mate with several different males near the time of conception, and primates cannot make use of phenotypic cues of relatedness, such as MHC. This conclusion is supported by negative results from a number of studies of paternal kin recognition (reviewed by Rendall 2004). However, the conventional wisdom must be reconsidered in light of recent evidence that shows that macaques and baboons can discriminate between paternal kin and non-kin.

Female macaques and baboons seem to recognize and interact preferentially with their paternal half-sisters. Female baboons in Amboseli groom, rest, and associate with paternal half-siblings as often as they do with maternal half-siblings, and prefer both types of siblings to non-kin (Smith et al. 2003). Similarly, in Cayo Santiago, female rhesus macaques associate and groom with maternal half-sisters at much higher rates than with paternal half-sisters, but they interact with paternal half-sisters at higher rates than with true non-kin (Widdig et al. 2001, 2002). Paternal kin discrimination does not extend to all behaviors. Females do not support paternal half-sisters at higher rates than non-kin, even though they selectively groom them (Widdig 2002).

This pattern might be a byproduct of early biases in association patterns if mothers selectively associate with the fathers of their offspring. In baboons, mothers of newborn infants form close associations with adult males (reviewed by Palombit 1999) who may be the fathers of their infants. If two females associate with the same male at the same time, their infants may develop close ties. However, this explanation does not seem to explain paternal kin recognition in rhesus macaques. Females' associations with one another do not map onto their mother's associations with one another when they were infants (Widdig et al. 2001).

In Amboseli, male baboons selectively support their own offspring in agonistic disputes (Buchan et al. 2003). Males provide a greater fraction of support to their own immature offspring than they do to unrelated immatures. Males are more likely to support their own offspring than they are to support unrelated offspring of former mates (females that they mated with, but whose infants were sired by other males). In contrast, males do not distinguish between unrelated offspring of former mates and unrelated offspring whose mothers they did not mate with. It is not entirely clear how males are able to recognize their own offspring, but it seems likely that behavioral cues may play an important role.

There is considerable variation in the extent to which males monopolize access to females near the time of conception. The proportion of a female's available consort time that a male monopolizes is linked to the probability that the same male will father her infant. The extent of male monopolization is also linked to the probability that males provide care for offspring. This suggests that males or females may make use of information about prior mating behavior to assess paternity. However, we cannot exclude the possibility that phenotypic cues play some role in paternal kin recognition.

Additional evidence for paternal kin recognition and paternal kin biases in behavior comes from the growing body of data on infanticide in primate groups. Infanticidal males selectively target other male's infants and avoid killing their own infants (reviewed by van Schaik 2000). Males also selectively defend their own infants from attacks by other males (Borries et al. 1999). These data imply that males are able to distinguish their own offspring from other male's offspring and channel investment selectively to their own offspring.

2.5.2.3 The distribution of unilateral altruism

Functional analyses of the deployment of social behavior are plagued by three problems. First, there are relatively few kinds of affiliative behaviors that occur often enough to quantify at the dyadic level (Cords 1997). Second, there are considerable uncertainties about fitness consequences of the behaviors that do occur often enough to analyze, i.e. grooming and proximity. Third, high rates of interaction within related dyads may be the product of kin selection or reciprocal altruism. To avoid these problems, we need to focus on unilateral behaviors with clear costs to actors and obvious benefits to recipients. There are at least two candidates for unilateral altruism in primate groups – food sharing and coalitionary support.

Females are generally more tolerant of kin than non-kin during feeding (de Waal 1986a, Stewart & Harcourt 1987, Bélisle & Chapais 2001) and more active forms of food sharing are often biased in favor of close kin (McGrew 1975). These sorts of data are often interpreted to mean that food sharing is favored by kin selection. However, these kinds of data are difficult to interpret because the costs and benefits of sharing are not easy to measure. If animals only give up food when they are satiated, then food sharing may not be altruistic. Moreover, food sharing may be the best response to scrounging in some situations (Blurton Jones 1984, Stevens 2004), not an unconstrained form of altruism (Chapais & Bélisle 2004).

In order to examine the evolutionary processes underlying food sharing more carefully, Bélisle & Chapais (2001) conducted a series of experiments on Japanese macaques in the laboratory. In these experiments, food could be monopolized by one individual. The frequency of co-feeding, or food sharing, increased as maternal relatedness increased (Bélisle & Chapais 2001), but the extent of nepotism was limited. Females shared selectively with their daughters, granddaughters and sisters, but not their aunts or nieces. However, when the costs of defending food were reduced by altering the experimental protocol, dominant females become more selfish (Bélisle 2002, cited in Bélisle & Chapais 2001). These data suggest that unilateral altruism is nepotistic, as Hamilton's rule predicts, and that females are sensitive to benefit-cost ratios.

Coalitionary support in agonistic disputes is often considered to be a form of altruism (e.g. Bernstein 1991, Silk 1992a, Bradley 1999). However, several authors have pointed out that most interventions are aimed at lower ranking opponents (Bernstein 1991, Chapais 2001). If subordinates are unlikely to retaliate against higher-ranking individuals, as is the case in groups with rigid dominance hierarchies, then intervention against lower-ranking targets may not be very costly to allies. In fact, the benefits gained by intervening against a lower-ranking opponent may outweigh the costs, making this type of intervention a form of mutualism (Datta 1983b, Chapais 2001). If intervention against subordinate opponents is shaped by mutualism, then we would not necessarily expect nepotistic biases in these interactions. However, intervention is strongly biased toward maternal kin in macaques (Kaplan 1977, 1978, Kurland 1977, Massey 1977, Silk 1982, Berman 1983a, 1983b, 1983c, Chapais 1983, Cheney 1983b, Colvin 1983, Datta 1983a, 1983b), baboons (Walters 1980, Pereira 1988, 1989, Silk et al. 2004), vervets (Hunte & Horrocks 1986), gorillas (Harcourt & Stewart 1987, 1989, Watts & Pusey 1993), Hanuman langurs (Borries et al. 1991) and ring-tailed lemurs (Pereira 1993). At the very least, these results suggest that kin selection supplements mutualism in the evolution of intervention behavior.

It is generally agreed that when individuals intervene against animals higher ranking than themselves, they risk retaliation and incur more substantial costs. Support against dominant opponents is strongly kin-biased (Kurland 1977, Watanabe 1979, Walters 1980, Silk 1982, Chapais 1983, Cheney 1983a, Hunte & Horrocks 1986, Netto & van Hooff 1986, Pereira 1988, 1989, Harcourt & Stewart 1989, Chapais et al. 1991). Moreover, monkeys are more likely to intervene on behalf of close kin against higher-ranking opponents than on behalf of distant kin (Datta 1983c, Chapais et al 1997). Much of this support is performed on behalf of immatures, who cannot effectively reciprocate.

2.6 Summary and conclusions

Kin biases in behavior are common among non-human primates, ranging from tiny grey mouse lemurs who forage alone but sleep in hollow trees with their mothers, sisters and nieces, to female baboons who selectively groom and associate with maternal and paternal kin, and red howler fathers and sons who jointly defend access to groups of females. The patterning of cooperative activities generally fits qualitative predictions derived from Hamilton's rule. That is, cooperation is more common among kin than non-kin, and the most costly forms of cooperation are reserved for close kin. But this does not mean that kin selection operates in every case or that kin selection is the only mechanism promoting cooperation. In cooperatively-breeding primates, not all helpers are closely related to the infants that they care for. In these species, individual benefits may favor cooperation, much as has been suggested for other cooperatively-breeding mammals. In many primate species, extended ties between mothers and their offspring may promote the development of social relationships among maternal kin, creating nepotistic biases which are sustained through reciprocity or mutualism. In addition, the inclusive fitness benefits derived from associating with kin may be offset by costs of competition. These factors must be taken into account when weighing the importance of kin selection in the evolution of cooperation in primate groups. However, there are certain forms of cooperation for which kin selection seems to be the only plausible evolutionary mechanism. These include consensual mother-daughter rank reversals in baboon groups, affinities among paternal half-sisters in macaque and baboon groups, and coalitionary support against higher-ranking opponents. Taken together, the evidence suggests that kin selection has played an important role in the evolution of cooperation in primate groups. Hamilton's insights about the evolution of cooperation have transformed our understanding of primate societies.

There are important gaps in our knowledge of nepotism in primate groups. We know a lot about kin biases in behavior in a small number of closely-related species, particularly baboons, macaques and vervets. This body of data provides us with a richly-detailed case study of how kinship is woven into the fabric of female monkey's lives. However, the cercopithecine primates share a common phylogenetic history (DiFiore & Rendall 1994) and represent a single branch in the evolutionary history of the primate order. Much less is known about the extent and pattern of nepotism in most prosimians, New World monkeys, colobines, or the great apes. What we do know suggests that maternal kin biases in behavior emerge whenever females form stable associations with their relatives. Even for solitary prosimians and orangutans, kinship structures the communities of females. However, the extent and consistency of kin biases across the primate order is difficult to evaluate because we know so little about so many species.

Most of our analyses are limited to maternal kinship relationships, but it is becoming clear that nepotism may extend to paternal kin. In multi-male groups of macaques and baboons, females preferentially affiliate with their paternal half-sisters. In red howlers, paternal ties between fathers and sons facilitate cooperation. In Amboseli, male baboons selectively support their own juvenile offspring in agonistic encounters. Male silverback gorillas in neighboring territories that maintain peaceful relationships are typically close paternal kin. In a range of infanticidal species, males protect their own offspring and launch attacks on unrelated infants.

There is reason to suspect that paternal kinship might facilitate nepotism in other species as well. Any primate species with high reproductive skew and well-defined birth cohorts are likely candidates for paternal kin recognition and paternal nepotism. Migratory alliances of age-mates are observed in a number of primate species with seasonal reproduction, including vervets, rhesus macaques, Hanuman langurs and red colobus. Affinities among age-mates, as seen in male squirrel monkeys in Costa Rica, male chimpanzees at Ngogo, and female red colobus monkeys in the Gambia, may reflect affinities for paternal half-siblings.

This work draws attention to the mechanisms underlying kin recognition in primates. There is a strong consensus that kin recognition in primates is based on familiarity, not on phenotypic cues such as MHC (e.g. Rendall 2004). This point of view is based partly on laboratory studies designed to decouple kinship and familiarity. In these studies, monkeys are generally unable to discriminate between unfamiliar kin and non-kin (Rendall 2004), suggesting that kin recognition is based on familiarity. The new data from the field suggest that monkeys may rely on multiple cues to assess relatedness in nature (Buchan et al. 2003). Familiarity, age-similarity, mating history, and phenotypic features may all contribute to kin recognition. Like other kinds of signals in nature, redundancy of signals about kinship may enhance their accuracy.

There is no reason to expect that kin selection will be the only process shaping the evolution of social behavior in primate groups. In cooperatively-breeding primates, related and unrelated helpers participate in infant care and other cooperative activities. Animals may gain direct benefits from helping others, or benefits may be reciprocated. Mutualism and reciprocity may often complement kin selection. If animals choose related partners for mutualistic interactions, they will gain direct benefits as well as inclusive fitness benefits. Similarly, kinship facilitates the initiation and maintenance of reciprocal interactions. Thus, kin selection, mutualism and reciprocity may jointly contribute to the high rates of cooperation among kin that we see in primate groups. This may create unwelcome complications when we try to identify the mechanisms that shape cooperation in primate groups, but nature is not arranged for our convenience.

To more fully understand the role that kin selection plays in the evolution of cooperation in primate groups, we will need to take advantage of all the tools at our disposal. Evolutionary theory arms us with quantitative and qualitative predictions about the dynamics of cooperative interactions that evolve through individual selection, kin selection, mutualism and reciprocity. Molecular genetics provides us with the methods to assess relatedness systematically, a critical parameter in Hamilton's rule. Fieldwork conducted on a range of primate species allows us to examine the effects of social structure, demography and ecological conditions on the dynamics of cooperation among males and females. Statistical methods, which allow us to cope with the complexities of dyadic interactions, will help us to unravel the causal processes shaping the patterns in our data. In carefully-designed experiments, we can examine the effects of manipulating benefit-cost ratios and evaluate the importance of previous interactions on future behavior. It will take careful, critical and creative work to make progress, but the first steps have already been taken.

Acknowledgments

I thank Peter Kappeler for the invitation to participate in the Fourth Göttinger Frielandtage and the opportunity to contribute to this edited volume. I thank Bernard Chapais, James Dietz, Ellen Kapsalis, Drew Rendall, and Jeff Stevens for providing me with copies of forthcoming papers. Bernard Chapais posed the challenge for interpreting evidence about nepotism in primate groups, and much of my thinking about this issue is derived from his thoughtful critique. I thank my colleagues, Susan Alberts and Jeanne Altmann, for sharing access to the Amboseli baboon groups, a venue in which these themes are played out on a daily basis. This project was funded in part by grants from the National Science Foundation (BCS-0003245), the LSB Leakey Foundation, the National Geographic Society, and the UCLA Academic Senate.

Chapter 3

Kinship, competence and cooperation in primates

BERNARD CHAPAIS

3.1 Introduction

In primate groups composed of several individuals varying in their degrees of relatedness to each other, cooperative activities, defined as interactions providing participants with direct benefits, are often expected, either implicitly or explicitly, to take place preferentially among closest kin (e.g. Silk, this volume). In contrast to this view, I will argue here that the role of kinship in the patterning of cooperation has probably been overestimated, that cooperation is expected to be kin-biased only under certain specific conditions, and that competence, rather than kinship, should drive the selection of partners for many cooperative activities. The widespread expectation that cooperative activities should be consistently kin-biased is based on solid empirical evidence, but on evidence which apparently underwent some important shifts in meaning. The expectation derives from two categories of observations: (1) some types of cooperative activities are indeed kin-biased, and (2) some types of non-cooperative social activities, namely altruistic ones, for which Hamilton (1964) proposed kin selection theory, may be extremely kin-biased, if not entirely restricted to kin when altruism is directed unilaterally.

When a mother protects her newborn against an aggressor or a predator, she does not gain any direct, personal benefits, whether immediate or delayed. She is altruistic, and unilaterally so. Unilateral altruism is not restricted to mother-offspring relationships in primate groups; it may be performed by other categories of kin, for example by grandmothers and sisters in the form of aiding in conflicts (Chapais et al. 2001). Inclusive fitness theory (Hamilton 1964), which posits that donors obtain indirect fitness benefits by contributing to the fitness of their kin, has proven especially useful to account for unilateral altruism, for which the theory of reciprocal altruism (Trivers 1971) is of no use, by definition. Accordingly, unilateral altruism is expected to be highly kin-biased, and kin selection has been consistently invoked to account for altruism in primate groups (Kurland 1977, Chapais & Schulman 1980, Silk 1982, 1987, 2002a, this volume, Walters 1987, Dunbar 1992, Maestripieri 1993, Schaub 1996, Chapais 2001, Chapais et al. 2001, Combes & Altmann 2001, Chapais & Bélisle 2004).

But the situation is different with cooperation. As mentioned above, the term cooperation subsumes various types of interactions that provide participants with direct, personal benefits (Pusey & Packer 1997). As a first possibility, referred to as mutualism, the benefits may be obtained concurrently by the participants in the course of the interaction itself. Examples include coalitions, group-hunting and food-sharing. In theory, the benefits could also be obtained through the reciprocation of altruistic acts that are immediately costly to the donor, but provide mutual benefits in the long run (reciprocal altruism, Trivers 1971). However, interactions that unambiguously meet the criteria of reciprocal altruism have barely been documented in non-human primates. Although the "donor" may incur a slight cost in terms of time and energy, the possibility of its simultaneously obtaining significant benefits cannot be eliminated (Dunbar & Sharman 1984, Bercovitch 1988, Chapais et al. 1991, Hemelrijk et al. 1992, Noë 1992, Chapais et al. 1994, Prud'homme & Chapais 1996, Widdig 2000, Chapais 2001). Rather, cooperation through reciprocation appears to involve actions which simultaneously provide both partners with a net benefit, but which are initiated or performed alternatively by each of them. Possible examples include reciprocal aiding against third parties, reciprocal grooming, and exchange of grooming for tolerance, access to food, or aiding (see below). Such interactions have much in common with mutualism, but differ in that partners take turns in initiating or performing the same or different behaviors.

Importantly, because partners in cooperative activities obtain direct (personal) benefits whether they are related or not, cooperation can take place between non-kin just as well as between kin. Nevertheless, kin selection theory may be used to predict that cooperation should be performed preferentially among kin because when kin cooperate together they benefit both directly, through the cooperative act itself, and indirectly, through the fitness benefits accrued via kin selection (Wrangham 1982; see below). Partly on this basis, it has become common to explain nepotism in general, whether altruistic or cooperative, in terms of kin selection (e.g. Gouzoules 1984, Walters 1987, Morin et al. 1994, Kapsalis & Berman 1996, Silk 2002a), as if kinship should have a similar impact on the two functional categories of interactions.

In this chapter, I argue that the impact of kinship on the patterning of cooperation should be much less extensive than its impact on unilateral altruism, and that in many circumstances, cooperation should not take place between closest kin in primate groups. I first describe two theoretical arguments for the existence of kin biases in cooperation. I then define the conditions under which one expects cooperation to be kin-biased, and those under which cooperation should not necessarily be kin-biased. I conclude that cooperative activities whose payoff is significantly affected by the partners' relevant qualifications, that is, by their competence, should not be consistently kin-biased. In reviewing the relevant empirical evidence, I focus on the striking difference in the degree of kin bias between male philopatric species, such as chimpanzees (*Pan troglodytes*), which exhibit relatively low levels of nepotism among matrilineal kin, and female philopatric species such as macaques, characterized by comparatively much higher levels of matrilineal nepotism. I argue that this comparison is deceptive, that a number of factors help considerably reduce the discrepancy, and that the role of kinship in shaping cooperation in primates has probably been overestimated.

Given the paucity of studies on the influence of patrilineal kinship on behavior in primates (reviewed by Strier 2004), let alone on the effect of patrilineal kinship on the patterning of cooperative activities, I test the present ideas on matrilineal kinship only. Nonetheless, I assess the possibility that patrilineal kinship might affect my interpretations.

3.2 Why should cooperation be kin-biased?

At least two different forces may generate kin biases in cooperation. The first requires kin selection and was proposed most explicitly by Wrangham (1982). When a female cooperates with a relative, she benefits in two ways. She obtains the direct (personal) benefits of the cooperative act B, and she also derives indirect fitness benefits that amount to a fraction r (degree of relatedness) of the direct benefits accruing to her kin Br. If the same female cooperates with a non-relative instead, she derives only the direct fitness benefits B of the cooperative act. Because $B + Br > B$, and considering only these factors, cooperation between kin pays more than cooperation between non-kin, hence cooperation should be kin-biased. In sum, when individuals have equal access to kin and non-kin (Wrangham 1982), those who choose kin as partners obtain a fitness bonus Br. Because the fitness bonus is obtained reciprocally between kin, kin partners are mutually dependent on two accounts: (i) to obtain the direct benefits of cooperation and (ii) to obtain its indirect benefits, which dictates their cooperating together. By comparison, non-kin are mutually dependent only to obtain B; they would not lose any fitness bonus by defecting. For this reason, kin would constitute more reliable partners compared to non-kin, and kin partnerships would be more stable as a result.

An alternative explanation for the occurrence of kin biases in cooperation is based essentially on the direct benefits of cooperation, and thus does not require kin selection (Chapais 2001). In group-living primates, maternal investment often extends throughout the lifespan, mothers maintaining long-term supportive and affiliative relationships with their daughters in female philopatric species (Fairbanks 2000), and with their sons in male philopatric species (Goodall 1986, Furuichi 1997). The very existence of lifelong bonds between mothers and offspring entails that siblings meet around the same mother on a regular basis and become disproportionately available and familiar to each other compared to non-kin. This bias is independent of any intrinsic attraction between the siblings themselves; it is a consequence of the siblings' common attraction to the same mother. Then, if siblings are suitable social partners, and if disproportionate availability and familiarity *per se* increase the chances of forming partnerships, it follows that siblings should cooperate preferentially with each other.

The two explanations differ fundamentally in that the second one does not require kin selection. It states that kin cooperate together for the same reason non-kin cooperate together, namely for the direct fitness benefits of cooperation, but that kin may cooperate more often than non-kin because they are more readily available as partners. While the indirect fitness bonus of kin cooperation is a central component in the first explanation, it is ancillary in the second (Chapais 2001). The two explanations should prove extremely difficult to differentiate because even if kin-biased cooperation was driven by the greater availability of kin, rather than by indirect fitness benefits, the fact is that kin obtain the indirect fitness bonus in any case. Fortunately, this problem does not matter for the present discussion.

3.3
The effect of competence

Whatever their relative merits, the two explanations have a central aspect in common; they do not take into account the qualifications and relative competences of partners as a criterion in the formation of cooperative partnerships. Not all types of activities call for competence. For example, suppose that two animals cooperate to keep warm in the context of huddling or co-sleeping (Anderson 1984, Takahashi, 1997). In this situation, the main qualification required from each partner is the ability to produce heat, which requires a minimal body size. Because kin easily meet this qualification, the two explanations for kin biases in cooperation should apply. First, the equation $B + Br > B$ is always satisfied; cooperation provides the same direct benefits B whether one cooperates with kin or with non-kin, but by cooperating with kin, individuals obtain additional indirect fitness benefits. Second, because close kin are no less valuable than non-kin for the task, but close kin are disproportionately available and familiar, they could be chosen as partners if only for this reason. I refer to cooperative activities such as social thermoregulation, whose payoff is little affected by variation in the partner's qualifications, as low-competence cooperation. The expression "attribute-independent cooperation" was used in a previous paper; Chapais & Bélisle 2004. Low-competence cooperation should be markedly and consistently kin-biased.

In contrast, competence differentials may be crucial in other situations. For example, suppose that an individual's goal is to gain access to resources monopolized by a high-ranking individual, and that this relationship translates into a grooming-for-tolerance cooperative partnership. The dominant partner's main qualification is its absolute power, determined to a large extent by its absolute rank. In this situation, kin are not necessarily the best partners. Which partner, then, should ego cooperate with? Let B represent the direct benefits of cooperation with a given kin, and q the ratio of competence between a potential non-kin partner and that kin, so that cooperating with the non-kin yields qB. For cooperation to be more advantageous with the non-kin, $qB > B + Br$, which reduces to $q > 1 + r$; that is, the ratio of competence between the non-kin and kin partners must be greater than one plus the degree of relatedness between ego and its kin. For example, a female having a choice between cooperating with a half-sibling ($r = 0.25$) or a non-kin, should choose the non-kin if its competence for the task is more than 1.25 times (or 25%) higher than the kin's competence. This condition may be easily satisfied considering that kin of the wrong age or rank may be considerably less competent than non-kin. I refer to such cooperative activities, whose payoff is markedly affected by variation in the partners' qualifications, as competence-dependent cooperation. The expression "attribute-dependent cooperation" was used in Chapais & Bélisle 2004. Competence-dependent cooperation should not be strongly and consistently kin-biased.

3.3.1 Low-competence cooperation

Table 3.1 classifies a sample of cooperative activities according to whether they belong to the low-competence category or to the competence-dependent category, specifying for each the nature of the partnership, the qualifications required, whether kin may meet these qualifications, and whether one expects kin biases as a result. Besides social thermoregulation, another possible form of low-competence cooperation is reciprocal grooming, assuming that grooming is performed for hygienic or comfort-related reasons (Hutchins & Barash 1976, Barton 1985, Schino et al. 1988, Boccia et al. 1989, Keverne et al. 1989, Tanaka & Takefushi 1993, Aureli et al. 1999). In this situation, a partner's qualification is its ability to reciprocate grooming, which only requires a minimal age. Because an individual's kin are likely to include such suitable grooming partners, one expects grooming to be kin-biased for the same reasons given in the case of social thermoregulation; that is, both because kin are readily available and because cooperating with them yields additional indirect fitness benefits. If, on the other hand, grooming is performed to obtain social benefits such as increased tolerance levels (Hemelrijk et al. 1992, Muroyama 1994, Henzi & Barrett 1999), access to food (de Waal 1997a), or coalitionary support (Seyfarth 1977, Seyfarth & Cheney 1984, Hemelrijk 1994), kin are not necessarily the most competent partners and grooming should not necessarily be kin-biased.

To test the present hypothesis about the differential impact of kinship on grooming distribution, one needs to differentiate and analyze separately the grooming episodes that individuals perform to obtain social benefits (competence-dependent cooperation), and those they perform to obtain grooming in return (low-competence cooperation). The difficulty of this task is commensurate to that of establishing clear causal relationships between behavioral categories, as exemplified by the relationship between grooming given and aiding received, first proposed by Seyfartyh (1977) and still debated 25 years later; e.g. contrast Schino (2001) with Henzi & Barrett (1999) and Henzi et al. (2003). But another way of testing the present hypothesis would be to compare whole grooming distributions in two situations: (i) when grooming is performed to obtain social benefits and (ii) when it is performed for its own value; only in the latter situation should grooming be reciprocal within dyads and markedly kin-biased.

Data on grooming among female chacma baboons (*Papio ursinus*) lend themselves to such a test. Henzi et al. (2003) compared the distribution of grooming in the same group between two periods, when ecological conditions favored contest competition for food, and later when this was not the case. When food competition was profitable, females had more diverse grooming partners in terms of rank distance, presumably because they sought to exchange grooming for tolerance at food sites with higher-ranking females. In contrast, when food competition was lower, females had a smaller number of partners, who ranked closer to themselves on average, presumably because they exchanged grooming only for

Table 3.1. Non-exaustive classification of cooperative activities in primates according to whether the activity's payoff is affected by the partners' qualifications (competence-dependent cooperation) or not (low-competence cooperation). See text for references.

Goal of cooperation	Nature of partnership	Qualifications required	Determinants of qualifications	Do kin meet qualifications?	Kin bias expected?
Low-competence cooperation					
Thermoregulation	Huddling/co-sleeping	Minimal heat produced	Minimal size	Yes	Yes
Receive grooming	Reciprocal grooming	Ability to groom	Minimal age	Yes	Yes
Gain maternal experience	Allomothering	Caring for an infant	Minimal age	Yes	Yes
Competence-dependent cooperation					
Gain access to resources[a]	Grooming for tolerance	Absolute power	High rank	Not necessarily	No
Obtain aid/rise in rank[b]	Grooming for aiding	Absolute power	High rank	Not necessarily	No
Rise in rank[c]	Mutual aiding	Relative power	Rank similarity	Not necessarily	No
Maintain one's rank[d]	Mutual aiding and grooming	Relative power	Rank similarity	Yes (matrilineal hierarchies)	Yes but amplified[e]
Gain access to resources	Mutual sharing	Relative power	Rank/age similarity	Not necessarily	No
Practice motor/social skills	Social play	Size similarity	Age similarity	Not necessarily	No
Catch preys/obtain meat	Group hunting/meat-sharing	Hunting experience	Absolute age	Not necessarily	No

[a] In exchange for grooming.

[b] Bridging alliance (Chapais 1995) formed between A and C in a A > B > C rank order.

[c] Revolutionary alliance (Chapais 1995) formed between B and C in a A > B > C rank order.

[d] In exchange for grooming and/or aiding

[e] In matrilineal hierarchies, kin rank close to each other, which may amplify nepotism. See text.

its own value (Henzi et al. 2003). Given that in baboons females close in rank are usually kin (Lee & Oliver 1979, Walters 1980, Hausfater et al. 1982, Johnson 1987, Silk et al. 1999), the data suggest that grooming was more kin-biased when food competition was lower and social benefits were less at stake. In a previous study, Barrett et al. (1999) also reported that when competition was lower grooming was exchanged reciprocally, presumably because females needed not exchange it for social benefits with high-ranking females. In sum, these data support the hypothesis that when grooming is performed in the context of a low-competence cooperative activity, it is more kin-biased, as predicted.

A third possible example of low-competence cooperation is the care of infants by individuals other than the mother, or allomothering. For allomothering to qualify as a low-competence cooperative activity, it must in the first place qualify as a cooperative one. Although not all allomothering is cooperative (Hrdy 1976, Nicolson 1987), Fairbanks (1990a) reported the existence of a system of mutualistic cooperation between helpers and mothers in vervet monkeys (*Cercopithecus aethiops*). Females who spent more time allomothering were more successful in keeping their first-born infant alive, presumably due to their greater maternal experience, and mothers using allomothers had shortened inter-birth intervals. Thus, both the helpers and the recipient mothers derived direct fitness benefits from allomothering. Second, for allomothering to qualify as a low competence activity, it should require few qualifications on the part of both the allomother and the infant and its mother. From the perspective of a helper seeking to gain maternal experience, the infant's qualifications and competence are probably irrelevant; any infant will do. But infant kin (e.g. sisters) are more available and familiar, and taking care of them provides additional inclusive fitness benefits (Fairbanks 1990a). From the infant's and mother's viewpoint, the same reasoning applies; older sisters are readily available to allomother their younger siblings, and allowing them to allomother provides the infant and the mother with inclusive fitness benefits (Fairbanks 1990a). Kin allomothers may also be less likely to harm the infant (Nicolson 1987).

Thus, even though allomothering may require some competence (e.g. experience), it is open to a large array of individuals, kin and non-kin, because specific qualifications which are often found among non-kin, such as a high rank or age similarity, are not required. In this sense, cooperative allomothering would be a low-competence activity, which could explain why it is often kin-biased (Nicolson 1987, Chism 2000).

3.3.2 Competence-dependent cooperation

Table 3.1 also lists possible examples of competence-dependent cooperative activities. In these examples, the partners' qualifications vary between absolute power (a correlate of high rank), relative power (a correlate of closeness in rank), experience (determined by absolute age) and size similarity (a correlate of age similarity among immatures). Because kin do not necessarily meet these criteria, the corresponding partnerships should not be consistently kin-biased. For example, as mentioned above, individuals might use grooming as a currency to obtain aid (Seyfarth 1977, Schino 2001) or tolerance at food sites (Henzi & Barrett 2003), from high-ranking individuals. Hence, such cooperative partnerships should not be markedly kin-biased. However, they could be among members of high-ranking matrilines because in this situation, the targeted partners are both related and high-ranking. Interestingly, rates of affiliative relationships were reported to be higher within high-ranking matrilines than within lower-ranking ones in baboons (Silk et al. 1999).

Social play provides another illustration of competence-dependent cooperation. Social play is hypothesized to provide mutual benefits to partners through its role in the development of motor and social skills (Fagen 1993). Thus, although social play is rarely viewed as a cooperative activity, it satisfies the criteria for mutualistic cooperation; partners gain direct benefits and they do so in the course of the interaction itself. When individuals have a choice between partners, age similarity proves to be a major determinant in the formation of play partnerships, presumably because age similarity correlates positively with peer familiarity and similarity in size and strength (Fagen 1981). Similar-age partners are often not available among close kin (e.g. siblings), even though older or younger close kin are often available. Thus, similar-age unrelated partners seem to be favored over dissimilar-age kin. In this sense, social play is a competence-dependent activity whose major qualification is age similarity. Accordingly, play is slightly or not kin-biased (Walters 1987, Janus 1989, Berman 2004).

Primates may also form partnerships on the basis of rank similarity. For example, among the chimpanzees (*Pan troglodytes*) of the Ngogo community (Uganda), Mitani et al. (2000) found that matrilineally-related males did not associate or cooperate preferentially. Cooperation was measured by mutual participation in alliances, meat sharing and boundary patrols. In a subsequent study, Mitani et al. (2002b) reported that males selected their partners on the basis of age similarity (same age class) and rank similarity (same rank class), and reasoned that the absence of nepotism reflected the non-availability of kin partners of the right age and rank due to the long birth interval of chimpanzees (5-6 years). Stated differently, chimpanzees gave priority to age and rank rather than to kinship; they were non-nepotistic because they engaged in competence-dependent cooperation. Chimpanzees are male-philopatric and live with several patrilineal relatives. I consider the possibility of patrilineal nepotism in a later section.

The effect of the rank similarity criterion on cooperation also operates among females, but then it may be confounded by kinship in situations where close-ranking females are also kin, as commonly happens in matrilineal dominance hierarchies. The observed high levels of attraction between female kin in matrilineal societies are commonly attributed to the effect of kinship *per se* because the respective effect of its correlate, rank similarity, is most often not assessed. But two studies that analyzed the separate influences of kinship and rank similarity revealed that both factors contributed to attraction and tolerance among close-ranking female kin. De Waal (1991b) found that closeness in rank significantly increased levels of proximity and co-drinking in two groups of rhesus monkeys (*Macaca mulatta*) independently of kinship, and, similarly, Kapsalis & Berman (1996) reported that rank similarity significantly increased levels of affiliation (approaches, proximity, contact and grooming) among free-ranging rhesus monkeys, independently of kinship. Both sets of results indicate that competence-dependent cooperative activities driven by rank similarity were probably underestimated in groups characterized by matrilineal dominance systems (see below). Both studies also found that kinship increased rates of behaviors independently of closeness in rank. Thus, in these studies and in others (e.g. Silk et al. 1999), the effect of kinship is real; it is not an artifact of the rank similarity correlate (see Silk, this volume).

3.3.3 The relationship between kinship and competence

The foregoing argument about the relative role of kinship and competence in patterning cooperation was framed in dichotomous terms. I argued that cooperative activities whose outcomes are minimally affected for competence differentials should be kin-biased, whereas cooperative activities whose outcomes vary substantially in relation with the partners' qualifications should not be kin-biased. At first sight, this dual classification may appear justified because any given cooperative activity can either require some well-defined qualifications (e.g. rank-based power and the capacity to offer help), or not (e.g. the capacity to provide heat). Although heuristically useful, the dichotomy is nonetheless somewhat simplistic because competence is a continuous variable.

As stated previously, ego should prefer a more competent non-kin over a less competent kin when $q > 1 + r$, where q reflects the competence ratio between the non-relative and the relative. When the competence ratio is higher than 1.5, ego should always choose the non-kin partner over all potential kin partners, even its closest kin because the maximal degree of kinship (in outbred populations) is 0.5. But when q is lower than 1.5, some kin could be advantageously chosen over the non-kin partner. For example, half-siblings ($r = 0.25$) should be chosen over non-kin when $q < 1.25$. Thus, competence-dependent cooperation could be kin-biased provided q is relatively low. In the previous discussion of competence-dependent cooperation, I assumed that high levels of competence were required for the activities considered (Table 3.1), and that there were no reasons to believe that kin partners were better qualified than non-kin. This assumption appears generally reasonable in light of the examples reviewed.

However, even when q is higher than 1.5, cooperation might be performed among kin for two reasons. First, kin might be the most competent partners. Two examples have already been mentioned. Assuming that the main qualification for a given cooperative activity is rank similarity, that criterion coincides with kinship in matrilineal hierarchies; hence, cooperation among females would be kin-biased because female kin have similar ranks. The other example concerned the high-rank qualification in matrilineal hierarchies. For members of the highest-ranking matriline, high-ranking partners are also close-ranking individuals so that cooperation among highest-ranking individuals could be kin-biased because it is rank-biased.

Second, competence-dependent cooperation could be kin-biased because competent non-kin are not available, even though they are present in the group.

In such a situation, individuals would have no choice but to cooperate with kin. For example, low-ranking individuals might be unable to interact with high-ranking non-kin if they are prevented from doing so by mid-ranking individuals, as in Seyfarth's (1977) grooming model. Such a situation would favor cooperation among kin.

I have hitherto discussed situations where competence-dependent cooperation could be kin-biased contrary to the main argument presented here. The reciprocal may also be true. Low-competence cooperation, which is expected to be kin-biased, could take place both among kin and non-kin if not enough kin are available. This is expected when matrilines are small as in decreasing populations (Dunbar 1988), or due to random demographic fluctuations.

I conclude that, in general, kinship should have little effect on the selection of partners for cooperative activities: (i) when the competence ratio (q) is high, or more specifically, higher than 1.5, and (ii) provided individuals have equal access to both kin and non-kin. Cooperation should be kin-biased when q is low and kin are available.

3.4 Matrilineal kinship and competence: the contrast between male and female philopatric societies

In this section, I pursue the investigation of the relative role of competence and kinship in the patterning of cooperation by examining an apparent contradiction between the importance of matrilineal kinship in male and female philopatric species. In female philopatric species, such as macaques and baboons, matrilineal kinship has a strong impact on the behavior of the philopatric sex (reviewed by Gouzoules 1984, Walters 1987, Bernstein 1991, Chapais 2001, Silk 2002a, Berman 2004, Kapsalis 2004). In contrast, available data on male philopatric species indicate that matrilineal kinship has little effect on the behavior of the philopatric sex (males). The male philopatric species for which we have the best data both on kinship and social interactions is the chimpanzee. Three studies carried out in three different populations of chimpanzees found that affiliation and cooperation among males were not biased towards matrilineal kin as assessed by mitochondrial DNA (mtDNA) haplotype sharing (Goldberg & Wrangham 1997, Mitani et al. 2000, 2002b, Boesch et al., this volume, see also Strier 2004). Similar results, but on smaller samples, were obtained for other male philopatric species: bonobos, *Pan paniscus* (Hashimoto et al. 1996) and muriquis, *Brachyteles arachnoïdes* (Strier et al. 2002). In light of these studies, the instances of cooperation between maternal brothers observed among the Gombe chimpanzees by Goodall (1986) would constitute the exception rather than the rule.

Why would matrilineal kinship promote cooperation among philopatric females, but much less so among philopatric males? One might think that this question is biased, or even irrelevant, because it does not take into account the other half of genetic relatedness, patrilineal kinship. Future studies might indeed reveal that male chimpanzees are nepotistic with their male patrilineal kin even though they are not with their matrilineal kin. I consider the issue of patrilineal kinship in a separate section (see below). However, treating matrilineal kinship separately in male philopatric societies, as was done by Hashimoto et al. (1996), Goldberg & Wrangham (1997) and Mitani et al. (2001, 2002b), is no less relevant than treating it separately in female philopatric societies, as was done in almost all studies on kinship in these species. If philopatric females are nepotistic with their matrilineal relatives, philopatric males should be as well. So, why do the available data apparently fail to support this prediction? I will examine three different reasons, dwelling on the last two.

First, the discrepancy might reflect a sex difference in the role of competence in the formation of cooperative partnerships. If cooperation in chimpanzees is not kin-biased because it is mostly competence-dependent, as argued above, conversely the high levels of nepotism of philopatric females might reflect a female bias for low-competence cooperative activities. For example, allomothering is practiced essentially by females and is markedly kin-biased (see above). Whether other cooperative activities display such a sex bias – for example whether grooming between males more often aims at obtaining social benefits, and grooming between females more often at obtaining further grooming – remains to be explored.

Two other factors help account for the greater importance of kin biases in female philopatric societies: (i) kin compositions and (ii) the confounding effect of rank similarity on kinship. I examine these factors in the next two sections.

3.4.1
Philopatry patterns and differences in kin compositions

Networks of matrilineal kin differ fundamentally between male and female philopatric societies. In female philopatric groups, females co-reside with several categories of matrilineal kin and discriminate some of them. For example, experiments on Japanese macaques (*Macaca fuscata*) revealed that the degree of kinship beyond which females treated their kin as non-kin was the same for three different categories of behavior: (i) aiding in conflicts (Chapais et al. 1997, 2001), (ii) tolerance at a monopolizable food source (Bélisle & Chapais 2001) and (iii) homosexual inhibition among females (Chapais & Mignault 1991, Chapais et al. 1997). Kin discrimination was manifest between mothers and daughters, grandmothers and grandoffspring, great-grandmothers/great-grandoffspring (tested for only one behavioral category: aiding), and between sisters. Kin discrimination was not manifest between aunts and nieces, and only inconsistently so between aunts and nephews. Using a different methodology, Kapsalis & Berman (1996) reported very similar results for free-ranging rhesus monkeys. Generalizing from these studies, philopatric females would discriminate a minimum of four or five categories of matrilineal kin among all those present.

The number of discriminated categories of matrilineal kin is substantially smaller in male philopatric species, such as chimpanzees, due to the dispersal pattern. Assuming that all males are resident and that most females disperse (Vigilant et al. 2001, Doran et al. 2002), male matrilineal kin reduce to a single category, maternal brothers, because female transfer entails that sons do not co-reside with their mother's kin. Even assuming that some females reproduce in their natal group so that these females' sons co-reside with their maternal uncles, given that kin discrimination is inconsistent between aunts and nieces in female philopatric societies (see above), the same may apply between uncles and nephews. Matrilineal kin also include mother-son dyads but I focus here on dyads of male matrilineal kin. Hence, to compare adequately the relative importance of matrilineal nepotism between male philopatric societies and female philopatric ones, one should focus on siblings, the only kin that are both available and discriminated on a regular basis in the two types of societies. The question then becomes: does the discrepancy in matrilineal nepotism remain, i.e. do maternal sisters in female philopatric societies cooperate more than maternal brothers in male philopatric societies?

It is not easy to answer this question because very few studies on female relationships partitioned data according to degree of kinship, and further differentiated between cooperative interactions and altruistic ones. But three sets of factors appear to reduce the discrepancy. First, sisters in matrilineal hierarchies have similar ranks and their cooperation may be rank-driven. In other words, if sisters ranked independently of each other, as do maternal brothers in chimpanzees, they might be less kin-biased (see below). Second, much of the cooperation between sisters may be of the low-competence type (e.g. reciprocal grooming, allomothering), which could account for a further portion of the kin bias. Third, it is noteworthy that levels of nepotism between sisters are much lower than between mothers and daughters (Kaplan 1977, Kurland 1977, Massey 1977, Glick et al. 1986, Kapsalis & Berman 1996, Chapais et al. 1997, Chapais & Bélisle 2004), and may even be lower than between grandmothers and granddaughters (Chapais et al. 1997) despite these two categories of kin sharing the same degree of kinship ($r = 0.25$), and sisters being even more closely related on average, if some sisters are full-siblings. Relatively low levels of nepotism between sisters probably reflect the intense and long-lasting dominance competition between them (Datta 1988, Chapais et al. 1994), whereas it is weak between grandmothers and granddaughters. Dominance competition was indeed found to act as a significant constraint on sister nepotism in Japanese macaques (Chapais et al. 1994).

The point here is that by limiting the comparison of the nepotistic tendencies of philopatric males and philopatric females to siblings, the discrepancy in the overall extent of kin biases between the two categories of species is reduced.

3.4.2
The amplifying effect of rank similarity on female nepotism

All female-philopatric societies for which we have good data on both kinship and behavior exhibit matrilineal hierarchies in which, by definition, kin rank close to each other. As mentioned above, both kinship and rank similarity contribute, independently, to increase attraction between individuals (de Waal 1991b, Kapsalis & Berman 1996). Thus, observed levels of nepotism in female-philopatric species are consistently amplified to a variable extent and this factor artificially increases the discrepancy in nepotism between these species and male-philopatric ones.

To better understand how rank similarity amplifies kin biases in cooperation, I have modeled the effect of rank similarity as if females chose their partners on that basis alone; that is, independently of kinship. Consider a matrilineal hierarchy composed of three matrilines (a, b and c) of four females each (a_1, a_2, and so on, in decreasing rank order). Suppose that any female cooperates preferentially with the two females ranking immediately below her, and the two ranking immediately above her. The female's four partners may be kin or non-kin. For example, of the four females that rank closer to b_1, two are kin (b_2, and b_3), and two are non-kin (a_3 and a_4); thus, 50% are kin. I calculated this percentage for each of the 12 females composing the hierarchy; the average is 75%. In other words, assuming that females in a nepotistic hierarchy choose their partners solely on the basis of closeness in rank, 75% of the partners nonetheless happen to be kin. If the rank order were not matrilineal (i.e. if female kin ranked independently of kinship in relation to each other), the average percentage of kin among a female's close ranking partners would be 27% (the proportion of kin dyads out of all dyads). The difference between the two percentages represents the maximal amplification of nepotism due to rank similarity *per se* in this particular hierarchy.

I calculated the percentages of close-ranking females that are kin for various matriline sizes, from three females per matriline (hierarchy of nine females) to seven females per matriline (hierarchy of 21 females), and for various definitions of closeness in rank, from one rank on each side of ego (two close-ranking partners) to three ranks on each side of ego (six close-ranking partners). The results are summarized in Fig. 3.1. The top curve defines closeness in rank as one rank on each side of ego and shows how the percentages of close-ranking females that are kin vary according to matriline size. The second and third curves define closeness in rank as two and three females on each side of ego, respectively. The bottom curve gives the percentage of close-ranking females that are kin when kinship and closeness in rank are decoupled in non-matrilineal hierarchies. The three curves are well above the baseline, indicating that nepotism is amplified for all three definitions of closeness in rank.

All three curves are ascending, indicating that for all three definitions of closeness in rank, the amplification of nepotism increases with matriline size. This is because in hierarchies composed of very small matrilines, close-ranking females are more likely to belong to different kin groups. But the larger the matrilines, the more likely close-ranking females are kin. The most detailed data on matrilineal kinship structures in primates come from provisioned populations of only two species, the rhesus macaques of Cayo Santiago (Rawlins & Kessler 1986) and various populations of Japanese macaques (Fedigan & Asquith 1991). Provisioned populations are often growing, and therefore have particularly extensive kinship structures (Dunbar 1988), which easily extend over four generations in macaques. Thus, given that our best data on the effect of matrilineal kinship on behavior come from provisioned groups composed of especially large matrilines, the amplification of the role of kinship in female cooperation, and our expectations of kin biases have been maximized.

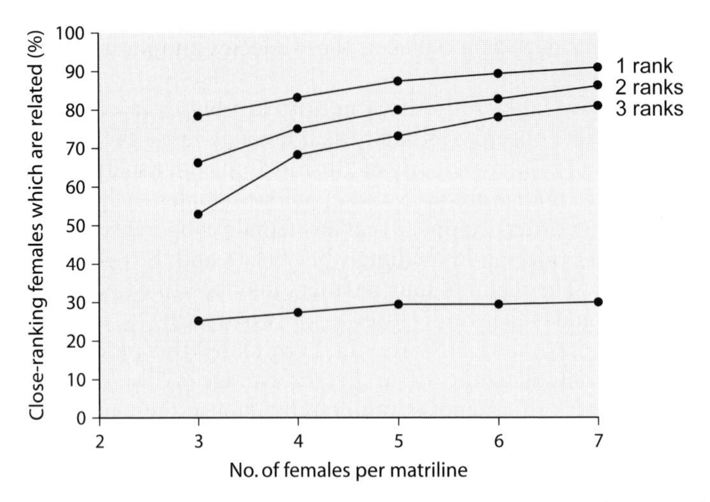

Fig. 3.1. Maximal amplification of nepotism due to kin ranking close to each other in matrilineal dominance hierarchies. The model assumes that females form cooperative partnerships solely on the basis of closeness in rank, which is defined in three ways: 1 rank: any female cooperates only with the female ranking immediately below her and the one ranking immediately above her; 2 ranks: the female cooperates with the two females immediately below her and the two above her; 3 ranks: the female cooperates with the three females immediately below her and the three above her. For any matriline size, the percentage of close-ranking females which happen to be kin increases the narrower the definition of closeness in rank. For any definition of closeness in rank, the percentage increases the larger the size of matrilines. The bottom curve (baseline) gives the percentage of close-ranking females that are kin when kinship and closeness in rank are decoupled in non-matrilineal hierarchies.

Fig. 3.1 also shows that for any matriline size, the amplification of nepotism increases as the definition of closeness in rank becomes narrower, being maximal for adjacent-ranking females (one rank position). In this situation, the closest-ranking female is almost always a relative, whereas in the case of larger rank distances, partners may belong to different matrilines. How close in rank to ego a female must be to be treated preferentially can only be determined through empirical studies. For example, data in de Waal (1991b, Fig. 3) show that a rank distance of one stands out from all other rank distances as having the maximal effect on attraction. If this result is representative of other groups and species, the one-rank curve in Fig. 3.1, nepotism maximally amplified, would be closer to reality than the other two curves.

In sum, both historical biases in the selection of populations for which we have the best data on kinship, and the meaning of closeness in rank from the animals' perspective, may have contributed in amplifying our perception of the importance of nepotism in female relationships. It follows that in order to assess the effect of kinship *per se* on female relationships, one needs data on female philopatric species that do not exhibit matrilineal hierarchies. Examples include Hanuman langurs, *Presbytis entellus* (Hrdy & Hrdy 1976, Borries 1993, Koenig 2000) and captive sooty mangabeys, *Cercocebus torquatus atys* (Gust & Gordon 1994), wherein females establish well-defined dominance relationships, but do not rank close to their kin. Kin biases in behavior should be weaker in such societies if only because nepotism is not amplified by closeness in rank. Unfortunately, detailed data on kinship are not yet available for these species. Even if they were, however, testing the prediction should prove difficult because non-matrilineal societies could exhibit lower levels of nepotism for another, confounding, reason. If female kin do not form alliances to transmit, acquire and maintain their birth rank, as they do in matrilineal hierarchies, levels of altruism and cooperation among female kin should be significantly lower. Thus, levels of nepotism in non-matrilineal societies could be lower both because they are not amplified by rank similarity and because females are socially less dependent on their kin. This is not to say that kin biases should be absent altogether. Kin biases should be clearly manifest in the females' low-competence cooperative activities.

In any case, when data on kinship and behavior become available for female-philopatric societies lacking matrilineal hierarchies, the discrepancy in levels of nepotism between maternal sisters in these groups, and maternal brothers in male-philopatric groups, should be much reduced, if still significant.

3.4.3
What about patrilineal kinship?

Although male chimpanzees do not cooperate preferentially with their matrilineal kin, they might do so with some of their patrilineal kin. One possible process of patrilineal kin discrimination derives from male reproductive skew. The fact that single males may fertilize several females over a limited period of time generates paternal sibships among the resulting offspring. The more pronounced male reproductive skew, the larger the resulting paternal sibships within a given birth cohort. Hence, age similarity might reflect paternal relatedness to some extent (Altmann & Altmann 1979, van Hooff & van Schaik 1994, Strier 2004). Interestingly, as mentioned above, age similarity was found to pattern cooperation among male chimpanzees, along with rank similarity (Mitani et al. 2002b). If age similarity does reflect patrilineal kinship, males might in fact be cooperating with their paternal half-brothers, a possibility evoked by Mitani et al. (2002b), Silk (2002a) and Strier (2004). To understand the implications of this idea, however, it is useful to distinguish clearly between two possibilities.

The first possibility is that males discriminate their paternal brothers among all their age mates, and cooperate preferentially with them through the operation of kin selection. No data on patrilineal kin recognition are available for male philopatric groups, but some data are available for female philopatric groups. Although a number of studies failed to find kin discrimination between paternal sisters (Fredrikson & Sackett 1984, Kuester et al. 1994, Ehardt et al. 1997), other studies reported kin discrimination between paternal sisters close in age in rhesus macaques (Widdig et al. 2001) and savanna baboons (Smith et al. 2003). On this basis, suppose that male chimpanzees do discriminate their paternal brothers among their age mates. Because males are expected to be equally familiar with all age mates, regardless of their degree of relatedness with them, patrilineal kin recognition could not be based on familiarity differentials. Another mechanism would be involved, for example phenotype matching, as suggested by Widdig et al. (2001) and Smith et al. (2003) for macaques and baboons. In male-philopatric societies, males co-reside with several categories of male patrilineal kin besides same-aged brothers, including their younger and older paternal brothers, their fathers, sons and uncles. Thus, if males use phenotype matching to recognize their similar-age paternal brothers, they should be able to use it to recognize their younger and older brothers as well as other categories of patrilineal kin, and they should cooperate preferentially with kin of various ages and ranks. This prediction is not easily reconcilable with the observation that male chimpanzees cooperate preferentially with males close in age and rank (Mitani et al. 2002b).

The second possibility is that males do not discriminate their paternal brothers among their age mates. Nonetheless, they would have been selected to cooperate preferentially with age mates because these are often paternal brothers. Thus, by cooperating with age mates in general, males would increase their inclusive fitness. In other words, males would use the age similarity criterion as a marker of potential paternal kinship, the age bias would be in fact a kin bias, and cooperation would be driven by kin selection. There are four problems with this hypothesis. First, only a fraction of same-age males are paternal brothers if male reproductive skew is weak. For example, in chimpanzees, Constable et al. (2001) reported that only 36% of paternities in the Kasakala community of Gombe could be attributed to the males that were in the alpha position at the time of conception (see also Strier 2004). Second, age similarity between males is a poor marker of patrilineal kinship in a male-philopatric society, as mentioned above. Only a fraction of all patrilineal kin would correctly identify each other through age similarity. Paternal brothers belonging to different age cohorts would not, nor would fathers and sons, for example. Note that age similarity is a much better marker of patrilineal kinship between females in a female-philopatric society because most patrilineal kin are paternal sisters. Third, if males use age similarity essentially as a marker of kinship, they should cooperate with all the kin that they can recognize, for example their maternal brothers. But this is apparently not the case, as mentioned above. Fourth, males also choose their partners on the basis of rank similarity. They do so independently of age similarity (Mitani at al. 2002b), and there is no reason to believe that rank similarity correlates with kinship. Thus, if the rank similarity qualification is used for its own sake, the age similarity criterion could be as well. The four arguments suggest that age similarity is not used as a marker of kinship.

Based on the available evidence, I conclude that it is unlikely that cooperation between age mates reflects patrilineal kinship in a male philopatric society. As argued above, if the cooperative activities of males are mostly competence-dependent, and age similarity a criterion of competence, partner selection may be driven by the direct benefits of cooperation, not by kin selection.

3.5
Summary and conclusions

The expectation that individuals should consistently cooperate with their kin implicitly assumes that kin and non-kin make equally valuable partners, or that individuals are unable to assess competence and its determinants, such as relative age and rank. Both assumptions are amply contradicted by empirical evidence, supporting the hypothesis that low-competence cooperation should be markedly kin-biased, but high-competence cooperation should not be, except when kin are more competent, or when competent non-kin are not accessible. Although much remains to be done to test this hypothesis, the evidence reviewed here provides preliminary support for it.

The expectation that cooperation should be kin-biased derives essentially from the role attributed to kin selection in its evolution. One important correlate of the preponderant role attributed to kin selection in the evolution of behavior in general is the idea that any kin-biased behavior is the likely product of kin selection. This correlate has also been re-examined recently. In a previous paper (Chapais 2001), I questioned the assumption that a number of kin-correlated behaviors, including certain forms of grooming, allomothering and aiding, were altruistic and the product of kin selection. Several such behaviors seem to have been forced into an altruism framework, with the emphasis being put on the costs to their performers rather than their benefits. If kin-correlated behaviors provide performers with direct benefits, they might evolve through mechanisms other than kin selection (see above), or kin selection might be involved, but secondarily (Chapais 2001; but see Silk, this volume). A similar line of argument has been made for cooperative breeding in vertebrates. Kin selection is widely invoked to explain why non-breeding individuals help others raise their young while apparently incurring a net cost, but a growing body of evidence reviewed by Clutton-Brock (2002) points to an underestimation of the direct benefits to helpers and the overestimation of the indirect benefits of helping. The role of mutualism in the evolution of cooperative breeding in vertebrates appears to have been relatively neglected (Clutton-Brock, this volume).

Due to its powerful logic and to its position as the unique evolutionary theory specifically devoted to kinship, inclusive fitness theory has had a deep impact on our view of kin-biased behavior. Perhaps as a consequence, we tend to attribute all forms of interactions between kin to the operation of kin selection, even though empirical tests of kin selection and alternative mechanisms are badly needed (Chapais 2001, Clutton-Brock 2002). By consistently equating nepotism with kin selection, we have come to view kinship as such a powerful determinant of behavior that we expect genetic relatedness to explain the distribution of cooperation as much as it explains the distribution of altruism. Such a view probably owes much to the fact that our knowledge about kinship in primates is predominantly derived from a few species and populations in which nepotism is particularly important, if not paramount (Chapais & Berman 2004). In particular, we should obtain a more balanced view of the explanatory value of kinship in female relationships when we obtain data on female-philopatric societies lacking matrilineal hierarchies.

Acknowledgments

I am greatly indebted to Peter Kappeler for inviting me to participate in the highly stimulating Göttinger Freilandtage on cooperation in primates. Carol Berman, Peter Kappeler, Jean Prud'homme, Joan Silk, Carel van Schaik, Shona Teijeiro and one anonymous reviewer provided most helpful comments on the manuscript.

Part III
Reciprocity

Chapter 4

Reciprocal altruism: 30 years later

Robert Trivers

"Two are better than one; because they have a good reward for their labour. For if they fall, the one will lift up the other: but woe to him that is alone when he falleth; for he hath not another to help him up. Again, if two lie together, then they have heat: but how can one be warm alone? And if one prevails against him, two shall withstand him; and a three-fold cord is not easily broken." (Ecclesiastes 4, 9–12; King James Version).

4.1 Introduction

A little over 30 years ago, I had the good fortune of publishing my first scientific paper on reciprocal altruism, a subject that had not yet been addressed from an evolutionary standpoint. Hamilton's (1964) great work on kinship and altruism made it clear that in humans there existed a major form of altruism that could not be explained by kinship. Its elaboration was responsible for the complex economic systems in which we now live and its regulation could plausibly be explained by a system of interconnected human emotions, including feelings of friendship, gratitude, sympathy, guilt, moralistic aggression, a sense of justice and (I would now add) forgiveness.

I brought no great talents to this enterprise, beyond a willingness to take the evolutionary problem seriously and to model evolutionary logic on easily inferred psychological facts regarding our own behavior (for a description of how the paper was written, see Trivers 2002). The paper was certainly timely. My 600 reprints were quickly exhausted and the evolutionary idea was off and running. There now exists a very large literature on the subject and many subareas have advanced far beyond my original paper.

The purpose of the present paper is to provide a personal review of some major developments since my paper. These include the Prisoner's Dilemma (PD) as a model for reciprocal altruism, other models and third-party observer effects. I concentrate on the human sense of justice and the selective forces likely to have molded it. In the process, I discuss recent empirical work (using economic games) that bears on our sense of fairness and what seems to me the most plausible way to interpret these results. I neglect many important topics, for example, discrimination against cheaters in symbioses (see Sachs et al. 2004).

4.2
What was accomplished in 1971

It is well to pause for a moment and remember what the pre-evolutionary period looked like. Although Charles Darwin and George Williams had devoted a sentence or two to the subject of reciprocal altruism, both anthropology and social psychology were thoroughly pre-Darwinian in their thinking. Anthropology recognized the importance of reciprocity but not the problem of the cheater within the system. Conceptual confusion was illustrated by the effort to define parental investment as an example of 'reciprocity between the generations'. You invest in your children and they in theirs. Social psychology saw that 'prosocial' tendencies were important in life but failed to see any problem in how they evolved (indeed, did not even raise the question) and therefore failed to differentiate between obvious subcategories such as kin-directed versus reciprocal altruism. Evolutionary theory had nothing to offer beyond group selection and general inattention.

In my paper, elementary distinctions were emphasized. Altruism is suffering a cost to confer a benefit. Reciprocal altruism is the exchange of such acts between individuals so as to produce a net benefit on both sides. Reciprocal altruism is one kind of return-benefit altruism. There can be a variety of ways in which an act of altruism can initiate a causal chain leading to a return benefit to the actor, of which reciprocal altruism is but one example. A warning call which keeps group members alive may bring an immediate return benefit and eating your cleaner may bring a cost in lost interaction with the now dead cleaner. I was primarily interested in reciprocal altruism and would probably have skipped these two examples of return-benefit altruism if I had any nonhuman examples of reciprocal altruism. At the same time, the cleaning symbiosis example demonstrated altruism without kinship and the bird example provided a host of return-effect alternatives to a kinship explanation. In any event, the larger area of return-benefit altruism has become much more important, although it is not always conceptualized this way. Thus, group selection language is often formally equivalent to the language of return effects, though this may be obscured or denied. A good recent review, especially in the context of symbioses, can be found in Sachs et al. (2004).

A second distinction of some importance concerns the cheater or non-reciprocator in the system. I deliberately chose the term cheater, even though the neutral term non-reciprocator (later, defector) was more precise, because the emotive and intuitive powers of 'cheater' were attractive to me. The key point is to figure out how the cheater may harm itself, often by the counteraction of others. For me, the simplest was simply to break off the relationship, thereby cutting one's own losses and coincidentally reducing the benefits of cheating. The second is direct punishment, but since this is itself costly, I imagined that it would evolve after mere non-reciprocation. I called it 'moralistic aggression' deliberately, since it had a moral flavor but was not necessarily moral.

I naturally wanted the domain of my paper to be as large as possible, so I avoided, wherever possible, any limiting assumptions. Both Darwin (1871) and Williams (1966) quickly restricted reciprocal altruism to species with complex cognitive powers, while I preferred to imagine that once reciprocity got underway, the requisite cognitive powers would quickly evolve. More recently, possible cognitive limitations to the evolution of reciprocity have been emphasized (Stephens & Hauser 2004).

Finally, the paper sketched out a few obvious ways in which the analysis could be extended to a complex, multi-party system, with norms, infractions of norms, observer effects, collective punishment, and so on. These are topics that have become very active (and, in some cases, contentious) areas of research in recent years.

4.3 Tit-for-tat

My approach began with reality and attempted to model its evolution. In particular, I turned to everyday life for an account of what seemed to me the key emotions regulating the system and I dressed these up with data from social psychology. This can be a very effective first step, and indeed a continuing source of ideas for theoretical development, but one soon wants to have theory that begins with fundamental assumptions and generates different possible worlds. The key first advance was provided by Axelrod & Hamilton (1981). Building on the results of Axelrod's computer tournaments to see who could devise the best strategy for playing iterated games of PD (Rapaport's tit-for-tat [TfT] was the winner), they were able to show that TfT is evolutionarily stable as long as there is sufficient probability of future interaction (always-defect is also stable). TfT was the simplest strategy introduced in the original tournaments and has only two rules: begin cooperative and then on the next move do whatever your partner did on the previous one (i.e. cooperate or defect).

The simplicity is itself beguiling. It immediately widens the range of situations in which to look for reciprocal relations. Why not bacteria? They certainly are capable of a contingent strategy with neighbors in which an individual produces a (cooperative) chemical first and then produces whichever chemical the neighbor does (cooperative or selfish). There is no question that a rich world of social interactions will be found in bacteria, sometimes unifying the behavior of millions or billions of separate bacteria, but disentangling kinship, green beard and reciprocal effects will not be easy. A defector mutant in a sea of cooperators will initially be unrelated to neighbors at the key locus, and *vice versa*. Exciting empirical work is now emerging on the PD and cooperation in viruses (Turner & Chao 1999), bacteria (Velicer & Yu 2003) and yeast (Greig & Travisano 2004).

The success of the TfT strategy also gives new meaning to the value in life of imitation or, at least, responding in kind. Imitation is classically viewed as a valuable form of learning but it may also often be the appropriate response in reciprocal relations. Do unto others as they have just done unto you. If someone is nice, be nice; if not so nice, not so nice; if nasty, be nasty, and so on. But there are limitations to the value of this rule, as seen in the large costs associated with outbreaks of reciprocal spite (see below).

TfT and Axelrod & Hamilton (1981) spread like wildfire through the behavioral literature, so that soon all reciprocal interactions seemed to be formulated in terms of the PD. The effect was to distance the analysis from subjective experience to the point where I had to translate papers back into English in order to understand what was being said; for example, the excellent work on predator inspection in sticklebacks (Milinksi 1987, Milinski et al. 1990). It soon seemed that theorists and empiricists alike were forgetting that iterated games of PD amount to a highly artificial model of social interactions; each successive interaction simultaneous, costs and benefits never varying, options limited to only two moves, no errors, no escalated punishment, no population variability within traits and so on. In fact, almost all of these simplifying assumptions have now been shown to introduce important effects.

4.4 Beyond tit-for-tat in the Prisoner's Dilemma

In a brilliant series of papers, Nowak, Sigmund and colleagues explored the consequences of relaxing the assumptions built into the original game of iterated PDs. In the process, they outlined a plausible account of the way in which new strategies may displace old ones in an evolving two-dimensional social world where the two dimensions are the probabilities that an individual will cooperate in response to either the partner's cooperation or defection on the previous move. Nowak (1990) first showed that the introduction of errors into the system brought forth the value of some degree of forgiveness by the tit-for-tatter and some relaxation in the rigidity of always doing what your partner just did. In a simple iterated PD between two tit-for-tatters, a mistaken move by one on any move (say, defect on the first) will put the two actors permanently out of synchrony with each other, one cooperating, the other defecting, then *vice versa*, and so on, never achieving a relationship of strong reciprocal benefits. Various strategies of partial forgiveness can cut through this dilemma and outcompete TfT. The first to be described was 'generous tit for tat' (gen TfT) in which a modified tit-for-tatter responds always to cooperation with cooperation but responds to defection some of the time with cooperation; in effect, forgiving the partner for that defection (Nowak & Sigmund 1992). If the partner is a tit-for-tatter (generous or original), cooperation will have been re-established, the error corrected. The expected frequency of forgiving is determined by the precise payoff matrix of the game (one-third in the original payoff conditions of Axelrod's tournament).

This work invites us to consider the many mistakes that can be made in life, not just accidental breakdowns in the underlying causal machinery. We can try to second-guess the situation and guess wrong, we can be in a negative state because of other interactions which spill over into this situation; there are 'innocent bystanders' (Nowak & Sigmund 1992) and so on. Error is intrinsic to life but perhaps more frequent with some kinds of strategies or in some kinds of settings than in others, an avenue of investigation that may prove fruitful.

Nowak & Sigmund (1993) then showed that yet another strategy did very well in a world of tit-for-tatters, generous tit-for-tatters and full-time cooperators (all-c). Called 'win-stay, lose-shift', it merely repeats the same move as on the previous if it was rewarded and changes its move if it was punished. Put another way, if your partner just cooperated, keep doing what you are doing; if the partner just defected, switch. This strategy has three benefits: (i) it protects against exploitation, (ii) corrects errors and (iii) exploits a naive cooperator. Note that it cannot invade in a world of defectors, because it switches every other interaction to cooperate, where it is exploited. But if TfT and gen TfT have driven out all-defect (all-d), then all-c will spread by drift, inviting the success of win-stay, lose-shift.

Nowak & Sigmund (1994) have also analyzed the alternating PD where, instead of acting simultaneously, players alternate roles of donor and recipient. This would seem to mimic real life more closely, though including some (non-zero) probability of repeating the same role would do even better. Generous TfT does relatively better than win-stay, lose-shift and the latter can only work well in the alternating game if memory extends back more than one move. When students play iterated PD games that are either simultaneous or alternating, they are more successful in the simultaneous game with win-stay, lose-shift strategies but, as expected, more successful in the alternating game with gen TfT (Wedekind & Milinksi 1996). In addition, the win-stay, lose-shift people adjusted their strategy to protect better against all-d and exploit better all-c.

The greater cognitive complexity of win-stay, lose-shift has also been nicely confirmed by Milinski & Wedekind (1998). They asked students to play the iterated PD either continuously or interrupted after each round by a memory task (playing the game 'Memory') that acts to reduce working memory capacity. Most people playing the regular game end up adopting win-stay, lose-shift strategies, while the remaining play gen TfT. With the memory task imposed, people play the cognitively simpler gen TfT relatively more often. In addition, students who continued to use complex win-stay, lose-shift strategies were successful doing so but became less good at playing the Memory game.

There have been additional developments along these lines, e.g. modeling the effect of the spatial structure of the social neighborhood in selecting for cooperation (Nowak & May 1992) or imagining different levels of satisfaction for win-stay, lose-shift strategies (Posch et al. 1999). More generally, it has been argued that most two-party games should be modeled as a series of interactions in which partners negotiate the final outcome (McNamara et al. 1999).

4.5 Expanding beyond the Prisoner's Dilemma

Important findings have appeared which evade the limitations of the PD (and my cursory treatment of the area is purely a function of my ignorance). Introducing individual variation alone tends to drive out the inflexible strategies we have reviewed above (Johnson et al. 2002). Cooperation can thrive through the appearance of a strategy that begins small and then, when matched, raises investment (Roberts & Sherratt 1998). Another way to make reciprocity more viable, is simply to break down a large transfer into a series of contingent small ones. If as a hermaphroditic sea bass I approach you laden with eggs, and give you all to fertilize, I have nothing left with which to bargain for the privilege of fertilizing your eggs (sperms being less expensive than eggs). You are free to swim away and seek sperm elsewhere. But if I offer you a few and then no more unless you offer a few in return, we can spend a happy afternoon in reciprocal egg exchanges, which eventually add up to what we could have achieved in one exchange (Fischer 1988). Similar behavior may have evolved in polychaete worms (Sella et al. 1997). And in impalas, mutual grooming is broken down into a series of short bouts which can be terminated on evidence of non-reciprocation (Hart & Hart 1992). Certainly in human life, we recognize the principle as well, much more willing to make a loan or extend some help, if it can be broken down into smaller parts, with evidence of some positive effect available from earlier dispensations before the later ones are extended. A general theory of when parceling is expected (and how) would be most welcome.

Recently, a new game has been introduced as a metaphor for cooperation, the snowdrift game. This is the same as 'hawk vs. dove' in aggressive encounters. Two cars are stuck in a snowdrift. Each driver can free its car by removing the drift, which also frees the other car, or the two can work together (and the relative fraction of effort can be made to vary). The key is that cooperative behavior produces direct benefits to another individual and to self simultaneously. Analysis of this model produces striking departures from findings in the PD. For example, in the PD, spatial structure often favors cooperation, but in the snowdrift, the opposite is true (Hauert & Doebeli 2004). Spatial structure reduces the frequency of cooperation for a wide range of parameters. This is because the payoff is greatest if one's strategy differs from that of one's partner or neighbor. Hauert & Doebeli (2004) show that (depending on assumptions regarding payoffs) many situations modeled already as PD games may better be modeled as snowdrift games. In the continuous snowdrift game in which a cost is incurred to transfer a benefit to self and other, a uniform population often gives way to a bifurcated one, with a stable proportion of individuals making large (cooperative) investments coexisting with a set of defectors who invest little (Doebeli et al. 2004).

4.6 Possible green beard effects

When I first tried to model reciprocal altruism, I gave each individual in a two-party interaction a different locus guiding their reciprocal tendencies. I did this to avoid a so-called green-beard effect (alleles at a locus able, in effect, to recognize themselves in another individual, even if the rest of the genome is uncorrelated between the two: Hamilton 1964) but I quickly became convinced (with Bill Hamilton's encouragement) that such complexities were better left to the future. Yet, they may be important. While kinship can be excluded as a factor in some cases, this is not so easy for green beard effects. Consider TfT models which conventionally assume genes at a single (haploid) locus, in which case, by definition, when two individuals hook up, they aid themselves but also identical copies at the same locus in another individual. Thus, by its very definition, reciprocal altruism, similarity between partners is expected, perhaps at the loci directing the altruism.

A start in this direction has been made by Riolo et al. (2001a) who claim to have shown that cooperation without reciprocity can evolve via tag-based altruism where a tag is any observable phenotypic trait (including behavioral) toward whom tag-bearers direct their altruism. The problem with such systems is that they are vulnerable to the evolution of individuals who display the tag but not the altruism, leading to the rapid collapse of tag-based cooperation (Roberts & Sherratt 2001). It is clear that any successful tag-based model requires a series of additional assumptions whose plausibility has not been addressed (Riolo et al. 2001b). For another attempt to model greenbeard effects in human reciprocal altruism, see Price (in press).

An interesting example of green beard cooperation has been described in some detail for the side-blotched lizard, *Uta stansburiana* (Sinervo & Clobert 2003). Three color morphs occur in males. The blue morph is cooperative and two such males will settle close to each other, without regard to relatedness, and cooperate in driving off yellow males who attempt to sneak copulations within the territories of blue males. The orange morph, in turn, dominates blue males. Although individuals assort themselves non-randomly by color, it remains unclear if this also truly occurs at an underlying switch gene controlling color and functionally-related behavioral characters. Another oft-cited example of greenbeard effects was described in fire ants but has not been confirmed in subsequent work (Keller & Ross 1998, KG Ross pers. com.). The only clear examples of genetic green-beard effects operate at the level of cell-cell interactions, as in gametophytic factors in plants, cell-adhesion molecules in cellular slime molds and, probably, in mother-fetal interactions in mammals (reviewed by Burt & Trivers 2006).

4.7 Intrapersonal reciprocity?

Let us briefly consider another genetic hypothesis. It is possible to imagine internal genetic conflict within an individual that may, in part, be ameliorated by reciprocal effects within the genome. This is a completely novel perspective that remains to be confirmed (or disconfirmed) but the logic is fairly simple to describe. The outbred, heterozygous genetic system can be conceived of as an example of reciprocal cooperation between (usually) unrelated genomic halves (maternal and paternal). A kinship-based system would show high internal relatedness (i.e. inbreeding between parents) but with the costly side-effect of increased homozygosity. Although the two halves are typically unrelated, no conflict is expected most of the time, because the fates of the two genomes are tightly bound together in the survival of the same individual. But at reproduction, the genes are separated and sent into new individuals and this gives rise to multiple avenues for conflict in the germline as genes are selected to increase their own replication at the expense of the larger organism (a massive topic, reviewed by Burt & Trivers 2006). Axelrod & Hamilton (1981) imagined that even here reciprocal relations could restrain selfish tendencies, if, for example, a chromosome in one cell could respond to evidence of transmission ratio distortion by its homolog in another cell by doing the same. This would sometimes lead to both chromosomes driving, producing a trisomic individual. In principle, this argument could partly explain the dramatic increase in trisomy 21 in humans associated with maternal age.

In humans, there is one striking exception to the rule that internal genetic conflict is limited to conflict over genetic transmission. Genomic imprinting permits parent-specific gene expression, so that maternal and paternal genes can act according to their exact degrees of relatedness to each parent (and relatives related through them; reviewed by Haig 2002). This conflict is now well established for early life in mice and humans, paternally-active genes (with the maternal copy silent) tending to extract more resources from the mother and *vice versa* for maternally-active ones. And there are strong hints of conflict later in life as well (e.g. effects on adult behavior and brain structure).

An interesting possibility is that within-individual reciprocal relations may develop between the maternal and paternal halves of our genomes (Haig 2003). There are many situations in which the two could benefit by foregoing the costs of opposing each other and agreeing on some fair bargain. Haig (2003) wonders whether oppositely-imprinted genes might, over time, evolve complex strategies of cooperation, defection and punishment that are conditional upon the expression of oppositely-imprinted genes. It would be very interesting if such genes could create stable paternal and maternal personalities that interacted as individuals. If so, interactions with parental molding would be expected, since one set of genes in the offspring, associated with one personality, has interests exactly aligned with one parent but not the other, and *vice versa*. Does the degree of reciprocity between a couple affect the degree of reciprocity within their children?

4.8 A false negative

Every living theory generates a backlash of some sort, often, new theoretical work claiming to undermine or sharply limit the original argument. Sometimes this leads to a lively exchange, which helps to clarify matters, as in the case of my theory of parent-offspring conflict (Trivers 1974), countered immediately by Alexander (1974; logic and evidence reviewed by Godfray 1995a). My paper on reciprocal altruism generated no such initial attack. Indeed, the paper was followed a decade later by Axelrod & Hamilton (1981), which only served to strengthen the underlying theory and point the way toward many future developments. It fell to Boyd & Richerson (1988) to publish the first serious challenge, not to the logic itself but to the notion that it could work in any but small groups (5 individuals is a large group, 25 very large). The paper is now often cited as a reason to skip quickly beyond reciprocal altruism (and the iterated PD) to a group-beneficial view of our sense of fairness (e.g. McElreath et al. 2003). Fehr (2004) goes the extra step and tells us that reciprocal altruism can only evolve in groups of two, a depressing fact, were it true.

Boyd & Richerson (1988) claim that their "model closely resembles models of reciprocity in pairs" and from it, one can safely conclude that "the conditions necessary for the evolution of reciprocity become extremely restrictive as group size increases". In fact, their model does not model reciprocity within pairs and their pessimism is entirely a result of artificial assumptions at the outset, primarily that "If an individual switches from defection to cooperation every other member of the group is better off". This strange assumption is justified because it "formalizes the idea that cooperation benefits other members of the group". But why insist that cooperation between two must benefit all? If I groom you (or feed you or protect you), why does every one else have to benefit? Boyd & Richerson (1988) are pretending to model dyadic relations but have forced on such interactions a large group effect, which inevitably grows in power with group size. They have created a world in which defectors automatically benefit from the trading of favors between reciprocators. This is an unnatural assumption.

Put differently, Boyd and Richerson seem to think that when we move to n-person groups, the natural extension of 2-person model is to impose the same kind of payoff structure on the entire n-person group. Indeed, a two-person interaction within an n-person group can not, by assumption, occur because it automatically affects all other members of the n-party group. This is highly unrealistic. In fact, the opposite is more likely, that most interactions within the group continue to be dyadic or triadic. Most interaction in real time is a sequence of fast decisions about how to behave toward a small number of individuals who are immediately present. What group size of n does is to affect whom one is likely to be interacting repeatedly with. It is a mistake to collapse all these interactions into one large n-adic interaction with the additional restriction near-universal benefit or cost for any act.

Elementary logic suggests that there can be no dramatic limitation to reciprocal altruism as groups include more than a few. Even in groups of 40, a tit-for-tatter should be able to learn in 40 costly interactions who are the fellow cooperators, hence capping its losses early, say after a week or a month, with months and perhaps years of benefits to flow from successful cooperation. Recent mathematical modeling, using the PD in finite populations, shows that TfT invades at very low frequencies in moderate-sized (N ~ 30) groups (Nowak et al. 2004). We now turn to the way in which reciprocal altruism can be extended to n-party interactions within such groups.

4.9
Indirect reciprocity, image scoring and reputation

What were just a few sentences in my original paper, the possibility of three-party interactions with important observer effects, has in the past 10 years become an entire sub-discipline. Although Alexander (1987) appreciated the importance of the subject, the key step was to model the evolution of 'image scoring', the tendency to assign an individual a positive image or a negative image, depending on whether that individual was seen to act altruistically or selfishly toward a third party (Nowak and Sigmund 1998a,b). Positive images induce altruism from others and negative images selfishness. Since in this model a strategy was also included of being indifferent to the images, it was possible to show that image-scoring itself could evolve. Observer effects, in turn, are strong enough to induce cooperation and fairness, along with punishment of non-cooperators (Sigmund, Hauert and Nowak 2001). This entire subject has now been beautifully reviewed by Sigmund and Nowak (*in press*), with numerous novel insights. A subtlety of some importance is that an actor who is seen to fail to give to another can be so acting because the actor is stingy or because the recipient is unworthy (i.e. itself stingy in interactions with others). Likewise, punishing an unworthy individual may improve your image, while punishing a worthy one will have the opposite effect. But how is the observer to have this kind of detailed knowledge? Opportunities for deception would seem to be rife, increasing furthe the cognitive complexity of reciprocal altruism, with associated selection on mental traits.

Once again, a pleasing feature of the explicit theory is that it has been coupled to laboratory experiments – games played for money – that explore the underlying dynamics with real people. For example, under experimental conditions of anonymity, in which individuals can respond to whether other have been generous or not, donations go more frequently to those who themselves have been more generous in earlier interactions (Wedekind and Milinski 2000). Also, while generosity induced generosity, it did not produce a net benefit, but this turned out to be an effect of how long the game was played (n=6). When the game is twice as long, a significant positive effect emerges (Wedekind & Braithwaite 2002). The work shows that cooperation through image scoring does occur even when individuals are known only by an arbitrary marker. It provides an easy way out of measuring the net effect, which turns out to be more positive the more rounds that are played. For additional work along these lines, consult chapter 14 (see also Milinski et al 2001, 2002).

4.10
A sense of justice

Perhaps the most important implication of my paper (for me, at least) was that it laid the foundation for understanding how a sense of justice evolved. At the time, the sense of justice in humans was usually considered a product of culture and upbringing with no biological component. I thought that grounding a sense of justice in biology would only strengthen our attachment to it and naively assumed that those with a self-professed interest in justice would greet the work warmly. This turned out to be true for the great philosopher of the subject, John Rawls (1971), while the pseudo-radicals of the 1970s tore after my work (and sociobiology more generally) like so many rabid dogs after a fleeing rabbit, missing the point entirely (Trivers 1981).

I was surprised on rereading my paper recently to find no reference to justice itself but only to the weaker term 'norms'. I apparently did not get around to using the proper term until Trivers (1981, 1985). The argument, in any case, is the same. Even in two-party interactions, but especially in multi-party ones, one needs a standard by which to judge deviations from symmetrical (or fair) interactions, the better to detect cheaters. Such cheating is expected to generate strong emotional reactions because unfair arrangements, repeated often, may exact a very strong cost in inclusive fitness. In that sense, an attachment to fairness or justice is self-interested and we repeatedly see in life, as expected, that victims of injustice feel the pain more strongly than do disinterested bystanders and far more than do the perpetrators. This is not to say that we have no response to injustice visited on people very far from us in space (or even time) but this does not imply that our sense of justice evolved with these distant events in mind.

The first appearance of the concern in children is thoroughly self-interested ('but that's not fair, mommy') and those who would push a group selection (or mere cultural diffusion) view of the sense of justice would do well to ask themselves when last they heard a child say, 'mommy, daddy, I am acting unfairly, please stop me'. Self-interest is often confused with selfishness, a mistake given wide prominence by Dawkins' (1976) misuse of the word 'selfish'. 'Enlightened self-interest', after all, is meant to call attention to the value for our own selves of funneling certain benefits through others.

It is easy (in the United States, at least) to underestimate the power of our sense of justice. For example, the neoconservative architects of the current U.S. bloodletting in Iraq are fond of saying that the United States is disliked in other countries not because its policies are perceived as unjust (and certainly not because they are unjust) but because the U.S. is envied for its size, power and wealth. Envy, however, is a trivial emotion compared to our sense of injustice. To give one possible example, you do not tie explosives to yourself to kill others because you are envious of what they have, but you may do so if these others and their behavior represent an injustice being visited on you and yours day after day and year after year, often with little alternative action available to you, as in Palestine, where the people suffer robbery of land, water and life and are denied little more than rifles in self-defense, while their next-door Israeli neighbors (and occupiers) are armed with nuclear weapons and the latest in U.S. lethal technology.

There is growing evidence that something like a sense of fairness has evolved in a range of non-human primates who, in turn, practice reciprocal altruism (see de Waal & Brosnan, this volume; for chimp reciprocity in the wild, see Watts 2002). Thus, there appears to be an aversion to inequity in both capuchins and chimpanzees (Brosnan & de Waal 2003, Brosnan et al. 2004). In chimpanzees, ungenerous individuals are attacked when they beg for food, as are those who do not support those who just supported them (de Waal 1992b). These observations are consistent with the view that a sense of fairness evolved slowly over a long period of time, first in dyadic relationships and then more widely. The number of dyadic relationships should not be underestimated; there may be a fair split between the interests of parent and offspring, sibling and sibling, near-neighbor and near-neighbor, and so on.

4.11 The experimental study of reciprocal altruism

One of the more welcome developments in the study of reciprocity, and the human sense of justice, is the development of economic games that attempt to mimic real life situations but can be played in the lab for real money. These games can be played cross-culturally and altered in a series of ways to explore relevant causal factors. This is a considerable advance over the world of social psychology, with its reliance on paper-and-pencil tests of human dispositions, artificial situations difficult to interpret and (sometimes) deception of the subjects under study. Indeed, Rapaport & Chammah (1965) saw this very clearly when they argued not only for the theoretical utility of the PD but its value in generating empirical data to test the theory. As we have seen, experiments in which people play iterated games of PD have provided valuable evidence on the cost-benefit ratio of such strategies as win-stay, lost-shift or gen TfT.

There is now a very large and excellent literature on economic games (see Roth 1995, Smith 2003). Many have been played in the laboratory under controlled conditions, games with names like the ultimatum game, dictator game, public goods game and so on. They can be played single shot or iterated, anonymous or non-anonymous, with and without onlookers, and they can be analyzed theoretically with simple models.

One game will serve as an example, the single-shot, anonymous ultimatum game. The game is played once by a proposer and a responder (Güth et al. 1982, Burnham 2002). The proposer is given an amount of money (by the experimenter) to split with an unknown responder. He (or she) proposes a split and if the responder accepts the offer, the two split the money accordingly, but if player two rejects the offer, neither player receives any money. No further interactions occur between the two. In the standard economic model of maximizing financial gain, responders are expected to take whatever they are offered, as long as it is not zero, and, thus, proposers are expected to make very low offers and to try to keep most of the money. But this is not what research shows (Forsythe et al. 1994). Offer modes and medians are 40-50%, offer means are 30-40%, and offers below 20% are usually rejected (Camerer 2003). This result now has been replicated, with some interesting variation, in scores of studies, with varying stakes, around the world (Henrich et al. 2005).

The problem arises in how to interpret these results. It is generally agreed that they show our attachment to a sense of fairness, even when this costs us money but why do we act this way? To many evolutionists, including myself, this sense of justice or fairness benefits us in everyday life by protecting us from unfair arrangements that harm our inclusive fitness. We are expected to react negatively to unfair offers by others, not out of envy of their extra portion, but because they chose to inflict this unfair offer on us and the unchallenged repetition of such behavior is expected in the future to inflict further costs on our inclusive fitness. Consistent with this, people accept lower offers from a computer than they do from another person, even though both offers offer identical payoffs to two humans (Blount 1995). This approach assumes that our responses were never selected to perform in the highly unusual, one-shot, anonymous interaction in a lab, with payoffs underwritten by a third party. Put another way, these experimental results seem on their face neither unexpected nor puzzling.

By contrast, some social scientists now playing these games have opted for a very different view (e.g. Fehr & Fischbacher 2003, Gintis et al. 2003). According to them, the results prove that our sense of fairness cannot have a self-interested function, all possibility of return effects having been removed. Instead, it must have been selected to benefit the group or appeared by some process of cultural diffusion. This they call 'strong reciprocity', to differentiate it from the 'weak' reciprocity of classic reciprocal altruism

4.12
'Strong reciprocity?'

In defense of this interpretation of the ultimatum game, Bowles & Gintis (2003) tell us "We do not think that subjects are unaware of the one-shot setting, or unable to leave their real-world experiences behind with repeated interactions at the laboratory door". Surely, awareness is irrelevant. You can be aware that you are in a movie theatre watching a perfectly harmless horror film and still be scared to death. As for leaving real-world experiences at the laboratory door, I know of no species, humans included, that leaves any part of its biology at the laboratory door; not prior experiences, nor natural proclivities, nor ongoing physiology, nor arms and legs, nor whatever. That is the whole point of experimental work. You bring living creatures into the lab (ideally, whole) to explore causal factors underlying their biology, the mechanisms in action. You do not imagine that you have thereby solved the problem of evolutionary origin; that is, that you can shortcut the problem of evolutionary function by simply assuming that the organism's actions in the lab represent evolved adaptations to the lab.

People do not leave their religion at the laboratory door and this alone can induce 'observer effects' for those who imagine that God is scrutinizing their every action (D. Stahl, pers. com.). Indeed, the impression that God will punish malefactors may have been promoted to encourage cooperation (Johnson & Krüger 2004). Nor do people leave their culture at the door, and it has been claimed cross-culturally that the higher the degree of market integration and cooperation in daily life, the more prosocial individuals act in anonymous games in the lab (Henrich et al. 2005). Nor do people leave their testosterone at the lab door and unpublished work shows that such levels are positively associated with rejection rates in men (T. Burnham unpubl. data). Apparently, those who promote 'strong reciprocity' believe that a little verbiage ('you know not who the other actor is nor will you ever know – and *vice versa*') is sufficient to put the human being into a state of suspended animation such that he or she automatically and appropriately adjusts to the full evolutionary implications of this novel, experimental situation.

This argumentation is supported, they argue, by the fact that humans do make discriminations in everyday life based on chance of future interaction, diminishing cooperation as frequency of interaction decreases (exactly as expected according to the theory of reciprocal altruism). But the fact that y, a function of x, is decreasing as x approaches zero, does not mean that the function passes through zero simultaneously on both axes. How do they handle this deficiency? With a little jargon, "individuals should be fully capable of taking a 'zero-baseline' of cooperation" (Fehr & Henrich 2003). This merely assumes what needs to be shown and is unlikely on its face.

These authors (and others) also claim that human evolution was characterized by frequent one-shot encounters with important fitness effects (Fehr & Henrich 2003). Leaving aside murder, this seems most unlikely, especially when one considers that what is meant is one-shot anonymous interactions of the form of an ultimatum game. As I have pointed out elsewhere, social interactions are intrinsically repeat interactions, certainly over very small time scales of seconds and minutes but almost always over longer time periods as well (Trivers 2004). What they are calling one-shot encounters were really repeat interactions lasting at least minutes and hours, if not days and months. Imagine I am walking through the woods with a man I have never met before and am unlikely to see after the walk, with not a soul in sight. He points to five low-hanging apples and suggests that he stand on my shoulders to pick them, after which he will give me one. I am likely to argue the point immediately (and a chance to communicate even once prior to an anonymous ultimatum game has been shown to affect the interaction; Rankin 2003). Even if I agree with him, as he lands on the ground, I may knock three apples out of his hand and run, or strangle him, taking apples and any other property he has (after all, observer and other return effects have been ruled out). This interaction all takes place over a span of several minutes. Thus, even an imaginary 'one-shot' encounter really consists of a series of interactions.

I have a close relative who, among other virtues, has a well-developed sense of spite. She has a long memory for slights and will repay in kind. As I described the ultimatum game to her, with her share of the pie (as recipient) now reduced to 30%, her fingernails began (literally) to dig into the table and her face contorted with anger (looks just like her everyday self to me, I thought). The intensity was remarkable. Those fingers were set to dig into someone's face, if need be. When Jamaican youngsters are given low offers, they often respond with a flash of anger and something like, 'this is all I get?'. Anger is not a mere emotion, it is (costly) physiological arousal for immediate aggressive action. Anger makes no sense without the possibility of future interaction. Neurophysiological work on the ultimatum game is also consistent with biological arousal for future interaction. Functional Magnetic Resonance Imaging (fMRI) work shows that unfair offers are met with activation of part of the recipient's brain (the anterior cingulate) involved in negative emotions, primarily anger and disgust, and control functions involving conflict, and the higher the activation, the greater the chance that such an offer will be rejected (Sanfey et al. 2003). In short, in our everyday behavior and neurophysiology we respond to so-called one-shot encounters as if they were the first in a chain of interactions.

The errors I have drawn attention to are but a few of many. An individual who turns down an 80:20 offer hardly benefits the group; 100% of resources disappear at once (Burnham & Johnson 2005). Indeed, moral philosophers have shown how the group-benefit approach to justice is inferior to 'justice as fairness', in which a fair arrangement is defined not by whether it maximizes group output, but by whether individuals would accept it if they did not know in advance which position they occupied (Rawls 1971). For those seeking a more in-depth treatment of the 'strong reciprocity' approach and its failings, see Burnham & Johnson (2005). These include idiosyncratic use of language. A lifetime of trading benefits with others in a fair manner is apparently 'weak reciprocity' while 'strong reciprocity' is a single, take-it-or-leave-it interaction with no possible reciprocity, underwritten by a third party. Likewise, punishment of unfair behavior is called 'altruistic punishment' on the assumption that such behavior has a net cost for the actor while benefiting others, something that needs to be shown, not assumed.

Finally, for me there is a feeling in all of this of déjà vu, all over again. There has been a long history in the social sciences of assuming whatever is necessary in order to make the argument that is desired. Is it necessary to assume that there is little or no genetic variation in human social traits in order to push an extreme environmental interpretation? So be it. Can we assume that humans are rational utility maximizers, where utility is anything the organism wishes to see maximized, and that on this foundation we can safely build social theory? Be our guest. May we assume that people's behavior in the highly unusual one-shot ultimatum game reflects how they were selected to act in precisely this situation? So let it be granted. On the bright side, there is actually some progress here. The first position denies both genetics and evolution. The second merely assumes that these subjects are irrelevant, while the third embraces evolutionary logic in principle but promptly gives it a truncated form.

4.13 Forgiveness and revenge

Forgiveness may play an important role in a system of reciprocal altruism, especially where the latter has degenerated into a system of reciprocal spite. Forgiveness tends to short-cut the spite, potentially saving enormous amounts of energy, both outward-directed and inner-consumed. There is strong evidence that positive emotions are associated with high immune performance (Rosenkranz et al. 2003) and that manipulations of mood which increase positive affect, e.g. through meditation, are themselves associated with an increase in immune response (Davidson et al. 2003). There is no direct evidence for similar benefits from forgiveness, but certainly many who have suffered grievous loss through the action of others have spoken of the corrosive internal effects of retaining the spiteful mentation of hatred and revenge (Cole 2004). As we saw earlier, the existence of mistakes automatically selects for a degree of forgiveness in everyday life. Forgiveness may also be efficient from a cognitive perspective, dropping a contradiction in one's mind between past losses, present failure to redress the injustice, which then requires planning for future activity. Letting go is letting go of a burdensome scheme of mentation. Set against this, is the impulse toward revenge and retribution. This seems most likely in social arrangements that endure over long periods of time, including several generations, so that long-delayed revenge is both possible and possibly instrumental in protecting later-borne relatives. A careful treatment of the evolutionary dynamics of forgiveness and/or revenge would be most welcome.

4.14 Justice and truth

Elementary logic suggests a possible connection in individuals between apprehension of the truth and commitment to justice in social relations. An immoral stance often requires deception, with its inevitable effects on self-deception (Trivers 2000). This tendency may be more pronounced at the top of a social hierarchy, where truth-telling by others is emphasized as a necessary virtue, while an illusion of the same by self is promulgated. There is a natural inclination to over-emphasize the justice of one's own position and thus to under-emphasize that of one's opponent, thereby underestimating the strength of their resistance. This asymmetry is especially striking in aggressive invasions of the land of others, where there ought to be a natural (moral and physical) presumption in favor of the prior occupants.

4.15 Torture

The demonstration that altruistic punishment (or moralistic aggression) is pleasurable to the punisher, as judged neurophysiologically (de Quervain et al. 2004), invites the speculation that this has made the appearance of torture more likely. Torture is the laser-like application of highly spiteful activities toward inducing pain, fear, shame, madness, and so on, in others. It may provide pleasure for the torturers, especially if they believe such action is morally justified, as in the words of U.S. Vice President Cheney, that the 'worst of the worst' are incarcerated in Guantanamo, Cuba and in Iraq. Recently, we have learned in the United States that internal government documents argue that torture is not torture if there is no organ failure or imminent death, a redefinition that eliminates nearly all non-fatal forms of torture. Such are the achievements of self-deception in the service of moralistic aggression.

4.16 Summary

Since 1971, an enormous and very sophisticated literature has grown up on reciprocal altruism and cooperation more generally, both in biology and economics. Noteworthy has been the success of generating a detailed series of theoretical findings, both within and outside the Prisoner's Dilemma paradigm. In the latter, we see conditions under which a series of strategies can displace each other or survive together: all-d, TfT, gen TfT, all-c and win-stay, lose-shift. We have seen alternatives to the PD such as the snowdrift game, which may produce very different effects (e.g. of spatial structure). And we have considered two genetic hypotheses of interest, green-beard altruism and within-individual genetic reciprocity, both of which await confirmation. We have outlined the way in which reciprocal altruism may generate a sense of justice as a means of guarding against cheaters and we have explored an alternative interpretation, popular in some circles, that our sense of fairness evolved without regard to return effects, a view for which there is as yet no useful, positive evidence. All of these areas of research are unusually vibrant and productive at the present time and much valuable new work is published literally every month. This work shows every promise of uniting, at last, important parts of economics with evolutionary biology.

Acknowledgments

I am grateful to Jim Bull, Terry Burnham, Peter Godfrey-Smith, Christoph Hauert, Marc Hauser, Peter Kappeler, Martin Nowak, Carel van Schaik, Dale Stahl, Claus Wedekind, Richard Wrangham and Darine Zaatari for many, helpful comments. I am grateful to Terry Burnham for first alerting me to the difficulty of interpreting the results of economic games.

Chapter 5

Simple and complex reciprocity in primates

FRANS B. M. DE WAAL, SARAH F. BROSNAN

5.1 Introduction

Ever since Kropotkin (1902), the proposed solution to the evolution of cooperation among non-relatives has been that helping costs should be offset by return-benefits, either immediately or after a time interval. Formalized in modern evolutionary terms by Trivers (1971), this principle became known as reciprocal altruism.

Reciprocal altruism presupposes that: (i) the exchanged acts are costly to the donor and beneficial to the recipient, (ii) the roles of donor and recipient regularly reverse over time, (iii) the average cost to the donor is less than the average benefit to the recipient, and (iv) except for the first act, donation is contingent upon receipt. Although the initial work on cooperation (especially from the prisoner's dilemma perspective) focused primarily on the payoff matrix to distinguish between reciprocity and mutualism, more recent efforts have included a significant time-delay between given and received services as an additional requirement for reciprocal altruism (e.g. Rothstein & Pierotti 1988, Taylor & McGuire 1988). Given that a distinction between immediate and delayed benefits is theoretically richer, we include a time-delay in our definition of reciprocal altruism.

The above considerations outline the steps of an evolutionary argument about how reciprocal cooperation may have come into existence. As such, it applies to organisms from fish to humans. This should not be taken to mean, though, that reciprocal help in human society is essentially the same as in guppies. This would be a fundamental error; the above theoretical framework only deals with the ultimate reasons for the existence of reciprocal exchange. That is, it provides an explanation for why animals engage in such behavior, and which fitness benefits they derive from it. It provides no explanation for how such cooperation is achieved, commonly referred to as the proximate explanation, as discussed by Brosnan & de Waal (2002). While it should be noted that, in the larger scheme of things, it is unlikely that human reciprocity deviates substantially from that of other animals, such as the apes, with which we share a long evolutionary history, humans probably have added unique complexities. The most parsimonious assumption with regards to recently-diverged species is that if they act similarly under similar circumstances, the psychology behind it is most likely similar too (de Waal 1991a).

One can imagine forms of reciprocal altruism in which the time-delay between the exchanged services is short, hence the need for record keeping minimal. Individual recognition is perhaps not necessary in such cases. This mechanism would approach mutualism as the time interval between exchanged favors becomes shorter. But also in the case of significant time-delays, exchanges are not necessarily based on give-and-take contingencies. They may simply reflect underlying characteristics of the relationship between individuals. If so, the role of memory would be minimal. This means that not all forms of reciprocal altruism require the cognition we tend to associate with it, such as scorekeeping, punishment of cheaters, attribution of intentions, and awareness of the respective costs of behavioral currencies.

Skeptics of reciprocal altruism in non-human animals sometimes fail to recognize this distinction between (i) the ultimate explanation, which merely postulates that the cost of help given be offset by the benefits of help received, and (ii) the proximate mechanism, which concerns the precise way in which benefits find their way back to the initial altruist. Satisfied with the most advanced mechanism only, they ignore simpler forms of reciprocity (Hammerstein 2003b, Stevens & Hauser 2004). Instead of wondering why reciprocity in animals is so rare, however, the real question is why we feel animals need to operate at the cognitive level that we are capable of, and even more pertinently, how we can be sure that we ourselves operate at that level most of the time? As soon as we move away from anthropocentric assumptions about the mechanism, reciprocity turns out to be widespread indeed (e.g. Dugatkin 1997).

This is not to say that determination of which behaviors evolved as reciprocal altruism, or not, is an easy task. Numerous examples have been posited, but often it is found that either the animals are related, and an alternative explanation for the observed altruistic exchange is kin selection (Wilkinson 1988), or else previously unnoticed benefits to the presumed altruist are found, indicating that the observed behavior is better described as byproduct mutualism (e.g. both animals benefit simultaneously: Koenig 1988, Clements & Stephens 1995; or pseudo-reciprocity: Connor 1986). Furthermore, it is difficult to assess reciprocity in situations in which the exchanged behaviors are dissimilar since the fitness value of different currencies is hard to compare (Seyfarth & Cheney 1988). Even within the same currency, fitness costs and benefits may vary for the parties involved due to individual differences in rank, size and age (Boyd 1992).

More than two decades ago, chimpanzee society was characterized as a 'marketplace' in which a variety of services are traded back and forth among individuals (de Waal 1982a). Here we will go into the quantitative details of this marketplace as expressed in coalitions, grooming and food-sharing among chimpanzees (*Pan troglodytes*). Chimpanzees, for instance, have a wide range of goods and services that can be exchanged, including coalitionary support, mating privileges, grooming and food-sharing.

The exchange of these commodities indicates a fairly high level of cognitive accounting in these marketplaces. Experiments on brown capuchin monkeys (*Cebus apella*) further illuminate the proximate side of cooperation and reciprocal altruism. Animals pursue immediate goals which, in the end, often beyond the cognitive horizon of the actors themselves, translate into benefits that form the material for natural selection. A study of proximate mechanisms helps to determine if evolutionary hypotheses are predicting behavior within the animal's range of abilities, because no matter how elegant or compelling an evolutionary scenario, it is useless if the organism lacks the capacity of behaving as the theory predicts (Stamps 1991).

5.2 Observational studies

5.2.1 Reciprocal coalitions and revenge

De Waal & Luttrell (1988) applied a matrix permutation technique to correlations between given and received agonistic support in over two thousand instances observed over a period of five years in the Arnhem Zoo chimpanzee colony as well as a large sample of interventions in mixed-sex groups of rhesus (*Macaca mulatta*) and stumptail macaques (*M. arctoides*) at the Wisconsin Primate Center. In all three studies, agonistic intervention was defined as a third individual responding with an aggressive act against one, and only one, of two participants in a dyadic confrontation. Interventions were recorded as triplets; individual A helps B against C. Reciprocity could occur in the domains of both *pro* (A helps B) and *contra* interventions (A goes against C), hence may reflect two kinds of *quid pro quo* as in "One good turn deserves another" and "An eye for an eye". The latter kind of punitive reciprocity has received far less theoretical attention than the first (but see Clutton-Brock & Parker 1995).

Table 5.1 presents Pearson correlations as well as partial correlations after statistical removal of the effects of symmetrical relationship characteristics. These effects are removed because any characteristic that is symmetrical between two individuals can be used to create reciprocal distributions of behavior if the characteristic causes both partners to show the behavior in question. The analysis controlled for symmetrical characteristics such as (i) time spent in proximity, (ii) matrilineal kinship and (iii) same-sex combination. The partial correlations resulted after correction for all of these characteristics at once.

The table confirms a significant level of reciprocity in *pro* interventions among adults of all three species, even after statistical adjustment for symmetrical relationships. The chimpanzees showed considerably higher reciprocity correlations than the macaques, however. An even more significant difference emerged with regards to harmful *contra* interventions. These interventions were significantly reciprocal in chimpanzees, but significantly anti-reciprocal in macaques. That is, if macaque A often intervenes against B, B will rarely do so against A, whereas in chimpanzees we find that if chimpanzee A often goes against B, B will do the same to A.

De Waal & Luttrell (1988) explain the absence of reciprocal *contra* interventions in macaques by their stricter hierarchy, which prevents subordinates from intervening against dominants. Most data in their study came from females, however. A similar analysis restricted to male bonnet monkeys (*Macaca ra-*

Table 5.1. Pearson reciprocity correlations (r) between given and received agonistic interventions for three primate species. The *pro* rate concerns beneficial interventions, the *contra* rate harmful interventions. Partial correlation coefficients (pr) have been adjusted for the effects of multiple symmetrical relationships characteristics. Probability levels (p) evaluate the partial correlations based on a permutation technique. From de Waal & Luttrell (1988).

Measure	Correlation	Rhesus	Stumptail	Chimpanzee
Pro rate	r	**0.36**	**0.35**	**0.61**
	pr	0.28	0.18	0.55
	p	0.005	0.025	0.005
Contra rate	r	**−0.17**	**−0.23**	**0.33**
	pr	−0.19	−0.29	0.32
	p	0.005	0.005	0.025

diata) did yield evidence for reciprocal *contra* interventions, perhaps reflecting a looser dominance structure among male than female macaques (Silk 1992b). There is also evidence for indirect retaliation among macaques, when defeated subordinates redirect aggression against their opponent's relatives (Aureli et al. 1992). The squaring of accounts in the negative domain, dubbed a revenge system by de Waal & Luttrell (1988), may represent a precursor to human justice, since justice can be viewed as a transformation of the urge for revenge, euphemized as retribution, in order to control and regulate behavior (Jacoby 1983, de Waal 1996b).

Symmetrical relationship characteristics are (or ought to be) an issue in every correlational approach to given and received acts of assistance across dyadic relationships, such as social grooming, food-sharing and agonistic support (e.g. Seyfarth 1980, de Waal & Luttrell 1988, de Waal 1989). Before concluding from a positive correlation that giving depends on receiving, the most obvious variable to control for is time spent in association; if members of a species preferentially direct favors to close associates, the distribution of favors will automatically be reciprocal due to the symmetrical nature of association. A similar argument applies to any symmetrical relationships characteristic (e.g. kinship, age or gender similarity). This mechanism for cooperation, dubbed symmetry-based reciprocity, needs to be distinguished from calculated reciprocity, which is based on mental score-keeping of given and received favors (de Waal & Luttrell 1988). In most species for which reciprocal altruism has been reported through observational methods, symmetry-based reciprocity has not been excluded and hence, remains the most likely mechanism (e.g. blood sharing in vampire bats, *Desmodus rotundus*, Wilkinson 1984; allogrooming in impala, *Aepyceros melampus*, Hart & Hart 1992).

This is not to say that uncorrected positive correlations are meaningless; obviously, symmetries are part of evolved social life. If they assist reciprocal relations that confer benefits, this is all that matters from an evolutionary perspective. Evidence limited to positive correlations, however, does not permit conclusions about contingencies between giving and received behavior. Although we know from experiments (see below) that monkeys are capable of contingent exchange, and although analyses that have gone beyond dyadic relationships, such as in biological markets (cf. Noë & Hammerstein 1994), show behavioral distributions that seem too finely tuned to the supply and demand of benefits as well as partners to be accounted for by symmetry-based reciprocity (e.g. Barrett & Henzi, this volume), we would still argue that correlations cannot reveal underlying processes and that it is best, therefore, to adhere to conservative interpretations.

In view of these problems, observational studies should add sequential analyses, which look at the unfolding of behavior over time. Does a beneficial act by individual A towards B increase the probability of a subsequent beneficial act by B towards A? These analyses get around the problem posed by symmetries. Preliminary sequential evidence for an exchange between affiliative behavior and agonistic support, and *vice versa*, exists for cercopithecine monkeys. De Waal & Yoshihara (1983) found increased post-conflict attraction and grooming between previous alliance partners in rhesus monkeys. Seyfarth & Cheney (1984) employed playbacks of calls that vervet monkeys (*Cercopithecus aethiops*) use to both threaten an aggressor and solicit support to gauge the reaction of individuals that had recently been groomed by the caller. They reported increased attention to previous grooming partners. Finally, Hemelrijk (1994) examined agonistic support after experimentally manipulating grooming among long-tailed macaques (*M. fascicularis*) and found indications that individuals supported those who had groomed them, i.e. individual A supported individual B more if B had groomed A, but not if A had groomed B.

The last study comes closest to demonstrating a temporal relation between one service and another, but what is still missing is evidence for partner-specificity, i.e. that the return service specifically targets the individual who offered the original service. The alternative is generalized reciprocity, or the 'good mood' hypothesis (see below), according to which the receipt of services leads to an indiscriminate increase in beneficial behavior. Our research on food-sharing in chimpanzees attempted to address this important distinction.

5.2.2
Food for grooming in chimpanzees

Although food-sharing outside the mother-offspring or immediate kin-group is rare in the primate order (Feistner & McGrew 1989), it is common in both capuchin monkeys and chimpanzees. Food-sharing lends itself uniquely to experimental research, because the quantity and type of food available, the initial possessor, and even the amount of food shared can be manipulated by the experimenter. Second, food-sharing provides a quantifiable currency. An observer can see exactly how many times the non-possessor obtains food and can estimate quantities shared. Finally, the observer can tell whether the sharing was active or passive.

Active food-sharing, a rare behavior, consists of one individual handing or giving food to another individual, while passive food-sharing, by far the more common type, consists of one individual obtaining food from another without the possessor's active help (Fig. 5.1). The sharing is selective, however, in that possessors are not equally tolerant of all individuals; only approximately half of the interactions between a possessor and an interested non-possessor resulted in an actual transfer of food.

There are three common hypotheses to explain food-sharing in primates: (i) the sharing-under-pressure hypothesis, (ii) the sharing-to-enhance-status hypothesis and (iii) the reciprocity hypothesis (reviewed by de Waal 1989, 1996b). The sharing-under-pressure hypothesis, similar to the tolerated-theft hypothesis of Blurton-Jones (1987), predicts that individuals will share in order to be left alone by potentially aggressive conspecifics (Wrangham 1975, Stevens 2004). This hypothesis is contradicted by the fact that the most generously sharing individuals are often fully dominant, hence have little to fear from anybody around them, and that most of the aggression in feeding clusters, rather than being by non-possessors against possessors, is directed the other way around (de Waal 1989). This confirms the remarkable 'respect of possession' (cf. Kummer 1991) already noted by Goodall (1971) in her first accounts of meat sharing among wild chimpanzees. The sharing-under-pressure hypothesis also fails to explain food transfers in experimental set-ups in which negative consequences of non-sharing are eliminated by physical separation (see below).

What remains, then, are the sharing-to-enhance-status hypothesis and the possibility of reciprocity. The first hypothesis predicts that generosity increases the altruist's standing in the community (Hawkes 1990), but there is as yet no evidence for this effect in non-human animals. The reciprocity hypothesis predicts that food is part of a service economy, hence that it is exchanged reciprocally for other favors. These two hypotheses are, of course, not mutually exclusive.

Our initial studies approached food-sharing by means of matrix correlations. This matrix approach yielded significant results in the predicted direction. However, food-sharing among chimpanzees correlates positively with proximity and grooming, hence the amount of time individuals spend together in non-food situations. As explained before, the effects of association must be removed before any explanation other than symmetry-based reciprocity may be invoked. When the matrix analysis was redone while statistically controlling for the effects of association, the correlation continued to be significant.

Statistical elimination of a variable is not as powerful as experimentally controlling for it, however. A new experiment was designed to measure temporal patterning within each dyad, thereby holding the effect of association constant. Partner specificity was addressed, i.e. whether a beneficial act by individual A towards B specifically affects B's behavior towards A (de Waal 1997a). The difficulty in measuring food-sharing across time is that after a group-wide food-sharing session, as used in these experiments, the motivation to share is changed (the animals are more sated). Hence, food-sharing cannot be the only variable measured. A second service that is unaffected by food consumption needs to

Fig. 5.1. A cluster of food-sharing chimpanzees at the Yerkes Field Station. The female in the top-right corner is the possessor. The female in the lower left corner is tentatively reaching out for the first time, whether or not she can feed will depend on the possessor's reaction. Photograph by Frans de Waal.

be included. For this, grooming between individuals prior to food-sharing was used. The frequency and duration of hundreds of spontaneous grooming bouts among the chimpanzees was measured during 90 minute observation sessions. Within half an hour after the end of these observations, the apes were given two tightly bound bundles of leaves and branches. Nearly 7000 interactions over food were carefully recorded by observers and entered into a computer according to strict empirical definitions described by de Waal (1989). The resulting database on spontaneous services exceeds that for any non-human primate.

It was found that adults were more likely to share food with individuals who had groomed them earlier in the day. In other words, if A groomed B in the morning, B was more likely than usual to share food with A later in the same day (Fig. 5.2). This result, however, could be explained in two ways. The first is the so-called 'good mood' hypothesis according to which individuals who have received grooming are in a benevolent mood leading to generalized reciprocity, i.e. increased sharing with all group members. The second explanation is the exchange hypothesis, in which the individual who has been groomed responds by sharing food specifically and only with the groomer. The data indicated that the sharing was specific to the previous groomer. In other words, each chimpanzee remembered who had just performed a service (i.e. grooming) and responded by sharing more with this particular individual. Also, aggressive protests by food possessors to approaching individuals were aimed more at those who had not groomed them than at previous groomers. All of this is compelling evidence for the reciprocal exchange hypothesis.

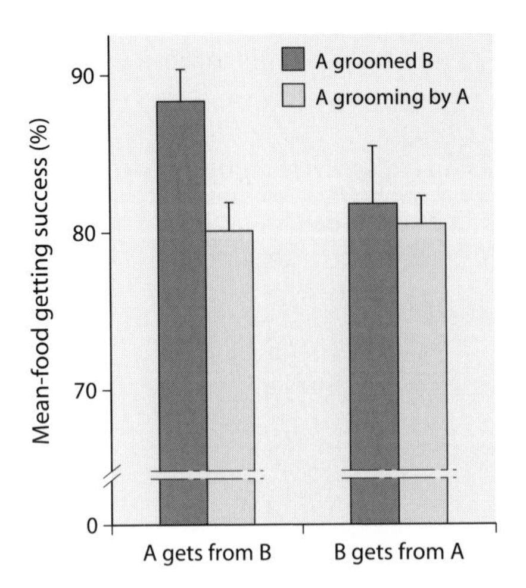

Fig. 5.2. Mean (+SEM) food-getting success per dyadic direction between adult chimpanzees during food trials. Two conditions are distinguished: either individual A had groomed B in the hours prior to the food trial, or no previous grooming by A to B had occurred. The left-hand side of the graph shows the success of A in obtaining food from B (A gets from B); the right-hand side shows the success of B in obtaining food from A (B gets from A). Success is defined as the percentage of approaches to a food possessor resulting in a transfer of food (regardless of whether the transfer is active or passive) from possessor to non-possessor. It was found that A's success in obtaining food from B increased significantly after A had groomed B, whereas B's success in obtaining food from A was unaffected by A's previous grooming. From de Waal (1997a).

It was further found that grooming between individuals who rarely did so had a greater effect on subsequent food-sharing than grooming between partners who commonly groomed. Among partners in which little grooming was usually exchanged, there was a more pronounced effect of grooming on sharing. There are several interpretations. It could be that grooming from a partner who rarely grooms is more noticeable, leading to increased sharing by the food possessor. Chimpanzees may recognize unusual effort and reward accordingly. Secondly, individuals who groom frequently tend to be close associates, and favors may be less carefully tracked in these relationships. Close friendships may be characterized by symmetry-based reciprocity, which does not have the high degree of conditionality found in more distant relationships. These explanations are not mutually exclusive; both will lead to a reduced level of conditionality the more common exchanges are in a relationship.

Of all existing examples of reciprocal altruism in non-human animals, the exchange of food for grooming in chimpanzees comes closest to fulfilling the requirements of calculated reciprocity. This study strongly suggests memory-based, partner-specific exchange in chimpanzees. It goes beyond symmetry-based reciprocity inasmuch as symmetry is a constant feature of relationships that cannot explain contingencies across time, as demonstrated here. There existed a significant time delay between favors given and received (from half an hour to two hours); hence, the favor was acted upon well after the previous positive interaction. Apart from memory of past events, for this to work we need to postulate that the memory of a received service, such as grooming, induces a positive attitude towards the same individual, a psychological mechanism described as 'gratitude' by Trivers (1971), and further explored by Bonnie & de Waal (2004).

5.3 Experiments on capuchin monkeys

Even though laboratory work on primate cooperation goes back to Crawford (1937), few experimental studies have been conducted since. What is especially lacking is the experimental manipulation of 'economic' variables, such as the relation between effort, reward allocation, and reciprocity. Recently, this situation has changed thanks to experiments on brown, or tufted, capuchin monkeys.

The *Cebus* genus seems particularly suited for cooperation research. These monkeys show high levels of social tolerance around food and other attractive items, sharing them with a wide range of group members both in captivity and in the field (Izawa 1980, Janson 1988, Thierry et al. 1989, de Waal et al. 1993, de Waal 1997b, Fragaszy et al. 1997). This level of tolerance is unusual in non-human primates, and its evolution may well relate to cooperative hunting. Perry & Rose (1994) confirmed reports by Newcomer & de Farcy (1985) and Fedigan (1990) that wild *Cebus capucinus* capture coati pups (*Nasua narica*) and share the meat. Since coati mothers defend their offspring, coordination among nest-raiders conceivably could increase capture success. This has also been suggested for hunting by capuchins on giant squirrels (*Sciurus variegatoides*; Rose 1997). Rose (1997) proposed convergent evolution of food-sharing in capuchins and chimpanzees based on group hunting. The precise level of cooperation of the hunt is not relevant for such evolution to occur; all that matters is that hunting success increases with the number of hunters. Under such circumstances, every hunter has an interest in the participation of others, something promoted by subsequent sharing.

5.3.1 Reciprocal food-sharing

In the delayed exchange test, or DET, a pair of monkeys is placed in a test chamber, separated from each other by a mesh partition that allows for food-sharing. Monkey A is given a bowl of cucumber pieces, placed well out of reach of monkey B. After 20 minutes, the cucumber is removed, and a bowl of apples is given to monkey B (second test phase). The same pair is given another DET later, on a different day, with the order reversed between the monkeys (Fig. 5.3).

In years of testing with this paradigm, our capuchins displayed an astonishing amount of social tolerance, sharing food on a reciprocal basis. Males tended to share more than females regardless of the sex of the partner. A matrix analysis found that, for the 14 female-female dyads in which the possessor was dominant, more sharing occurred between partners who in the group in which they lived

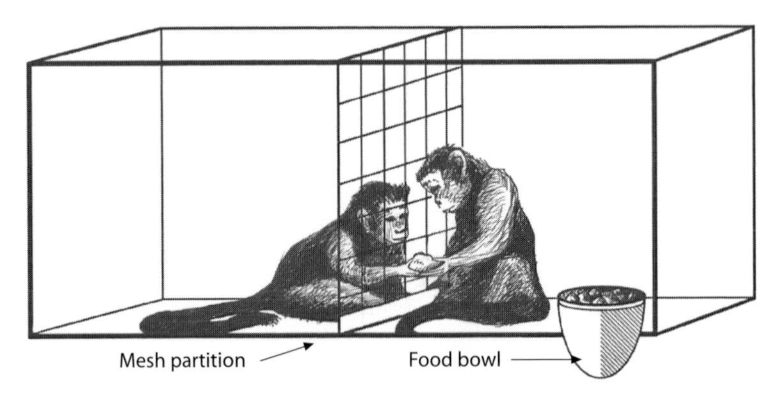

Fig. 5.3. Sketch from an actual video still showing active food-sharing in a pair of capuchin monkeys. The monkeys are separated by a mesh partition, and the monkey on the right has access to a food bowl containing apples. Active food-sharing is rare, but facilitated taking, in which the food possessor drops pieces by the mesh and allows the other monkey to take them, is common. Drawing by Frans de Waal, from de Waal (1997b).

had: (i) fewer agonistic interactions, (ii) shorter rank distances (i.e. were close in the dominance hierarchy) and (iii) higher levels of proximity and grooming. Furthermore, the number of tolerant food transfers in the first test phase was significantly correlated with the number of tolerant food transfers in the second phase (de Waal 1997b). The most parsimonious explanation of this result is symmetry-based reciprocity, i.e. reciprocity based on the symmetrical nature of relationships. The capuchins were already familiar with each other (pair members lived in the same group), and food-sharing might have resulted from a combination of affiliation and tolerance towards conspecifics. Our next concern was whether or not reciprocity could be attributed to anything besides the symmetry inherent in the relationship.

For this, changes within each relationship over time were examined. The test was similar to the previous one, but incorporated six DETs on each pair. For each DET, individual A was given apple pieces for 20 minutes, then these were removed and individual B was given carrot pieces for 20 minutes. The roles between individuals remained the same over the six tests. The results were compared across tests to see how sharing in the second test phase was affected by sharing in the first phase between the same two individuals. This approach allowed us to correlate events over time, rather than across relationships, tightening a possible argument for causality between the behaviors in both dyadic directions. Sharing rates were found to significantly covary over time within each pair of individuals, indicating something more than symmetry-based reciprocity (de Waal 2000c).

Calculated reciprocity, or mental scorekeeping, however, may still be too complex a mechanism. To explain these results, de Waal (2000c) proposed 'attitudinal reciprocity', that is, each individual's behavior mirrors the partner's attitude in close temporal succession. Instead of the monkeys keeping careful track of how much they gave and received, they may merely have responded positively (i.e. with proximity and tolerance) to a positive attitude in their partner. Such mirroring of social predispositions might explain the reciprocal distribution of food-sharing without the requirement of scorekeeping.

5.3.2 Cooperation

Despite indications of cooperation among wild capuchins, tests of their cooperative abilities in the laboratory initially failed. Early tests used electronically-mediated or other complex devices that were beyond the monkeys' comprehension (Chalmeau et al. 1997, Visalberghi et al. 2000, Brosnan & de Waal 2002). Adoption of the paradigm pioneered by Crawford (1937), on the other hand, quickly led to success. This paradigm, in which two individuals pull food towards themselves, is entirely mechanical. As such, it is intuitive; the monkeys can see how their pulling causes the food to move towards themselves and they also immediately feel the effect of their partner's pulling.

In our case, two capuchin monkeys had to work together to pull in a counterweighted tray, at which point one or both of them would be rewarded (Fig. 5.4). They were placed in the test chamber separated from each other by a mesh partition, giving them the option to share food. Each monkey had its own bar to pull in the tray, although these bars could be removed for control tests. Food was placed in transparent bowls so that each monkey could see which one was about to receive the food.

Initially, monkeys were taught to pull in the tray individually, which they quickly learned. At this point (and throughout the experimental period, which lasted three years) each monkey was given regular strength tests to determine how much weight he or she could pull in individually. For trials in which only one monkey pulled, the tray was weighted just under what this individual could pull. For trials in which both monkeys pulled, the tray was weighted more heavily than the strongest individual could pull alone, but somewhat lighter than their combined strengths. Each test consisted of four 10-minute trials conducted on seven same-sex pairs of adult capuchins. The five test conditions were:

- Solo effort test (SOL), in which only one monkey had a pull-bar and only this individual received food, although both monkeys were present in the test chamber. This required no cooperation.
- Mutualism, or double test (DBL), in which both monkeys were required to pull together and both cups were baited.
- Cooperation test (COP), in which both monkeys were required to pull together but only one food cup was baited. This represented altruism on the part of the helper.
- Obstructed view test (OBS), which was the same as the above COP test except that the mesh partition was replaced by an opaque one. This eliminated visual communication between the monkeys, but they still could both see both cups on the tray, and that only one was baited.

Fig. 5.4. The test chamber used for the cooperative pulling task in capuchin monkeys inspired by Crawford's (1937) classical study. Two monkeys are situated in adjacent sections of the test chamber, separated by a mesh partition. The apparatus consists of a counter-weighted tray with two pull bars, with each monkey having access to one bar. The bars can be removed. In the solo effort test, two monkeys were in the test chamber, but only one monkey had a pull bar and only this individual's food cup was baited. In the mutualism test, both monkeys were required to pull their respective pull bars, and both food cups were baited. In the cooperation test depicted here, both monkeys were required to pull, but only one individual's food cup was baited. Drawing by Sarah Brosnan.

- Unrestricted cooperation test (UCP), which was the same as the COP test, except that the partner was free to move in and out of the test chamber, which had an open connection to part of the group cage. This meant that the helper, needed for successful pulls, was not always at hand.

As expected, the success rate of cooperative trials was significantly lower than that of mutualism tests or solo efforts. In the unrestricted cooperation tests, bar-pulling attempts by the food possessor significantly decreased when the partner left the test chamber, indicating that the monkeys had learned to associate their partner's presence with successful pulling. They might even have made the more complex association that they could succeed only with their partner's help (Mendres & de Waal 2000).

Bar-pulling success also decreased significantly in the obstructed view tests as opposed to the cooperation tests. In the obstructed view test, vocal communication was still possible and the monkeys continued to make pulling efforts at the same rate they did in the cooperation tests. Since both monkeys could see the food cups, their success rates should not have decreased if the impetus to pull simply stemmed from seeing food. What changed was their ability to see each other's behavior, indicating that success was at least partially dependent on visual coordination with the partner. The failure to succeed when visual access was cut off indicates that the monkeys were paying attention to each other's actions and coordinating their efforts. This result countered the claim of Chalmeau et al. (1997) that capuchins do not understand the need for a partner in cooperative tasks (Mendres & de Waal 2000).

5.3.3 Sharing following cooperation

The central question underlying this project was whether food-sharing would increase in the context of a cooperative enterprise. In a service economy, food can be exchanged for assistance in cooperation, or the converse. Our analyses of the amount of food-sharing indicated that capuchins share significantly more in successful cooperative trials than in solo effort trials, in which the partner is present, but does not, and actually cannot, assist (de Waal & Berger 2000).

Furthermore, the partner pulled more frequently after successful trials. Since 90% of successful trials included food transfers to the helper, capuchins are assisting more frequently after having received food in a previous trial. The simplest interpretation of this result is that motivational persistence results in continued pulling after successful trials. But a causal connection is also possible, i.e. that pulling after successful trials is a response to the obtained reward and the expectation of more.

The most cognitively-demanding interpretation of these results is that the food possessor understands that its partner has helped and that the partner must be rewarded for cooperation to continue. This would represent calculated reciprocity, in which the exchange of favors on a one-on-one basis drives reciprocal altruism. Each individual understands the other's costs (assistance in pulling or loss of food) and out of gratitude returns the favor.

However, a simpler explanation of the cooperation and food-sharing in these trials is a variation on attitudinal reciprocity (cf. de Waal 2000c), in which the possessor and partner feel closer after a coordinated effort. The attention and coordination that cooperation entails may induce a positive attitude in the partner, which is expressed in social tolerance and mutual attraction, which translate into food-sharing. After a food-sharing episode, similar mechanisms lead to increased pulling by the partner and hence further cooperation.

The conclusion from these experiments is that capuchins are quite good at performing, and probably also understanding, cooperative tasks. The mechanism most likely to underlie cooperation and sharing in these monkeys is attitudinal reciprocity in which cooperation partners mirror the attitude shown by their partner. This is different from symmetry-based reciprocity in that reciprocation is not induced by symmetrical relationship characteristics but by attitudes that vary over time. This rather conservative explanation does not preclude the possibility of more complex processes, though. Indeed, the results of a recent experiment on cotton-top tamarins (*Saguinus oedipus*) indicate that monkeys are sensitive to benefits received from others, and that they may even recognize whether or not these benefits were intended (Hauser et al. 2003). If confirmed, these capacities have the potential of adding considerable complexity to attitudinal reciprocity.

5.3.4 Cooperation based on projected returns

Group hunting is characterized by a phase of coordination followed by a phase in which the parties collect around the captured prey. The latter phase decides who gets what for their efforts. In a variation on the above cooperation paradigm, de Waal & Davis (2003) mimicked this situation by allowing individuals to move around freely during the pulling task instead of being confined to separate areas, as done previously. This way, cooperation partners could compete over the acquired resource. We further manipulated (i) opportunities for competition by presenting the resource in clumped versus dispersed distribution (i.e. the cups with food could be placed far apart at both extremes of the tray or side-by-side, touching each other, in the center), and (ii) the tendency for competition by comparing unrelated pairs and adult mother-daughter pairs. Numerous primate studies indicate greater tolerance and more co-feeding among kin than non-kin (Yamada 1963, Feistner & McGrew 1989, de Waal 1989, 1991b, Schaub 1996).

In investigating how cooperative tendency varied with the potential for competition (clumped versus dispersed rewards), we were particularly interested in the speed of the decision-making process. Monkeys may need to learn incrementally which specific conditions are favorable for cooperation, or they may be able to make instantaneous adaptive decisions. In the first case, the pros and cons of each specific condition need to be learned through direct experience; hence, behavior will gradually change in response to any new condition. In the second case, there is a fast adjustment to new conditions since decisions are based on generalization from pre-existing knowledge.

Questions regarding the role of food distribution and the speed of adjustment are relevant to models of the evolution of cooperation. Imagine a genetic variant that cooperates readily with any member of its group to obtain resources, yet is a slow learner. The variant would have enormous trouble distinguishing profitable from unprofitable partnerships; it would need to go through many reiterated interactions before it understands which partners and situations provide optimal payoffs. Each time a new situation arises it would need to go through this learning process. Unless the cooperative tendencies of this individual selectively favor kin, they would impose serious costs. On the other hand, a variant that could quickly distinguish profitable from unprofitable partnerships would minimize costs in any interaction and hence enjoy higher fitness.

After pre-training, each of eleven pairs of monkeys was subjected to multiple tests consisting of fifteen 2-minute trials each, with rewards available to both parties. Clumped reward distribution had an immediate negative effect on cooperation, which was visible from the first trial (Fig. 5.5). Even in tests in which we alternated clumped and dispersed conditions across trials, there was an adjustment on each trial. The drop in cooperation under the clumped condition was far more dramatic for non-kin than kin, which was explained by the tendency of dominant non-kin to claim most rewards. The immediacy of responses suggests a decision-making process based on predicted outcome of cooperation rather than the totality of rewards available. Decisions about cooperation thus take

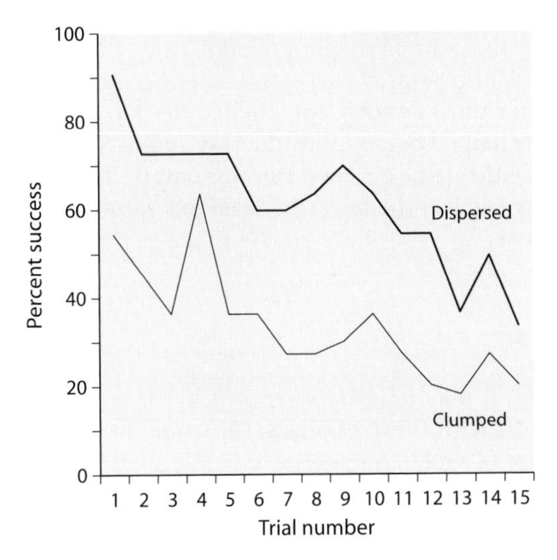

Fig. 5.5. Percentage of trials with successful cooperation for 15 consecutive trials per test in which the two food bowls are far apart (dispersed) or close together (clumped). The latter condition makes monopolization of the food by the dominant partner easy since, in these tests, there is no mesh between the partners. Since the data on all 11 pairs are pooled for each trial number, this graph provides no error data. The graph shows that the rate of success is lower right from the start in the clumped condition.

into account both the opportunity for and the likelihood of subsequent competition over the spoils.

The decisions observed probably reflected a lifetime of exposure to a variety of partners (e.g. dominant versus subordinate, or kin versus non-kin) under competitive conditions. The monkeys thus showed an ability to generalize previous knowledge to the novel test condition. It is particularly important to stress the generalizability of knowledge and the complexity of the variables that enter into decision-making given that social considerations are almost entirely absent from traditional learning research. For example, there is no mention of cooperation or almost any other socioemotional skills (e.g. conflict resolution, alliance formation, empathy) in a recent 700-page book on human and animal cognition (Shettleworth 1998). Many biologists, in contrast, believe that the social milieu has provided the main impetus for the evolution of intelligence in the large-brained order of primates (Humphrey 1976, Byrne & Whiten 1988). 'Planning' and 'foresight' are terms used in relation to chimpanzee power struggles (de Waal 1982a), and social intelligence is accorded special status in these highly social animals (Gigerenzer 1997, Dunbar 2001).

The study by de Waal & Davis (2003) supports the assumption that primates are extraordinarily sensitive to the reactive social field within which they operate. An anecdote (transcribed from a videotaped test) helps show how this sensitivity sometimes expresses itself. In a cooperation task on two female capuchins, Bias was paired with higher-ranking Sammy. Both females pulled in the tray together. Sammy quickly grabbed all of the food on her side, and released the tray without locking it into place, so that the counterweight pulled it away. Bias was left with her food cup out of reach. While Sammy was consuming her rewards, Bias started screaming at her partner until Sammy approached her bar again. While looking at each other, Sammy helped Bias pull in the tray again. Sammy did not do this for herself, because by this time her own cup was empty. This incident suggests protest by Bias for having lost the rewards 'deserved' for the first pull, and Sammy's corrective response.

5.3.5 Expectations about reward division

During the evolution of cooperation, it may have become critical for actors to compare their own efforts and payoffs with those of others (Brosnan, in press). Negative reactions may ensue in case of violated expectations. A recent theory proposes that aversion to inequity can explain human cooperation within the bounds of the rational choice model (Fehr & Schmidt 1999). Similarly, cooperative non-human species seem guided by a set of expectations about the outcome of cooperation and access to resources. De Waal (1996b, p. 95) proposed a "sense of social regularity", defined as: "A set of expectations about the way in which oneself (or others) should be treated and how resources should be divided. Whenever reality deviates from these expectations to one's (or the other's) disadvantage, a negative reaction ensues, most commonly protest by subordinate individuals and punishment by dominant individuals".

The sense of how others should or should not behave is essentially egocentric, although the interests of individuals close to the actor, especially kin, may be taken into account (hence the parenthetical inclusion of others). Note that the expectations have not been specified; they are species-typical (de Waal 1996b). Our experiment on clumped versus dispersed rewards (above) supports the role of expected returns in that it shows that cooperation disappears when subordinates anticipate a disadvantageous outcome. To further explore expectations held by capuchin monkeys, we made use of their ability to judge and respond to value. The ability to notice and respond when either reward value or efforts vary promotes cooperation by allowing individuals to recognize beneficial interactions. We knew from previous studies that capuchins easily learn to assign value to tokens, both through direct interaction with the items and through social learning (Brosnan & de Waal 2004a,b). Furthermore they can use these assigned values to complete a simple barter (Brosnan & de Waal 2004b). This allowed a test to elucidate inequity aversion by measuring the reactions of subjects to a partner receiving a superior reward for the same tokens.

We paired each monkey with a group mate and watched their reactions when their partners got a better reward for doing the same bartering task. This consisted of an exchange in which the experimenter gave the subject a token that could immediately be returned for a reward (Fig. 5.6). Each session consisted of 25 exchanges by each individual, and the subject always saw the partner's exchange immediately before their own. Food rewards varied from lower value rewards (i.e. a cucumber piece), which they are usually happy to work for, to higher value rewards (i.e. a grape), which were preferred by all individuals tested. All

Fig. 5.6. A monkey in the test chamber returns a token to the experimenter with her right hand while steadying the human hand with her left hand. Her partner looks on. The capuchin does not see the reward she is to receive prior to successful exchange. Drawing by Gwen Bragg and Frans de Waal after a video still.

subjects received four tests, including: (i) an Equity Test, in which subject and partner did the same work for the same lower-value food, (ii) an Inequity Test, in which the partner received a superior reward (grape) for the same amount of effort, (iii) an Effort Control Test, designed to elucidate the role of effort, in which the partner received the higher-value grape without any task-performance, and (iv) a Food Control Test, designed to elucidate the effect of the presence of the reward on subject behavior, in which grapes were visible but not given to another capuchin.

Fig. 5.7 shows that individuals who received lower value rewards showed both passive negative reactions (e.g. refusing to exchange the token, ignoring the reward) and active negative reactions (e.g. throwing out the token or the reward). Compared to tests in which both received identical rewards, the capuchins were far less willing to complete the exchange or accept the reward if their partner received a better deal (Brosnan & de Waal 2003). Capuchins refused to participate even more frequently if their partner did not have to work (exchange) to get the better reward, but was handed it for 'free'. Of course, there is always the possibility that subjects were just reacting to the presence of the higher value food, and that what the partner received (free or not) did not affect their reaction. However, in the Food Control Test, in which the higher-value reward was visible but not given to another capuchin, the reaction to the presence of this high-valued food decreased significantly over the course of testing, which is the opposite change from that seen when the high value reward went to an actual partner. In the latter case, the frequency of refusals to participate rose over the

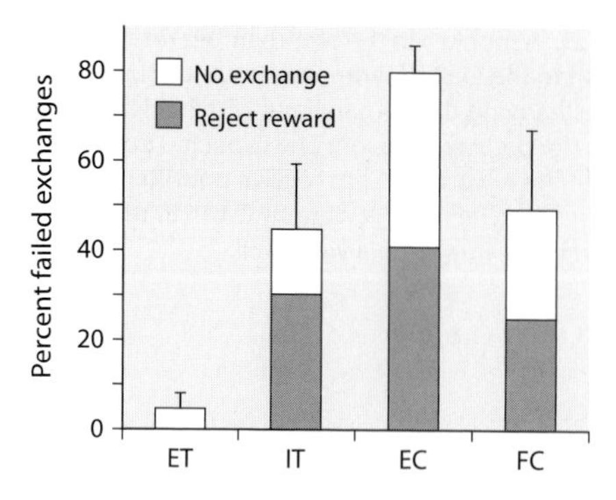

Fig. 5.7. Mean percentage + SEM of failures to exchange for females across the four test types. Black bars (RR) represent the proportion of non-exchanges due to refusals to accept the reward, white bars (NT) represent those due to refusals to return the token. SEM is for combined non-exchanges. ET = Equity Test, IT = Inequity Test, EC = Effort Control, FC = Food Control. The *y*-axis shows the percentage of non-exchanges. From Brosnan & de Waal (2003).

course of testing (Brosnan & de Waal 2004). While it has been suggested that these differences are not significantly variable (Wynne 2004), it is important to note that: (i) some reaction is always expected to the mere presence of a higher value reward, as it is inherently more desirable and (ii) the decrease in the level of response when no partner receives the reward is significantly different than when the partner does, demonstrating that the capuchins make the distinction between the two situations (Brosnan & de Waal 2004c).

Whereas the capuchins' reactions to this situation may not be identical to those of people (Henrich 2004), they fit well with the proposed evolutionary trajectory of inequity aversion (Brosnan & de Waal 2004). In fact, like humans, capuchin monkeys seem to measure reward in relative terms, comparing their own rewards with those available, and their own efforts with those of others. Although our data cannot elucidate the precise motivations underlying these responses, one possibility is that monkeys, like humans, are guided by social emotions. These emotions, known as 'passions' by economists, guide human reactions to the efforts, gains, losses and attitudes of others (Hirschleifer 1987, Frank 1988, Sanfey et al. 2003). As opposed to primates marked by despotic hierarchies, tolerant species with well-developed food-sharing and cooperation, such as capuchins, may hold emotionally-charged expectations about reward distribution and social exchange that lead them to dislike inequity.

5.4 Summary and conclusions

Although theories about the evolution of cooperation and reciprocal altruism are well established, proximate mechanisms have been little studied. There probably exist several levels of reciprocity, ranging from the more complex end of the spectrum, such as the kind originally proposed by Trivers (1971), which

Table 5.2. Three poximate mechanisms proposed by de Waal & Luttrell (1988) and de Waal (2000) to explain reciprocal distributions of benefits over dyadic relationships. The mechanisms are arranged from the least to the most cognitively demanding.

Mechanism	Catch phrase	Definition
Symmetry-based reciprocity	'We're buddies'	Symmetrical relationship characteristics prompt similar behavior in both dyadic directions: low degree of contingency in close relationships
Attitudinal reciprocity	'If you're nice, I'll be nice.'	Parties mirror each other's social attitudes: high degree of immediate contingency
Calculated reciprocity	'What have you done for me lately?'	Scorekeeping of given and received benefits: high degree of delayed contingency

involves obligations and punishment of cheaters, to reciprocity merely reflecting social symmetries. The evolutionary perspective simply postulates that the cost of help given be offset by the benefits of help received, which can be achieved in multiple ways, all of which fall under the general rubric of reciprocal altruism. Two decades of research on coalitions, grooming and food-sharing in macaques, chimpanzees and capuchin monkeys has allowed us to gauge the cognitive level of cooperation. Table 5.2 proposes three potential mechanisms, but we cannot exclude the possibility of more.

The cognitively least demanding explanation of reciprocal altruism is that individuals interact based on symmetrical features of dyadic relationships, which cause both parties to behave similarly to each other (de Waal & Luttrell 1988, de Waal 1992a). This mechanism requires no scorekeeping since reciprocation is based on pre-existing features of the relationship, such as kinship, mutual association, and similarities in age or sex. It produces reciprocity without a strong contingency between given and received behavior. A certain mutuality in the exchange of benefits is probably required for the stability of any social relationship, but this can be achieved without careful record keeping. All that is required is an aversion to major, lasting imbalances in incoming and outgoing benefits. We believe that such moderately conditional mutual aid is common in primates, including people, not only among kin but also among close friends and associates. The prediction, then, is that the contingency between given and received benefits decreases with closeness of the relationship. Conversely, the impact of a single act on future exchanges will be greatest in more distant relationships, as found by de Waal (1997a) in chimpanzees. Similar issues have been addressed in close versus distant human relationships by Clark & Mills (1979) and Clark & Grote (2003).

The second proposed mechanism is attitudinal reciprocity in which an individual's willingness to cooperate cofluctuates with the attitude the partner shows or has recently shown (de Waal 2000c). This 'If you're nice, I'll be nice' principle divorces cooperative interactions from the symmetrical state of the relationship, making them contingent upon the partner's immediately preceding behavior. The principle appears to approximate mutualism, but with the difference that both parties do not benefit at the same time. The involvement of memory and scorekeeping seems rather minimal, as the critical variable is general social disposition rather than specific costs and benefits of exchanged behavior.

The third and final mechanism is calculated reciprocity, in which individuals reciprocate on a behavioral one-on-one basis with a significant time interval. This requires memory of previous events, some degree of scorekeeping, partner-specific contingency between favors given and received, and perhaps also punishment of cheaters. The best evidence for this 'What have you done for me lately?' principle of reciprocity in non-human animals concerns, perhaps not coincidentally, our closest relative, the chimpanzee (de Waal 1997a). Whereas active punishment was not demonstrated in chimpanzees, we did find aggressive protest against partners trying to obtain services without previous payment. In addition, the demonstrated principle of exchange entails passive punishment in that it predicts forfeited services for those who fail to provide services themselves.

It is logical to expect that calculated reciprocity, with its higher cognitive requirements, will be found only in a few species whereas cognitively less demanding forms will be more widespread. For any species for which reciprocal exchange is reported, we suggest that the default mechanism is symmetry-based. The burden of proof rests on those who assume more complex mechanisms. With respect to primates, it could be argued that we know enough, based on studies such as those reported here, to consider complex exchange within their capacity. This sounds reasonable, but should never be taken to mean that these animals necessarily rely on these capacities all the time. So as to reduce memory overload, non-human primates, and probably humans as well, can most of the time be expected to follow processes simpler than calculated reciprocity. We therefore recommend that correlational studies on primate behavior always be complemented with sequential ones, in which behavior is tracked over time. Such studies allow a more careful monitoring of exchange, including the establishment of contingency between given and received behavior. Such monitoring is necessary given that seemingly complex levels of reciprocity can easily be explained by a combination of symmetry-based and attitudinal reciprocity.

One factor that has made scholars skeptical about reciprocity among unrelated individuals has been a concern about how such behavior could possibly have evolved in the face of its high initial costs. It has recently been suggested, however, that cooperation could evolve if the initial investment were minimal after which cooperators increased their investment contingent upon increasing confidence in the relationship (Roberts & Sherratt 1998). All of the above forms of reciprocity have been found in relatively low-cost exchanges and may well have provided the evolutionary starting point for more risky and costly forms of reciprocity. Our findings suggest that primates keep a close eye on exchanges, respond immediately if the outcome of cooperation risks being asymmetrical, and react negatively if they receive less than others. These findings are consistent with the view that cooperation is not pursued purely for the probability that it will be rewarded but rather as a social enterprise in which payoffs are compared between individuals and decisions are based on the likelihood of equitable outcomes. The study of proximate mechanisms thus enriches our view, adding a cognitive component that seems more variable than commonly assumed.

Acknowledgments

We are grateful for constructive comments from John Mitani, Peter Kappeler and one anonymous referee. The authors thank the technicians, animal care staff and undergraduate students who helped during their projects. The capuchin work was supported by the National Science Foundation (IBN-0077706) to the first author, and a grant from the National Institutes of Health (RR-00165) to the Yerkes Regional Primate Research Center. The Center is fully accredited by the American Association for Accreditation of Laboratory Animal Care.

Chapter 6

Reciprocal exchange in chimpanzees and other primates

JOHN C. MITANI

6.1 Introduction

A central problem in the study of animal behavior concerns why individuals cooperate and exchange altruistic acts (Dugatkin 1997). Considerable theoretical attention has focused on explaining the evolution of cooperation and altruism in taxa as diverse as insects and primates (Hamilton 1964a, 1964b, Trivers 1971, Brown 1983). Empirically, cooperation among males generates substantial interest because the resource over which they primarily compete, females, is not easily divided and shared (Trivers 1972).

Chimpanzees (*Pan troglodytes*) have played a significant role in discussions of cooperative behavior among males. Chimpanzees live in 'unit groups' or 'communities' whose membership is open due to female dispersal (Nishida 1968, Nishida & Kawanaka 1972, Pusey 1979, Goodall 1986, Boesch & Boesch-Achermann 2000). Philopatric male chimpanzees form strong social bonds and cooperate to compete with conspecifics between and within communities. Males engage in boundary patrols to defend their territories against neighbors (Goodall et al. 1979, Boesch & Boesch-Achermann 2000, Watts & Mitani 2001). Interactions between individuals of different communities are typically hostile, and during these males some times join forces to kill others (Wilson & Wrangham 2002). Within communities, male chimpanzees cooperate by grooming and by forming short-term coalitions and long-term alliances (Simpson 1973, Nishida 1983, Nishida & Hosaka 1996, Watts 1998, Watts 2000b). In coalitions, two or more individuals direct aggression toward others, while alliances involve enduring cooperative relationships between animals (de Waal & Harcourt 1992). Male chimpanzees also hunt vertebrate prey and frequently share meat, a scarce and valuable resource, with conspecifics (Boesch & Boesch 1989, Nishida et al. 1992, Mitani & Watts 2001).

The extent and diversity of cooperative behavior displayed by chimpanzees provide the basis for investigating reciprocity in acts given and received between pairs of individuals. Studies conducted in captivity reveal that chimpanzees exchange commodities that are both similar and different in kind (de Waal & Brosnan, this volume). For example, chimpanzees groom each other reciprocally and exchange grooming for food at the Arnhem Zoo and Yerkes Primate Center, respectively (de Waal 1989, 1997, Hemelrijk & Ek 1991). Until recently, evaluating these observations has been difficult given the absence of comparable data from chimpanzees living in their natural environmental and social settings. Our recent observations of an unusually large community of chimpanzees at Ngogo, Kibale National Park, Uganda, have furnished an ideal opportunity to investigate cooperation and reciprocity between males (Mitani & Watts 1999, 2001, Mitani et al. 2000, 2002b, Watts 2000b, 2002). In this paper, I present new analyses that indicate male chimpanzees at Ngogo trade services in the contexts of grooming, coalition formation, and meat-sharing. I compare these findings with those reported earlier from this same community and from other anthropoid primates. I conclude by considering the processes that might account for the evolution of reciprocal exchange in wild chimpanzees.

6.2 Methods

6.2.1 Study site and subjects

I observed chimpanzees at the Ngogo study site in Kibale National Park, Uganda, for 13 months over four years between 1999 and 2002: June–August 1999; May–August 2000; June–August 2001; May–July 2002. Ngogo lies at an interface between lowland and montane rain forest and is covered primarily with moist, evergreen forest. Ghiglieri (1984), Butynski (1990) and Struhsaker (1997) provide detailed descriptions of the Ngogo study area.

The size of chimpanzee communities ranges from 19–95 individuals at other sites (mean = 42, SD = 26, n = 11 communities; data from Wrangham 2000). In contrast, the Ngogo chimpanzee community is unusually large and contained approximately 150 individuals between 1999 and 2002. I restricted observations and analyses to the 22 adult male chimpanzees that were present throughout the four years of study. We have worked with the Ngogo chimpanzees since 1995 and do not know the precise age of any of the adult males. To derive estimates of the ages of males, I used morphological and behavioral criteria to classify them into three categories originally established and defined by Goodall (1968): young adult males (16–20-years-old); prime and middle-aged adult males (21–33-years-old); and old adult males (≥33-years-old).

6.2.2 Behavioral observations and statistical analyses

I sampled the behavior of target males during hour-long periods. Targets were selected on a pseudorandom basis, with priority given to those individuals who had been sampled infrequently. While following targets, I recorded their associations with other males and conducted scan samples at 10-minute intervals. During scans, I noted grooming activity between targets and other males. I recorded coalitions between males *ad libitum* whenever two or more individuals directed aggression toward third parties. I also recorded meat-sharing between males *ad libitum* during hunting events involving mammalian prey. Subordinate chimpanzees give a distinctive call, the pant grunt, to higher-ranking individuals (Bygott 1979, Hayaki et al. 1989). I recorded these calls as they occurred and used them to rank males. The analyses presented here are based on 1681 hours of observations with each of the 22 males followed a minimum of 68 hours (mean ± SD = 76 ± 8).

I employed K_r matrix correlation tests (Hemelrijk 1990a) to investigate reciprocity in grooming given and received between males. I performed similar tests to examine reciprocity in coalition formation and meat-sharing. Matrix correlation tests furnish information regarding the pattern of interaction between male chimpanzees at a group level. To examine whether acts were exchanged evenly between individuals within dyads, I constructed reciprocity indices for grooming, coalition formation and meat-sharing. Reciprocity indices were created by computing the cumulative binomial probabilities that two males initiated interactions with each other in samples of dyadic grooming, coalition formation and meat-sharing events. An index for each pair was calculated as the ratio of the two probabilities, with the probability of the individual who initiated fewer events placed in the numerator (Silk et al. 1999). The reciprocity index ranges from zero to one. Zero values reflect cases where interactions are always initiated by one individual of the pair, while values of one indicate that interactions are perfectly balanced.

I used the tau K_r matrix partial correlation procedure (Hemelrijk 1990b) to test the null hypothesis that correlations between the number of acts given and received were unaffected by the amount of time males spent together. For these tests, I constructed a third matrix of association frequencies between dyadic pairs of males based on observations made during following episodes of target individuals. I performed two comparisons, testing reciprocity in grooming given and grooming received, and support given and support received while controlling for association frequency. I conducted an additional test examining the relationship between meat given and meat received while controlling for the number of times males participated together in hunts.

I carried out three additional sets of matrix partial correlation analyses to investigate reciprocity in behavior while controlling for the effects of other potentially confounding variables. In the first set of analyses, I examined the relationship between the number of acts given and received while holding the effect of male age constant. For these analyses, I used an age matrix containing values of ones and zeros that represented pairs of males of the same and different age classes, respectively. I employed the distribution of pant grunts to create a 22 × 22 dominance matrix and used the MatMan software package (Version 1.0; de Vries et al. 1993) to assign ranks to males. In a second set of analyses, I built a male dominance rank matrix in which the ranks of males were entered in columns (Hemelrijk 1990a). I then calculated correlations between grooming, support and meat-sharing while controlling for male rank. In a final set of analyses, I performed similar tests of reciprocity controlling for male genetic relatedness. For these, I created a matrix consisting of ones and zeros that reflected mitochondrial DNA (mtDNA) haplotype identity and non-identity between males, respectively. MtDNA haplotypes were constructed using information derived from sequences of a 1038 bp region of the mitochondrial D loop (Mitani et al. 2002b). The 22 males possessed 11 different haplotypes. Six males had unique

Table 6.1. Matrix correlation results of reciprocity in grooming, coalitionary support, and meat-sharing. Two-tailed p values derived from 10000 permutations of the data are shown.

	Tau K_r	p
Grooming given and received	0.35	0.0002
Support given and received	0.44	0.0002
Meat given and received	0.28	0.0002
Grooming given and received, association controlled	0.33	0.0002
Support given and received, association controlled	0.34	0.0002
Meat given and received, joint hunting participation controlled	0.25	0.0002

haplotypes. Two other haplotypes were shared by pairs of males, while three more males displayed another haplotype. A group of six males shared the eleventh haplotype.

In the following, I use the same data in matrix correlation tests to perform multiple tests. To reduce the probability of committing Type I errors, I employed a sequential Bonferroni procedure adjusting the criteria of significance downward for each family of tests (Holm 1979). For k tests, I set the adjusted alpha level, α', to α' = α / / (1 + k – i), where α = 0.05 is the overall experiment-wise error rate and i is the ith sequential test from first to last. I considered a family of tests to involve a set of comparisons that evaluated the same general hypothesis. For example, I regarded tests of reciprocity in grooming, meat-sharing, and coalition formation as one set of tests, while another family of tests investigated whether acts given and received were affected by association frequency.

6.3 Results

6.3.1 Reciprocity in grooming, coalition formation and meat-sharing

Adult male chimpanzees at Ngogo groomed each other reciprocally at a group level. There was a significant correlation between the amount of grooming given and received between males (*Tau* K_r = 0.35, p = 0.0002, α' = 0.02; Table 6.1). Ngogo males also displayed reciprocity in coalition formation and meat-sharing (coalitions: *Tau* K_r = 0.44, p = 0.0002, α' = 0.025; meat-sharing: *Tau* K_r = 0.28, p = 0.0002, α' = 0.05; Table 6.1). The pattern of reciprocity in grooming, coalition formation and meat-sharing characterized the majority of all indi-

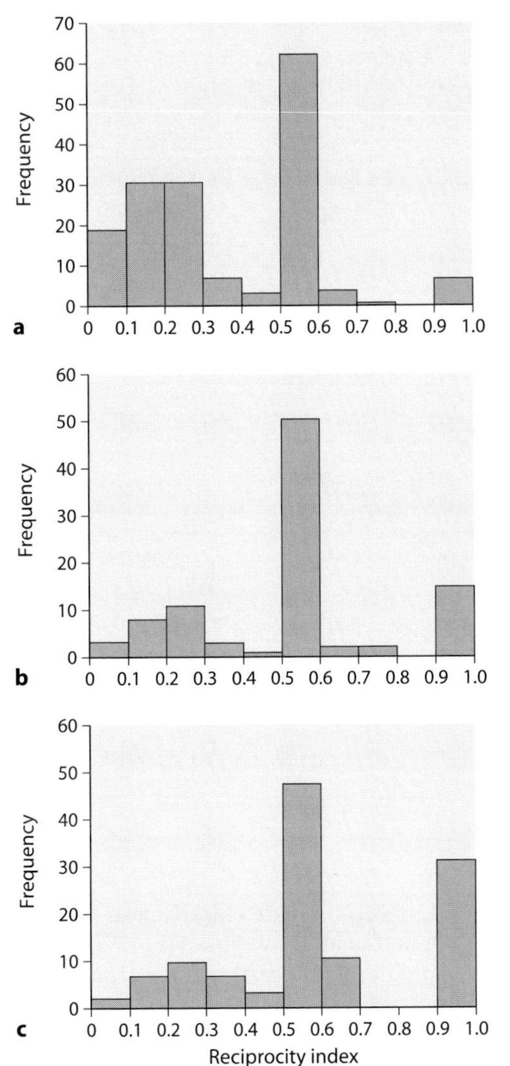

Fig. 6.1. Frequency distribution of reciprocity indices for (**a**) grooming, (**b**) coalition formation and (**c**) meat-sharing. Values near zero indicate that one male initiated more acts than the other within dyads, while values near one indicate that the exchange of behaviors was evenly balanced between males.

vidual males in the group, with most (18–21 of 22) showing positive row-wise correlations.

While results of matrix correlation tests reveal that male chimpanzees reciprocated grooming, coalition formation and meat-sharing at the group level, they do not indicate whether acts were balanced within dyads. Additional analyses failed to reject the hypothesis that individuals exchange all three behaviors evenly within dyads. Grooming took place between 166 of 231 possible dyads (72%). Reciprocity indices in these 166 pairs ranged from zero to one, with many clustering around the middle of the distribution (Fig. 6.1a). Only in a very few cases (14/166 = 8%) did one male initiate significantly more than one-half of

Table 6.2. Matrix partial correlation results of reciprocity in grooming, coaltionary support, and meat-sharing. Two-tailed p values derived from 10000 permutations of the data are shown.

	Tau K_r	p
Grooming given and received, age controlled	0.34	0.0002
Grooming given and received, rank controlled	0.34	0.0002
Grooming given and received, mtDNA haplotype controlled	0.34	0.0002
Support given and received, age controlled	0.41	0.0002
Support given and received, rank controlled	0.39	0.0002
Support given and received, mtDNA haplotype controlled	0.41	0.0002
Meat given and received, age controlled	0.27	0.0002
Meat given and received, rank controlled	0.21	0.0034
Meat given and received, mtDNA haplotype controlled	0.27	0.0002

all dyadic grooming bouts (Binomial test, $p < 0.05$). Male chimpanzees formed coalitions in 96 dyads or slightly less than half of all those possible (42%). Reciprocity indices again clustered in the middle of the distribution around 0.5 (Fig. 6.1b). Coalition formation was balanced between males in the vast majority of all dyads. One male initiated significantly more than one-half of all coalitions in only one of the 96 dyads (Binomial test, $p < 0.5$). Males shared meat between themselves in about one half of all possible dyads (120/231 = 52%). Meat was typically shared evenly between males within each dyad, with several reciprocity indices at or near one (Fig. 6.1c). In a single case, one male shared with another individual significantly more than one half of all times meat was swapped between the two (Binomial test, $p < 0.05$).

Given the fission-fusion nature of their society, chimpanzees do not associate with all members of their community equally often. Thus, reciprocal exchanges could result as byproducts of association if males directed behaviors toward those with whom they remained in contact frequently (de Waal & Luttrell 1988, Hemelrijk & Ek 1991). Additional analyses did not support this suggestion. Partial K_r tests revealed that correlations between grooming given and grooming received, and support given and support received remained significant after controlling for association frequency ($p = 0.0002$ for both tests; Table 6.1). Similarly, the correlation between meat given and meat received held after controlling for the number of times individuals were together at hunts ($p = 0.0014$; Table 6.1).

Before concluding that giving depends on receiving, it is important to exclude other variables that might confound the relationship (de Waal & Luttrell 1988, deWaal 1997, Hemelrijk & Ek 1991). In rhesus macaques, for example, similarity in age and rank affect association patterns (de Waal & Luttrell 1986) leading to the expectation that reciprocity in behavior may result from exchanges between closely bonded individuals who share common characteristics. In fact, recent observations indicate that at Ngogo males who are similar in age and rank tend to engage in coalitions and to share meat more than one would expect by chance (Mitani et al. 2002b). Nevertheless, results of partial K_r tests revealed that reciprocity in grooming, support and meat-sharing persisted after controlling for the effects of male age and rank ($p \leq 0.003$ for all six tests; Table 6.2). In addition, reciprocity in all three behaviors was still evident after holding the effects of genetic relatedness between males constant ($p = 0.0002$ for all three tests; Table 6.2).

6.3.2 Reciprocal exchange of different behaviors

The preceding results indicate that male chimpanzees at Ngogo exchange services that are similar in kind. Additional analyses show that males at Ngogo also trade services in different currencies. A positive and significant correlation existed between grooming given and support received (*Tau* $K_r = 0.26$, $p = 0.0006$; $\alpha' = 0.002$; Table 6.3). This relationship persisted after controlling for the effects of association frequency, male age, rank and genetic relatedness ($p \leq 0.0056$ for all four tests; Table 6.3). Males also traded grooming for the meat that they obtain in hunts. A significant correlation existed between grooming given and meat received (*Tau* $K_r = 0.30$, $p = 0.0002$, $\alpha' = 0.025$; Table 6.3). The correlation between these two variables held after controlling for the effects of association frequency, joint participation in hunts, male age, rank and genetic relatedness ($p \leq 0.001$ for all five tests; Table 6.3). Male chimpanzees at Ngogo also exchanged meat for support (*Tau* $K_r = 0.27$, $p = 0.0002$, $\alpha' = 0.05$; Table 6.3). As in the prior examples, this correlation persisted after controlling for the effects of several other potentially confounding variables ($p \leq 0.0008$ for all five tests; Table 6.3).

6.3.3 Comparisons with previous studies at Ngogo

The results presented here are consistent with those obtained from previous studies at Ngogo. Using independently-collected observations, Watts (2000b, 2002) has shown that the Ngogo males reciprocate grooming and coalitionary support. These correlations hold after controlling for the effects of male dominance rank and association frequency (Watts 2000b, 2002). In other studies, we have reported that males at Ngogo share meat reciprocally (Mitani & Watts 1999,

Table 6.3. Matrix correlation results of reciprocal exchange of different behaviors. Two-tailed p values derived from 10000 permutations of the data are shown.

	Tau K_r	p
Grooming given and support received	0.26	0.0006
Grooming given and support received, association controlled	0.21	0.0056
Grooming given and support received, age controlled	0.25	0.0008
Grooming given and support received, rank controlled	0.21	0.0030
Grooming given and support received, mtDNA haplotype controlled	0.25	0.0008
Grooming given and meat received	0.30	0.0002
Grooming given and meat received, association controlled	0.27	0.0006
Grooming given and meat received, joint hunting participation controlled	0.28	0.001
Grooming given and meat received, age controlled	0.29	0.0002
Grooming given and meat received, rank controlled	0.23	0.0028
Grooming given and meat received, mtDNA haplotype controlled	0.29	0.0004
Meat given and support received	0.27	0.0002
Meat given and support received, association controlled	0.27	0.0002
Meat given and support received, joint hunting participation controlled	0.24	0.0008
Meat given and support received, age controlled	0.26	0.0004
Meat given and support received, rank controlled	0.24	0.0002
Meat given and support received, mtDNA haplotype controlled	0.26	0.0004

2001a). Here I show that this correlation cannot be attributed to the effects of association frequency or joint male presence at hunts.

The preceding results also replicate previously reported observations of reciprocal exchange of different behaviors between male chimpanzees at Ngogo. Watts (2000b, 2002) has found that males at Ngogo exchange grooming for coalitionary support. This result holds after controlling for several potential confounds, including reciprocity in grooming, and in support and relationships between grooming, support and male rank (Watts 2002). Elsewhere, we have shown that the Ngogo males trade meat for agonistic support (Mitani & Watts 2001). Based on the preceding analyses, I now demonstrate that this relationship is not the byproduct of association frequency between males or their joint presence at hunts. Given these findings, it is perhaps not surprising that males at Ngogo additionally exchange grooming for meat. The observations that I present here are also consistent with our prior studies that indicate genetic relatedness as assayed by mtDNA haplotype sharing does not have any consistent effect on patterns of social behavior and relationships among the Ngogo chimpanzees (Mitani et al. 2000, 2002b).

6.4 Discussion

These new observations and analyses confirm those obtained from prior research and show that male chimpanzees living in an unusually large community at Ngogo cooperate in several behavioral contexts. Cooperation involves the reciprocal exchange of acts in similar and different currencies. At Ngogo, males reciprocate grooming, support and meat-sharing at the group level. The exchange of grooming, support and meat was evenly balanced within dyads; only rarely did one individual in a pair initiate more interactions than the other. Male chimpanzees at Ngogo also reciprocally exchange grooming for support, grooming for meat, and meat for support. These exchanges are unaffected by several potentially confounding variables that include association frequency, male age, rank and genetic relatedness.

6.4.1 Comparisons with other primates

Individuals in many Old World anthropoid primates reciprocally exchange behaviors that are similar in kind. The evidence is strongest and most extensive for reciprocity in coalitionary support and in grooming. Like male chimpanzees, individuals in other primate species engage in coalitionary behavior (Harcourt & de Waal 1992). In several species of macaques, individuals provide coalitionary support reciprocally (de Waal & Luttrell 1988, Silk 1992b, Widdig et al. 2000). Reciprocity in support has also been reported in vervet monkeys and gorillas (Hunte & Horrocks 1986, Watts 1997). Primates spend an extraordinary amount of time grooming each other (Dunbar 1988), and reciprocity in grooming has been documented in several cercopithecine primates, including Japanese macaques, blue monkeys and chacma baboons (Muroyama 1991, Rowell et al. 1991, Barrett et al. 1999, Silk et al. 1999). Food-sharing between adults occurs only rarely in primates, limiting the possibilities for reciprocal food exchange in this taxon. Some New World monkeys regularly share food with conspecifics, but investigations have produced mixed results with respect to reciprocity in this behavior. Captive golden lion tamarins share food spontaneously with others but do not do so reciprocally (Rapaport 2001). In contrast, recent experiments with cotton-top tamarins indicate that animals give food to genetically unrelated conspecifics, who have consistently shared with them in the past; these same animals fail to give food to those who have behaved selfishly by not sharing (Hauser et al. 2003). Similarly, brown capuchin monkeys share food reciprocally under captive experimental conditions (de Waal 2000c).

Some of the best evidence that primates reciprocally trade services in different currencies comes from studies of captive chimpanzees. Chimpanzees at the Yerkes Primate Center were more likely to share food with others after being groomed by them (de Waal 1989, 1997a). There is a paucity of other examples that demonstrate primates other than chimpanzees reciprocally exchange behaviors differing in kind. Studies of captive bonnet macaques reveal that males support those who groom them and intervene against those who intervened against them (Silk 1992b). Likewise, individuals trade coalitionary support for grooming in captive long-tailed macaques (Hemelrijk 1994). In the wild, unrelated female patas monkeys appear to exchange grooming for allomothering (Muroyama 1994). Considerable controversy exists over some examples of primates trading services in different currencies. Seyfarth (1980) reported that female vervet monkeys exchange grooming for agonistic support. The apparent pattern of exchange, however, may be due to relationships between these two variables and additional confounds. For example, Hemelrijk (1990b) showed that the relationship between grooming given and support received in vervets vanished after controlling for the effect of dominance rank. Female vervets directed their grooming toward high-ranking individuals preferentially, while these same high-ranking animals supported others frequently. Similarly, Vervaecke et al. (2000) reported a significant relationship between grooming given and support received in a captive group of bonobos, but this correlation disappeared after controlling for the effect of dominance rank.

Several other studies do not support the hypothesis that primates reciprocally exchange different behaviors (Henzi & Barrett 1999). Female white-faced capuchin monkeys fail to exchange grooming for coalitionary support (Perry 1996). In captive golden lion tamarins, animals exchanged grooming for coalitionary support, but this occurred in only one of three groups (Rapaport 2001). Among free-ranging and captive macaques, females do not provide agonistic support to those who groom them (Silk 1982, de Waal & Luttrell 1988, Kapsalis & Berman 1996, Matheson & Bernstein 2000). Even in captive chimpanzees, evidence for the reciprocal exchange of different behaviors is not always consistent. Hemelrijk & Ek (1991) found a significant positive correlation between grooming given and support received during periods with a clear alpha male, but these relationships disappeared after controlling for other potentially confounding associations. In sum, there is good evidence that primates reciprocally exchange behaviors that are similar in kind. It remains unclear, however, whether primates regularly engage in the reciprocal exchange of different behaviors.

6.4.2
Evolutionary processes underlying reciprocity in wild chimpanzees

Male chimpanzees are thought to derive important fitness benefits by developing social bonds with each other and cooperating (Riss & Goodall 1977, Nishida 1983, Nishida et al. 1992, Nishida & Hosaka 1996, Watts 1998, Mitani & Watts 2001). For example, coalitionary support is often necessary for males to achieve and maintain a high dominance rank (Riss & Goodall 1977, Nishida 1983, Nishida et al. 1992). High male rank in turn is related to mating success (Nishida 1983, Watts 1998, Watts & Mitani 2001). Given the importance of these coalitions, male chimpanzees appear to cultivate social relationships to obtain this valuable social service. The results presented here are consistent with our previous findings that indicate male chimpanzees exchange meat, a scarce and valuable resource, for support in agonistic contests (Mitani & Watts 2001). To develop and maintain their social relationships, male chimpanzees cooperate by exchanging behavioral acts that not only differ in kind but that are also similar. For instance, male chimpanzees display well-differentiated grooming relationships, and the preceding analyses are consistent with prior studies that show males at Ngogo groom each other reciprocally (Watts 2000b, 2002). These results also accord with observations of chimpanzees at other sites; reciprocity in grooming between males has been documented in the Taï chimpanzees (Boesch & Boesch-Achermannn 2000).

While reciprocal exchanges of grooming, support and meat-sharing are likely to have important fitness consequences for male chimpanzees, the evolutionary processes that account for them remain unclear. The preceding results indicate that male chimpanzees balance the number of behavioral acts that they give and receive, but they do not reveal the processes that have produced these reciprocal exchanges over evolutionary time. Three well-known evolutionary processes that have been used to account for reciprocity include kin selection, reciprocal altruism and mutualism (Dugatkin 1997).

Kin selection has historically been invoked to explain the evolution of cooperative relationships among philopatric male chimpanzees (e.g. Morin et al. 1994). Thus far, our findings from Ngogo are not consistent with this explanation. Using mtDNA haplotype sharing to assay genetic relatedness between individuals, we have previously shown that kinship is a poor predictor of who cooperates with whom (Mitani et al. 2000, 2002b). The results presented here indicate that reciprocity in grooming, support and meat-sharing persist after taking into account the effect of mtDNA genetic relatedness. These mtDNA data are admittedly only crude assays of kinship, and we are currently working to obtain better estimates of relatedness using data derived from nuclear DNA and Y chromosome markers (Langergraber et al., in prep.). With these data in hand, we will be in a better position to examine the role of kin selection in the evolution of male chimpanzee social behavior at Ngogo.

Reciprocal altruism occurs if individuals restrict their help to those who aid them in return (Trivers 1971). Though sometimes cited as an evolutionary mechanism leading to cooperation (e.g. Packer 1977, Seyfarth & Cheney 1984), potential examples of reciprocal altruism are often shown to involve mutualism where both partners benefit (Bercovitch 1988, Noe 1992, Clements & Stephens 1995). Theoretical analyses of reciprocal altruism emphasize the contingent nature of interactions; in situations where partners defect, reciprocity dissolves (Axelrod & Hamilton 1981). The results presented here document reciprocity using correlational procedures applied to the behavior of several individuals at a group level over the course of four years. Additional analyses indicate that both individuals typically contribute, with exchanges only rarely unbalanced within dyads. Though consistent with reciprocal altruism, these results cannot be used to test the hypothesis that reciprocal exchanges between male chimpanzees at Ngogo have evolved as a result of this process. A rigorous test would involve assessing the contingent nature of a male's interactions with his partners. Such an analysis is more likely to be derived from carefully-designed experiments in the laboratory where the temporal sequence of interactions between specific pairs of males can be monitored in detail. Experiments conducted by de Waal (1997a) and Hauser et al. (2003) provide good models of this kind of work and some of the strongest evidence for the evolution of cooperative relationships through reciprocal altruism.

Mutualism occurs in situations where actors and receivers benefit immediately through interaction and has been proposed to explain patterns of coalitionary behavior, grooming and meat-sharing in primates. Male barbary macaques show reciprocity in coalition formation, but most interventions are by males who outrank both opponents and risk few costs by doing so (Widdig et al. 2000). In this case, interveners reinforce their own dominance ranks, and coalitionary behavior represents a low cost behavior in which both partners obtain immediate benefits. Henzi & Barrett (1999) have suggested that female primates trade grooming for the mutual benefits that grooming itself provides over the short term and that time matching and reciprocation during the same grooming bout will be common. Recent observations of female chacma baboons furnish support for this latter hypothesis (Barrett et al. 1999). Finally, Stevens & Gilby (2004) argue that meat-sharing in primates frequently involves coercion; here sharers are forced to accept an immediate cost such that the net benefit of cooperating exceeds that of behaving selfishly by not sharing.

Recent observations support the hypothesis that male coalitions at Ngogo are relatively low risk behaviors; the vast majority of coalitions there form between males who outrank their opponents (Watts 2002). Additional work will be necessary to examine whether male chimpanzees at Ngogo time match and reciprocate their grooming efforts. Further study is also required to investigate the extent to which meat is shared under pressure. In sum, investigations into the evolutionary mechanisms underlying reciprocal exchanges in chimpanzees and other primates remain a fertile ground for future research.

Acknowledgments

I am grateful to Peter Kappeler for inviting me to participate in the Göttinger Freilandtage Conference and to contribute to this volume. My fieldwork has been sponsored by the Ugandan National Parks, Uganda National Council for Science and Technology, and the Makerere University. I thank G. I. Basuta, J. Kasenene and the staff of the Makerere University Biological Field Station for logistic support. Comments by R. Noe, C. van Schaik, and an anonymous reviewer improved the manuscript. S. Amsler, C. Businge, K. Langergraber, J. Lwanga, A. Magoba, G. Mbabazi, G. Mutabazi, L. Ndagizi, H. Sherrow, S. Teelen, J. Tibisimwa, A. Tumusiime, M. Wakefield and D. Watts provided help in the field. My field research at Ngogo has been supported by grants from the Detroit Zoological Institute, L.S.B. Leakey Foundation, National Geographic Society, the NSF (SBR-9253590 and BCS-0215622), University of Michigan, and the Wenner-Gren Foundation to J. C. Mitani and L.S.B. Leakey Foundation and National Geographic Society to David Watts.

Chapter 7

Causes, consequences and mechanisms of reconciliation: the role of cooperation

Filippo Aureli, Colleen Schaffner

7.1 Introduction

The term reconciliation was first used by de Waal & van Roosmalen (1979) to describe a specific pattern of post-conflict behavior of the chimpanzees housed at the Burgers' Zoo in Arnhem, the Netherlands. Reconciliation was used as a heuristic label for the friendly exchanges witnessed between former opponents in the aftermath of an aggressive conflict. There had been earlier descriptive accounts of such friendly reunions in various primate species (see de Waal 2000a for a historical review), but the Arnhem study was the first systematic investigation. Since then, post-conflict friendly reunions between former opponents have been demonstrated in over 30 primate species and a few non-primate species as well (Aureli & de Waal 2000, de Waal 2000b, Schino 2000, Aureli et al. 2002, Arnold & Aureli, in press). The function of restoration of the damaged relationship implied by use of the term reconciliation has also been demonstrated in all studies that tested for this function (see below). Thus, for the rest of the chapter, we use the term reconciliation to refer to post-conflict friendly reunions between former opponents.

The aim of the chapter is to explore the role of cooperation between partners in the occurrence of reconciliation after their conflicts. We start by reviewing the assumptions at the basis of the concept of reconciliation, the evidence supporting these assumptions, and the implications of accepting these assumptions for the role of cooperation in reconciliation as cause and consequence. We then focus on the cause-consequence relationship linking cooperation and reconciliation by reviewing the evidence for higher rates of reconciliation after conflicts between individuals with more cooperative relationships. We also present cases that qualify the set of circumstances in which a high degree of cooperation between partners is associated with a high likelihood of reconciliation after their conflicts. Next, we focus on the role of cooperation as a mechanism, i.e. in the act of reconciliation itself. Here, two aspects are explored in detail: the act of reconciliation as a tool for relationship negotiation and as a means of communication about the relative value each partner attaches to the social relationship. We conclude by suggesting emotional mediation as a proximate mechanism that can be at the basis of the role of cooperation in reconciliation patterns.

7.2 Assumptions underlying the concept of reconciliation

The fact that reconciliation may take place after an aggressive conflict implies that the latter produces a situation that requires resolution. Thus, the first assumption in the concept of reconciliation is that aggressive conflict produces disturbance in the social relationship between the two opponents. The second critical assumption is that the post-conflict friendly reunion between the two opponents functions in restoring their social relationship disturbed by the previous conflict. We first review the evidence for these two assumptions.

Post-conflict disturbance can affect the degree of tolerance between two partners and, as a consequence, the degree of uncertainty in interaction and the emotional state of each partner. Recipients of aggression are more likely to be attacked again by the original aggressor in the period following the attack relative to control periods (York & Rowell 1988, Aureli & van Schaik 1991, Aureli 1992, Cords 1992, Watts 1995, Silk et al. 1996, Castles & Whiten 1998, Schino 1998, Kutsukake & Castles 2001). In contrast, there is no evidence that former aggressors are targets of elevated rates of aggression following the conflict (Castles & Whiten 1998, Das 1998). Recipients of aggression may also experience negative socioecological consequences as a result of increased post-conflict uncertainty. Wild long-tailed macaques (*Macaca fascicularis*) reduce their foraging time after receiving aggression, possibly because of the need for social vigilance to keep a close watch on the former aggressor's actions and avoid further attacks (Aureli 1992).

A critical test of the damage that aggressive conflict can inflict upon social relationships was carried out in an experimental setting by Cords (1992). Following aggressive conflict, pairs of long-tailed macaques showed reduced tolerance around a limited resource relative to baseline conditions. Thus, aggressive conflict is likely to produce uncertainty about the future of the relationship and the potential loss of benefits (Aureli et al. 2002).

Data on self-scratching and other self-directed behavior have been used to investigate the possible increase of anxiety-like emotion in post-conflict periods. There is behavioral and pharmacological evidence that some self-directed behavior is a reliable indicator of emotional states similar to human anxiety associated with uncertain situations (Maestripieri et al. 1992, Schino et al. 1996, Troisi 2002). Recipients of aggression increase the rates, relative to baseline, of self-directed behavior following a conflict (Aureli et al. 1989, Aureli & van Schaik 1991, Aureli 1992, Castles & Whiten 1998, Schino 1998, van den Bos 1998, Kutsukake & Castles 2001, but see Manson & Perry 2000 and Arnold & Whiten 2001). Similarly, former aggressors also increase post-conflict rates of self-directed behavior (Aureli 1997, Castles & Whiten 1998, Das et al 1998, Schino 1998). The post-conflict increase in anxiety cannot be explained only in terms of immediate potential negative consequences, as no increased risk of post-conflict aggression is present for aggressors. It is likely therefore that at least some of the post-conflict anxiety in both former opponents is due to the uncertainty about their relationship (Aureli 1997).

The evidence presented above supports the first assumption in the concept of reconciliation, that is, that aggressive conflict produces disturbance in the social relationship between opponents. To provide support for the second assumption, we review studies that explored the role of post-conflict friendly reunions between former opponents in restoring their social relationship disturbed by the previous conflict.

Post-conflict reunions reduce the hostility between former opponents by reducing the rates of attacks to baseline levels (Aureli & van Schaik 1991, Cords 1992, de Waal 1993, Silk et al. 1996, Castles & Whiten 1998, Koyama 2001, Kutsukake & Castles 2001, Wahaj et al. 2001, see also Bshary & Wurth 2001 for interspecific evidence). Post-conflict reconciliation also restores tolerance between former opponents, which normally is reduced following aggressive conflict (Cords 1992; Wittig & Boesch 2005). Interestingly, post-conflict interspecific reconciliation in the form of tactile stimulation functions in re-establishing interaction between cleaner fish and client reef fish (Bshary & Wurth 2001).

The occurrence of reconciliation reduces self-directed behavior of both opponents compared to post-conflict periods without reconciliation (Aureli & van Schaik 1991, Castles & Whiten 1998, Das et al. 1998, Schino 1998, Arnold & Whiten 2001, Kutsukake & Castles 2001). Support for an uncertainty-reduction function of reconciliation also comes from playback experiments on wild chacma baboons (*Papio ursinus*), which use soft grunts during post-conflict reunions (Cheney et al. 1995, Silk et al. 1996). After the playback of grunts of former aggressors, recipients of aggression approached them and tolerated their approaches more often than during periods without playbacks (Cheney & Seyfarth 1997).

The reduction of post-conflict uncertainty and anxiety after reconciliation does not seem to be due simply to the calming effect of friendly behavior in general. Post-conflict friendly contacts with third parties do not reduce the rate of self-scratching of the original aggressor (Das et al. 1998). Similarly, reconciliation reduces post-conflict heart rates of former opponents to baseline levels faster than post-conflict friendly contacts with third parties (Smucny et al. 1997, Aureli & Smucny 2000).

The two assumptions find further support from a study that investigated the long-term effect of aggressive conflict. If reconciliation does not take place, Japanese macaques (*Macaca fuscata*) experience long-term negative consequences (Koyama 2001). In the ten days following conflicts, affiliation rates between former opponents are lower and aggression rates higher than at baseline. By contrast, if reconciliation takes place no such negative changes occur, suggesting that the post-conflict friendly reunions reconciled the opponents and prevent disruption of their interaction patterns and thus of their relationship.

7.3 Cooperation as cause and consequence of reconciliation

Based on the evidence for the first assumption in the concept of reconciliation, it is easy to identify a potential role of cooperation in the causation of reconciliation. The disturbance in the social relationship between the two opponents produced by the aggressive conflict would certainly entail a post-conflict disruption of cooperation between such partners. Thus, the post-conflict anxiety associated with the potential loss of benefits due to the disruption of cooperation is expected to serve as a strong incentive for each former opponent to pursue reconciliation (Aureli 1997, cf. the 'relational model' in de Waal 1996a, 2000b).

Similarly, the evidence for the second assumption in the concept of reconciliation provides the background to view cooperation as a natural consequence of reconciled conflicts. The restoration of the social relationship between the two opponents obtained through post-conflict friendly reunions is expected to include the re-establishment of cooperation between such partners, which was disrupted by the previous conflict.

The potential roles of cooperation as a cause and consequence of reconciliation are clearly not independent and stem from the very same function at the basis of reconciliation implied in the two assumptions. Thus, we can review support for such interlinked roles together by reviewing evidence for a relation between the degree to which partners cooperate and the likelihood of reconciliation after their conflicts. Given the beneficial outcomes following reconciliation reviewed in the previous section, we can predict that conflicts between partners with more cooperative relationships would be reconciled at higher rates. This prediction fits well with both the view of cooperation as a cause (i.e. stronger interest in reconciling of partners with highly cooperative relationships given the greater loss of benefits and post-conflict anxiety due to the post-conflict disruption of the relationship) and the view of cooperation as a consequence (i.e. greater benefits to be regained by the restoration of the relationship between partners usually involved in highly cooperative actions). Below, we review the evidence available to test this prediction.

Experimental evidence supports the prediction. Pairs of long-tailed macaques were trained to depend on each other in order to obtain food in a cooperative task (Cords & Thurnheer 1993). The training aimed to make the relationship more cooperative and thus increase the value of the partner. Reconciliation was measured before and after the training, and was found to have increased dramatically after the training. Thus, the monkeys adjusted their conciliatory tendency to the increased cooperative value of the partner.

The relative cooperative value of social relationships can also be measured by directly recording naturally-occurring behavior and then calculating the rates of certain interactions exchanged between group members. Obvious candidates for such interactions are coalitions in which a third party provides support to one of the two combatants in an ongoing aggressive interaction (Harcourt & de Waal 1992). A reliable supporter would be a great asset in competition for any resource and thus individuals are expected to be highly interested in restoring such relationships after potentially damaging conflicts. Early attempts to find support for a positive relation of reconciliation levels and frequencies of agonistic support failed. Kappeler & van Schaik (1992) investigated whether there was a correlation between reconciliation frequencies and rates of polyadic conflicts (i.e. conflicts in which multiple opponents are involved) across species, but no such relationship was found. However, one could argue that the overall rate of polyadic conflicts in a group or species is not the fine-grained measure needed to investigate the potential correlation between reconciliation and aggressive support at the relationship level.

Two studies used rates of support between group members, but they did not find any relation between reconciliation and coalition rates. The first one focused on juvenile long-tailed macaques and the authors argued that a possible cause of failing to find such a relation was due to the asymmetry in the value of the social relationship between the juveniles and the adult supporters (Cords & Aureli 1993). Such relationships may not be very cooperative because juveniles cannot return effective agonistic support, and the primary interest of supporters might be to direct aggression at the other opponent rather than to aid the juvenile. The second study was on a captive group of chimpanzees, and the authors argued that their negative finding might be due to the low overlap between affiliation and support networks and to the highly asymmetrical support relationships (Preuschoft et al. 2002).

More recently, three studies found support for the prediction. Wild female Assamese macaques showed higher reconciliation rates with females with whom they exchanged higher rates of agonistic support (but this was not the case for males: Cooper et al. 2005). A similar result was found in Japanese macaques, even when the effect of kinship was controlled (Koyama in prep.). Contrary to previous studies, social relationships characterized by high exchanges of affiliative interactions were those in which agonistic support was given. In a study on wild chimpanzees, dyads that supported and shared food with one another reconciled more often than other types of dyads, although analyses of the effect of either of these factors alone did not yield significant results (Wittig & Boesch 2003). It seems therefore that coalition and reconciliation rates are positively correlated only when agonistic support is exchanged between partners for support on other occasions or for other services (e.g. food-sharing or grooming). This finding, of course, supports our prediction because those relationships in which the exchange is more symmetrical are the most cooperative.

There is also indirect evidence for the prediction of more reconciliation in more cooperative relationships from studies that inferred the relative cooperative value of relationships from the general pattern of interaction of the group or species. For example, in those macaque species, in which matrilineal kin cooperate closely, reconciliation occurs more often after conflicts between kin than non-kin (reviewed by Aureli et al. 1997, Demaria & Thierry 2001, cf. Thierry 1990). In chimpanzees, in which males typically form coalitions with one another for within- and between-group conflicts, male-male conflicts are reconciled more often than female-female conflicts (de Waal 1986b, Goodall 1986, Arnold & Whiten 2001, Wittig & Boesch 2003, but see Kutsukake & Castles 2004). In mountain gorillas, where females receive support from adult males during female-female conflicts as well as protection against potentially infanticidal males from outside, females only reconcile conflicts with adult males (Watts 1995). Variation in reconciliation between groups of the same species can also be interpreted in support for the prediction. In a study of two groups of pig-tailed macaques (*Macaca nemestrina*) living in very similar enclosures, the average reconciliation frequency in the well-established group was twice as high as the frequency in the newly-formed group (Castles et al. 1996). This finding suggests that reconciliation is influenced by the possibility for group members to form reliable cooperative relationships with one another over several years.

Interspecies comparisons of rates and patterns of reconciliation can also be useful to examine possible links with cooperation. For example, there is variation in average reconciliation rates across macaque species (Thierry 1986, 2000, de Waal & Ren 1988, Aureli et al. 1997, Petit et al. 1997, Demaria & Thierry 2001). One possible reason for this variation can be related to the relative value of the average group member that may vary depending on the need of cooperative actions requiring several group members, such as the protection from external threats (e.g. predators, other groups of the same species, or infanticidal males). In species where such need is high, subordinates have leverage over dominant individuals and the latter are expected to be more tolerant towards other group members in exchange for support in such cooperative actions (e.g. van Schaik 1989). This would result in variation in the degree of despotism and tolerance towards unrelated group members across macaque species. In particular, in species in which more cooperative actions with the average group member are needed, a weaker power asymmetry (i.e. a less strict effect of the dominance hierarchy) is expected in contests.

Using a combined data set of eleven groups of nine macaque species, we have recently tested whether the variation in reconciliation rates is associated with interspecific differences in power asymmetry (Thierry et al., in prep.). We found a positive correlation between conciliatory tendency and the percentage of counter-aggression in contests (i.e. a proxy of power asymmetry). We also found that in species with a higher percentage of counter-aggression (i.e. lower power asymmetry) there was a smaller difference in the conciliatory tendencies between kin and non-kin. Both findings suggest that reconciliation rates are higher and less kin-biased in macaque species with lower power asymmetry. If our reasoning about the higher need of cooperation among group members in less despotic species holds true, then these findings support the prediction of more reconciliation in more cooperative relationships.

Based on the evidence reviewed above, we can conclude that there is clear support for a relation between the degree of partners' cooperation and the likelihood of reconciliation after their conflicts. Specifically, conflicts between partners with more cooperative relationships are reconciled at higher rates. Although this relation is true in most cases, there are some caveats that we address in the next two sections.

7.3.1 The case of callitrichids

Obvious test cases for the prediction of a positive association between the degree of cooperation and the level of reconciliation are cooperatively-breeding species. Among primates, the best examples of such species are callitrichids, i.e. marmosets and tamarins (Solomon & French 1997). In these species, individuals live in extended family groups in which older offspring delay dispersal from the natal groups and help parents to raise their younger siblings (French 1997). Group members cooperate in virtually all tasks: (i) rearing of young, (ii) travel coordination, (iii) food-sharing and (iv) anti-predator surveillance (reviewed by Caine 1993, Schaffner & Caine 2000). Marmosets and tamarins also pack together during sleep (Caine et al. 1992) and affiliate with each other at high rates (Price 1992a, Wormell 1994). Given the highly cooperative nature of their relationships and the high degree of tolerance, callitrichids are expected to reconcile at high rates.

Two studies have been carried out on reconciliation in callitrichids. In red-bellied tamarins (*Saguinus labiatus*), no evidence for reconciliation was found (Schaffner & Caine 2000). In common marmosets (*Callithrix jacchus*), the occurrence of reconciliation was demonstrated, but only if cases of mere proximity (i.e. without affiliative overture) between former opponents were included in the analysis (Westlund et al. 2000). Based on evidence from other primate species and given the highly tolerant nature of callitrichids, post-conflict reunions in these species are expected to include overt friendly behavior, such as grooming or special contacts not usually common outside the post-conflict context (cf. 'explicit reconciliation' in de Waal 1993). Thus, post-conflict proximity in marmosets may have a different meaning (see below), and Westlund et al.'s (2000) findings are probably better interpreted as another failure to demonstrate reconciliation in callitrichids. This is clearly surprising given the highly cooperative nature of their relationships.

One possible explanation for the surprising results is that in the above study groups, the first assumption of the concept of reconciliation (see above) does not hold. Aureli et al. (2002) emphasized that the critical issue is whether aggressive conflicts disrupt cooperative relationships and jeopardize the benefits associated with such valuable relationships. Thus, it is paramount to investigate whether there is evidence for post-conflict disturbance of the social relationships between the two opponents in callitrichids.

Using the same data set of Schaffner & Caine's (2000) study on red-bellied tamarins, we tested for evidence of post-conflict disturbance of social relationships (Schaffner et al. 2005). First, we did not find evidence for a reduction of post-conflict tolerance between former opponents as their proximity levels after aggression did not decrease compared to baseline levels. Second, this finding held even in those cases in which post-conflict renewal of aggression between opponents took place. Third, there was no evidence for a post-conflict disruption of opponents' mutual activities as they resume the activity (e.g. feeding in close proximity grooming, or playing) they were engaged in before the conflict as often as they would in control periods.

Close inspection of the results of Westlund et al.'s (2000) study on common marmosets reveals striking similarities with the findings for our study on red-bellied tamarins. First, there was no evidence for a post-conflict decrease in tolerance between opponents as their proximity levels were actually higher in post-conflict periods than in control periods (see above). Second, there was circumstantial evidence for rapid post-conflict resumption of the same activity former opponents were engaged in prior to the conflict. Third, the authors reported that the marmosets hardly showed any behavioral sign of distress in the aftermath of aggressive conflicts (Westlund et al. 2000).

The findings of the two studies suggest that reconciliation does not take place because there is no post-conflict disturbance of social relationships, probably due to their high predictability and resilience (i.e. highly secure relationship; cf. Cords & Aureli 2000; see below). The lack of disruption may be the result of alternative conflict management mechanisms that reduce the frequency and intensity of aggressive conflicts (Schaffner & Caine 2000). The findings, however, are important because they point out that it is critical to test for the assumptions underlying the reconciliation concept. Thus, the case portrayed by the two studies has more general implications for other species as well; it is the exception that confirms the rule. If there is no post-conflict disruption of cooperation, there is no need to restore it with reconciliation.

7.3.2 Biological market effects

Reconciliation frequency within a social relationship may change over time when changes in the way partners value each other occur. The experimental study by Cords & Thurnheer (1993) reported above is a brilliant illustration of these interlinked changes. Reconciliation frequency within a social relationship may also change depending on modifications in the social environment. In particular, partner value (relative to others) is expected to change as a function of the availability of alternative partners. This dynamic view of partner value is best interpreted within a biological market framework in which animals are considered as traders engaging in exchanges of commodities (Noë et al. 1991, Noë & Hammerstein 1995). As in human economic markets, the animal traders would select the social partner that offers the best value on the basis of the supply and demand of the commodity. Thus, the levels of supply and demand as well as competition among traders determine the value of commodities and ultimately the relative value of potential partners (Barrett & Henzi 2001). Several empirical studies confirmed the predictions of the framework for exchanges of grooming and other commodities in various animal species (Barrett & Henzi, this volume, Noë, this volume). Thus, if the cooperative value of group members varies depending on local biological markets, reconciliation patterns are expected to vary accordingly. Reconciliation frequency within a social relationship is likely to change depending on the availability of alternative partners and their overall relative value. Below, we illustrate this rationale with three possible scenarios.

In pair-living species, the pair members have a valuable relationship with each other as they rely on the other for sharing tasks and cooperative actions. As they depend so highly on each other, aggressive conflicts are expected to be rare. In addition, post-conflict reconciliation may not be needed as their relationships are valuable, but also highly secure (Cords & Aureli 2000). The situation can change if there is an external threat that undermines the pair bond. We would predict higher investment in the pairmate including conciliatory overtures if an alternative partner appears on the scene. Although no data on reconciliation are available to test this prediction, experimental data of challenges of established pairs with intruders suggests that this could be possible. For example, pair-living titi monkeys (*Callicebus moloch*) spent more time in close proximity to each other when experimentally exposed to conspecific strangers than during baseline conditions (Anzenberger et al. 1986). The increase in overall proximity included a higher proportion of time spent huddling with each other. In addition, the pairmates joined in agonistic displays toward the strangers, and these displays often culminated in mutual vocal calls (i.e. duets), which were never performed in other circumstances (Anzenberger et al. 1986). These findings suggest that behavioral changes elicited by the presence of strangers may serve to strengthen the bond between pairmates when exposed to a potential challenge to their relationship security. Under this scenario, reconciliation is also expected to occur to maintain the valuable relationship between pairmates.

Another possible scenario is the case of cooperative breeders. As discussed above, there is no evidence for post-conflict relationship disturbance in the two studies of callitrichids carried out so far. Similarly to pair-living species, group members in cooperatively-breeding species are strongly interdependent on one another for survival and reproductive success. Their relationships are therefore valuable and secure, and aggression is rare (Schaffner & Caine 2000). Variation in relationship quality across group members can however be expected with changes in group size. When groups are small, every group member is needed to perform the various cooperative tasks for successful rearing of the young. The overall positive disposition toward other group members when group size is small can be illustrated by the high degree of tolerance the breeding female shows toward strangers when no subadult or adult helpers are present in the group (Schaffner & French 1997). However, when group size is larger, tolerance towards strangers is absent. Furthermore, in large family groups, severe instances of aggression against some subadult or adult helpers are observed (Schaffner & Caine 2000). This form of aggression is likely to disrupt opponents' relationships. It is, however, unlikely that reconciliation would take place in these cases because the most likely outcome of the prolonged hostility is the eviction of the targeted individual, indicating that such relationships are no longer valuable (and it is difficult to reintroduce ejected individuals into the group; see Inglett et al. 1989). In support of this view, the breeding female of large groups shows high intolerance towards intruders when tested in experimental settings (Schaffner & French 1997). Thus, a possible scenario for the occurrence of reconciliation in callitrichids is when in middle-sized groups mild aggression results in a loss of benefits because individuals can 'shop around' for alternative cooperative partners (cf. Lazaro-Perea 2001). Under these circumstances, former opponents are likely to be interested in reconciling their conflicts to reaffirm the co-dependency to each other and thus maintain the benefits associated with their relationships.

A third scenario to illustrate the effect of biological markets on reconciliation can take place in all those species in which cooperative actions, such as coalitions, may involve different partners. For example, in many macaque species, coalitions occur between matrilineal kin, but also between individuals belonging to different matrilines (Chapais, this volume, Silk, this volume). In most despotic macaque species, there is however a bias toward matrilineal

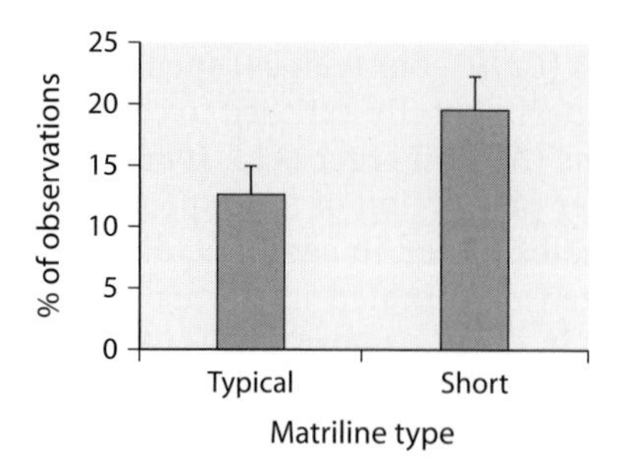

Fig. 7.1. The mean (+SE) percentage of post-conflict observations in which a post-conflict friendly reunion between non-kin opponents occurred during the first three minutes in two groups of long-tailed macaques differing in their matrilineal structure. Typical: a group with the typical matriline structure; Short: a group with short matrilines.

kin to serve as coalition partners (Chapais, this volume, Thierry 1990, 2000). This is usually the case in long-tailed macaques (de Waal 1976, 1977) in which matrilineal kin reconcile more often than non-kin (Aureli et al. 1989, 1997). The relative preference for matrilineal kin as coalition partners can also be affected by their relative availability. We could thus hypothesize that in groups where only a few matrilineal kin are available relatively more non-kin serve as coalition partners. Under these circumstances, we would expect non-kin to have relatively more valuable relationships and therefore to reconcile relatively more often. We tested this prediction with data from two captive groups of long-tailed macaques; one group had typically well-developed matrilines, whereas the other group contained only a few matrilineal kin (the group was formed by splitting a large group to minimize the number of kin for unrelated study). As predicted, we found that reconciliation between non-kin was higher in the group with short matrilines than in the group with the typical matriline structure (Fig. 7.1). Although we need to be cautious in our conclusion as we did not test directly for differences in non-kin cooperation between the groups, this finding suggests that reconciliation is affected by what is available in the market place.

7.4 Cooperation in the act of reconciliation

Reconciliation is usually a dyadic phenomenon. Individuals may differ in their interest in reconciling, but both former opponents need to participate for the post-conflict friendly reunion to occur. One of the former opponents may take the initiative for the reunion, but the other needs to accept or reciprocate the friendly overture for reconciliation to really take place. One-sided attempts may occur, but they are not expected to function as reconciliation (i.e. fulfilling the second assumption of the concept of reconciliation). Thus, it seems that cooperation between former opponents in the very act of friendly reunions can be at

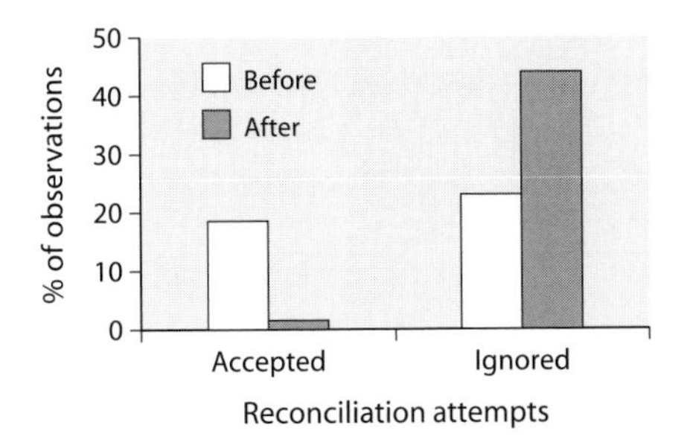

Fig. 7.2. The mean percentage of post-conflict observations in which aggressive behavior between former opponents occurred before and after accepted and ignored reconciliation attempts in 4 to 6-year-old boys (modified from Ljungberg et al. 1999).

the basis of the resumption of pre-conflict interaction and restoration of their social relationship.

A critical test of the functional importance of cooperation in the act of reconciliation is to examine whether only reunion attempts that are accepted by the partner restore their social relationship. No published study on nonhuman primates examined the functional consequences of attempts to reconcile depending on their success, probably due to the difficulty for researchers to recognize failed or aborted attempts (Cords & Aureli 2000). There is at least one study on children that addresses this issue (Ljungberg et al. 1999). There are many similarities in the principles and actual patterns of conflict resolution between nonhuman primates and children (Cords & Killen 1998, Butovskaya et al. 2000), and the use of language can facilitate the researchers' discrimination between accepted and ignored attempts to reconcile in children. Ljungberg et al. (1999) found that among 5-year-old boys when reconciliation attempts were accepted, both post-conflict stress-related behavior and hostility were reduced (Fig. 7.2). In contrast, reconciliation attempts that were ignored did not reduce post-conflict stress-related behavior and even increased post-conflict hostility between former opponents (Fig. 7.2). Here, we need to be cautious about the direction of causality as the boys may ignore reconciliation attempts due to persistent hostility or instead become more hostile as a result of the rejected reconciliation. Hence, this study supports the view that the cooperation required in accepted reconciliation attempts may convey the mutual interest of former opponents to end the hostility and start the process of relationship repair.

Evidence for such a role of cooperation in the act of reconciliation may be gathered in nonhuman primates by distinguishing the patterns of friendly reunions. For example, attempts that are reciprocated by the partner (e.g. mutual grooming) or that require direct participation from the partner (e.g. the 'hold-bottom' ritual in stumptailed macaques, Fig. 7.3) can be more effective in terms of restoring the relationship than attempts that are characterized only by unidirectional friendly contacts (e.g. brief gentle touch by one individual without any overt response from the partner). If this prediction is supported by empirical data, it will point out that the degree of cooperation involved in the act of recon-

Fig. 7.3. The 'hold-bottom' ritual in stumptailed macaques, which is often used for reconciliation. Photograph by Josep Call.

ciliation can be directly related to its effectiveness in restoring social relationships. After all, what can be more convincing for former opponents about the reciprocal interest in repairing their social relationship than an act of cooperation taking place right after an aggressive conflict?

7.5 Reconciliation as a tool for relationship negotiation

Given the need for both former opponents to participate at least to some extent in the post-conflict friendly reunion, it is possible that the acceptance of reconciliation attempts is open to negotiation. The likely asymmetry in the benefits obtained from their social relationship is expected to underlie differential interest between former opponents in repairing the disrupted relationship and thus in initiating and accepting reconciliation attempts (Cords & Aureli 2000). Within a dyad, the individual with less at stake in the social relationship is in a more powerful bargaining position for negotiation about the future course of their interactions (cf. Dunbar 1988). Accepting or not accepting a reconciliation attempt can, therefore, be used as a tool for relationship negotiation (de Waal 1996a) and may affect future cooperation.

The implications of the differential bargaining power between former opponents for the process of conflict resolution were clearly recognized by de Waal (1986b). He suggested that aggressors could make conditions for reconciliation to take place and labeled the process as 'conditional reassurance' (de Waal 1986b). Reconciliation can serve as a reassurance mechanism when it alleviates distress (see above for evidence) and can be conditional when it is granted depending on whether the partner exhibits a wanted behavior or stops an unwanted behavior. De Waal (1986b) proposed a mother's behavior toward the offspring during weaning as a clear illustration of conditional reassurance in species where the mother-offspring relationship extends beyond the weaning period. After having pushed the offspring away from her breast, a chimpanzee mother, for example, allows it to return in contact with her, so reducing the distress caused by the forced separation, on condition that the offspring refrains from reestablishing nipple contact. This is a case of conditional reassurance in which the partner needs to stop an unwanted behavior. An example of the condition in which the partner needs to display a wanted behavior is the granting of post-conflict reconciliation only after the partner gives a signal of submission. De Waal (1986b) illustrated this case by reporting on dominance struggles between male chimpanzees. During the last phase of the struggles, the winner of a conflict avoided to accept conciliatory overtures from the loser until the latter displayed a submissive greeting.

A similar interpretation of conflict resolution in terms of relationship negotiation was proposed by Aureli et al. (1989). They emphasized the emotional and physiological costs of receiving aggression and the role of reconciliation in reducing such costs. Then, they predicted reconciliation patterns according to the relationship value from the winner's point of view because the loser is expected to be the one more likely to seek reconciliation due to the higher post-conflict costs. Winners would offer or withhold reconciliation as a means to increase or decrease the post-conflict costs to the loser depending on whether it is in the winners' interest to maintain a positive relationship with the loser (Aureli et al. 1989). Thus, according to this view, the granting of reconciliation is ultimately a means for manipulation of the loser's post-conflict costs depending on its value to the winner.

The possibility of post-conflict relationship repair through reconciliation may also strengthen the feasibility of using aggression as punishment to manipulate others (cf. Clutton-Brock & Parker 1995). Aggression can be used more effectively as a tool for relationship negotiation because mechanisms for conflict resolution, such as reconciliation, can undo the damage to the social relationship (de Waal 1996a, 2000a). Based on the views outlined above (de Waal 1986b, Aureli et al. 1989), we would like to take this perspective one step further and relate the relationship negotiation more directly to cooperation.

Individuals not only can aggressively punish partners that are unwilling to cooperate, but they can withhold reconciliation as a potentially additional punishment. It is likely that this tactic also entails some short-term costs for the individuals withholding reconciliation as they cannot count on the benefits of their relationship with the former opponent in the aftermath of a conflict (this claim is supported by increased post-conflict rates of self-directed behavior of former aggressors; Aureli 1997, Castles & Whiten 1998, Das et al 1998, Schino 1998). However, additional punishment through withholding reconciliation is possible when the short-term costs are outweighed by the benefits of successful instigation of future cooperation. As the benefits may be substantial, such a tactic is expected to occur regularly to negotiate cooperation (i.e. altruistic punishment; Gächter & Herrmann, this volume).

7.6 Reconciliation as a means of communication

The need for cooperation between former opponents in the act of reconciliation opens the possibility that the act itself can be used as a means of communication about their relationship. In particular, the behavior of each opponent during the reconciliation process can convey information about the relative value each individual attaches to the relationship. Because each opponent needs to attend and respond appropriately to the partner's behavior to achieve successful reconciliation (see above), the sequence and patterning of the exchange can have an important communicative value. The readiness to attempt reconciliation and the willingness to accept it are obviously related to each partner's interest in repairing the relationship. The mutual behavior of the two partners during such a process can, therefore, communicate their relative interest in the relationship and possible future cooperation.

The communicatory value of the act of reconciliation has been emphasized by Silk (1996, 1997). She suggested that post-conflict reunions may communicate that the signaler's future behavior will be non-aggressive and perhaps benign. This is clearly important in reducing uncertainty about future interactions between former opponents. When former opponents are both interested in coordinating their post-conflict interactions for mutual benefit, they can use signals of benign intent that are inexpensive and inconspicuous (Silk 1997). A good example of such signals is the use of soft grunts by baboons during post-conflict reunions (Cheney et al. 1995, Silk et al. 1996).

Our suggestion for a communicatory value of the act of reconciliation is complementary to Silk's interpretation, but it focuses on post-conflict exchanges that require more coordination between former opponents. Many behaviors exchanged during post-conflict reunions are conspicuous and involve the participation of both partners (de Waal 1993, Aureli et al. 2002; e.g. Fig. 7.3). In addition, many of these reunions involve close body contact between the former opponents. Thus, whereas vocalizations, such as the baboon grunts, are given at a distance and entail little cost, contact reunions can be highly risky. After all, they require close proximity between individuals that have just exchanged hostility. The relative readiness of former opponents to exchange post-conflict friendly contacts despite the risk of renewed aggression can therefore convey information about their interest in the future of the relationship.

Risky post-conflict reunions could also serve as tests of the strength of the bond between former opponents with insecure relationships (i.e. relationships in which the partner's behavior is rather unpredictable; Cords & Aureli 2000). Zahavi (1977b) proposed that such tests require imposing a stressful stimulus on the partner. Such a stimulus typically has a strong sensual component that can be perceived by the recipient as pleasant if the bond is strong (possibly leading to reciprocation) or unpleasant if the relationship is weak (likely leading to withdrawal). The recipient's response then provides reliable information about the recipient's current interest in the relationship (Zahavi 1977b). The potential close body contact with a former opponent in the aftermath of a conflict is likely to be perceived as stressful by partners with insecure relationships. The partner's response to engage in such contact or to avoid it could then serve as a signal of the partner's interest in the relationship and of the partner's reliability in future cooperative actions.

7.7 Summary and conclusions

Our review of the role of cooperation in the occurrence of reconciliation reveals a close link between the two phenomena in terms of causes, consequences and mechanisms. The potential disruption of cooperation due to aggressive conflict is probably one of the most important factors at the basis of the post-conflict emotional changes (e.g. post-conflict anxiety) that mediate reconciliation and the restoration of cooperative interaction (cf. Aureli & Smucny 2000). This cause-consequence relation linking cooperation and reconciliation is supported by several studies showing higher rates of reconciliation after conflicts between individuals with more cooperative relationships.

Whether the pattern of reconciliation across group members is based on long-term planning by single individuals for future cooperation is an issue still open to investigation. Although the possible involvement of long-term planning has intriguing implications for the cognitive abilities underlying reconciliation, the influence of relationship quality on the conciliatory process may be based more on the history of previous interactions between former opponents than on an evaluation of potential future exchanges. Relationship quality could influence reconciliation patterns via emotional mediation (Aureli 1997, Aureli & Smucny 2000) when emotional experiences with partners are viewed as 'summaries' of the history of previous interactions with such individuals (Aureli & Schaffner 2002, Aureli & Whiten 2003).

Involvement of long-term planning is not necessary from a functional point of view. The function of reconciliation in relationship repair and restoration of cooperation can occur through short-term changes that bring the interaction quality between former opponents back to what it was before the aggressive conflict (Cords & Aureli 1996). This short-term view of the conciliatory process does not imply that reconciliation does not have medium- or long-term effects. There is actually evidence for reconciliation functioning in undoing long-term negative damage caused by aggressive conflicts (Koyama 2001).

The relation between a high degree of cooperation between partners and a high likelihood of reconciliation after their conflicts is not unconditional. We presented cases that qualify which sets of circumstances modulate such a relation. The case of callitrichids pointed out that if aggressive conflict does not disrupt cooperation between former opponents, there is no need for reconciliation (cf. Aureli et al. 2002). The framework of biological markets (Barrett & Henzi, this volume, Noë, this volume) helped us to interpret cases in which the likelihood of reconciliation may vary as a function of changes in partner value relative to other individuals. If the cooperative value of group members varies depending on local biological markets, reconciliation frequency within a social relationship is likely to change depending on the availability of alternative partners and their overall relative value. A focused research effort is needed to explore the role of biological markets in modulating the relation between cooperation and reconciliation.

Cooperation also plays an important role as a mechanism in the conciliatory process. Both former opponents need to participate in a post-conflict friendly reunion for such an act to function as reconciliation. This critical role is nicely supported by the finding of a study on children that only accepted attempts at reconciliation reduced post-conflict stress and hostility (Ljungberg et al. 1999).

The role of cooperation in the conciliatory act also has implications for the possible use of reconciliation as a means of relationship negotiation and as a means of communication about the relative value each partner attaches to the social relationship. Further investigation of these two aspects is critical because both are expected to play a role in future cooperation between the partners. If empirical evidence supports these views, cooperation and reconciliation will be even more interlinked than presented in this chapter.

Acknowledgments

We thank Peter Kappeler and the German Primate Center for organizing the conference, funding the first author's attendance and inviting us to write the chapter; Frans de Waal, Peter Kappeler, Joe Manson, Carel van Schaik and other participants of the Freilandtage in Göttingen 2003 for their comments; and Josep Call for permission to use his photograph in Fig. 7.3.

Part IV
Mutualism

Chapter 8

Cooperative hunting in chimpanzees: kinship or mutualism?

Christophe Boesch, Hedwige Boesch, Linda Vigilant

8.1 Introduction

Cooperation characterizes human societies today and is thought to have been important in our evolutionary past as one of the main characteristics that allowed humans to dominate the planet (Isaac 1978, Mithen 1996). The fundamental role of cooperation is, for example, evident in hunter-gatherer societies, where the prevalence of food-sharing combined with a sexual division of labor is the basis of the economic system (Kaplan et al. 2000, Heinrich et al. 2001, Hill 2002). Analysis of the factors favoring the evolution of cooperation is, therefore, vital for the understanding of the course of human social evolution.

Cooperation can be defined as two or more individuals acting together to achieve a common goal. Cooperation can evolve when two basic conditions are fulfilled. First, the benefits of the common action must be shared sooner or later between the participants, and second, the benefits to the participants must exceed the costs of the common action. However, cooperation can be rather unstable, as it is very susceptible to cheaters, individuals that are not investing in the cooperative task, but are nonetheless trying to gain access to the benefits. The greater the extent of cheating, the less likely cooperation is going to occur (Axelrod & Hamilton 1981, Maynard Smith 1982). Therefore, theories of the evolution of cooperation rely on four different mechanisms to explain its evolution. First, kin selection can facilitate cooperation among related individuals, as each agent will not only benefit directly, but also indirectly from the gains of the others the more they are related to them (Hamilton 1964). Second, in a mutualistic scenario, all partners in a cooperative act would directly profit from the outcome. Third, reciprocity in repeated interactions could allow each partner in turn to obtain the benefit of the cooperative act (Trivers 1971). Finally, the byproduct of sociality scenario suggests that individuals have been selected to live together for other reasons and, by virtue of their proximity, may happen to engage in cooperative acts even when not directly profiting from them (Mesterton-Gibbons & Dugatkin 1992). It is important to note that these explanations for cooperation are not mutually exclusive and several mechanisms may operate simultaneously. Kin selection theory has convincingly explained the evolution of some dramatic forms of cooperation, such as worker sterility in eusocial insects (Reeve et al. 1998). However, it is possible that the indirect benefits of cooperative behavior have been overestimated and little consideration given to direct benefits in other animal societies. This is illustrated by the exemplary work of Clutton-Brock and colleagues, in which careful field observation and experimentation using large numbers of wild meerkats has led to a greater appreciation of direct benefits as an explanation for helping in that species (Clutton-Brock et al. 2001b, 2001c, Clutton-Brock, this volume).

An important distinction between mutualism and reciprocity is that in a mutualistic scenario, the benefits to actors are immediate, whereas the benefits to one partner are deferred in a reciprocal interaction. This is important because some observations have shown that most animals appear wholly or mostly incapable of keeping track of past or future exchanges and so mutualism rather than reciprocity might be more likely to be observed in nature (Clements & Stephen 1995; but see de Waal & Brosnan, this volume).

Many carnivores hunt in groups, but even studies of the same species differ with regard to explanations of the mechanisms promoting this possibly cooperative behavior. In some populations of lions and cheetahs, group hunts have not been observed to provide a direct benefit to all hunters, so that in many instances it would seem better for individuals to hunt alone (Packer et al. 1990, Caro 1994). Accordingly, byproduct mutualism has been suggested to be the main mechanism explaining the observed group hunting in those populations. In other populations living in more difficult or less prey-rich habitat, lions and cheetahs were observed to hunt in groups more frequently and it could be shown that under these conditions, individual benefits increased with group size (Cooper 1991, Stander 1992, Creel & Creel 1995c). In carnivores, group hunting and the benefit extracted from it is not a fixed characteristic of a species, as it is the conditions of the habitats where the hunt take place that will determine the outcomes for the different participants in the hunt. Thus, viewing cooperation in hunting as a constant in a species living in different populations in different habitats is misleading and additional studies may be required to understand the features of the cooperative abilities of a particular species.

We are going to concentrate in this chapter on group hunting in wild chimpanzees (*Pan troglodytes*) and discuss which mechanisms could be responsible for its evolution and maintenance.

8.2 The puzzle of chimpanzee hunting

Wild chimpanzees live in large multi-male, multi-female communities that may contain from 10 to more than 100 individuals. Membership in these communities is very stable, except for females that normally transfer once, upon reaching maturity. Males are philopatric, remaining in their natal group their entire lives (Goodall 1986, Nishida 1990, Boesch & Boesch-Achermann 2000, Mitani & Watts 2002). Cooperation in chimpanzees has been observed in two main contexts. First, the territory of one community is defended by macro-coalitions of adult and adolescent males that regularly patrol the borders, repelling all intruders they see or hear (Goodall 1986, Boesch & Boesch-Achermann 2000). Encounters between communities are normally aggressive and in some instances, have been seen to lead to the deaths of individuals (Goodall et al. 1979, Goodall 1986, Boesch & Boesch-Achermann 2000, Wilson & Wrangham 2003). Hunting of small mammalian prey is the second typical context in which cooperation among varying numbers of males has been observed (Boesch & Boesch-Achermann 2000). Hunting by wild chimpanzees has been observed throughout the area of distribution of the species and it represents a special challenge, as it involves the pursuit and capture of small monkeys fleeing through the forest canopy. As in many apparently cooperative activities, the outcome of a hunt is dependent upon the spontaneous participation of different actors. In fact, hunting success increases with the number of individuals actively hunting (Boesch 1994, Boesch & Boesch-Achermann 2000).

Hunting seems to be a universal behavior in chimpanzees as it has been observed in all populations subject to long-term study (Boesch & Boesch-Achermann 2000). However, detailed observations have been limited to only a few of the long-term studies on that species. First, Geza Teleki (1973) described hunting behavior of the Gombe chimpanzees and detailed some of the tactics used by the male hunters. Subsequent studies at Gombe have complemented our understanding of the hunting behavior in that population (Busse 1978, Goodall 1986, Stanford et al. 1994, Stanford 1998). Studies of hunting behavior have also been conducted on the chimpanzees in Mahale Mountains National Park, some 200 km south of Gombe, and revealed a rather similar picture of the hunting behavior (Nishida et al. 1983, 1992, Uehera et al. 1992). Our understanding of this behavior in chimpanzees was broadened when new observations of the behavior of the Taï chimpanzees in Côte d'Ivoire became available (Boesch & Boesch 1989), revealing surprisingly large variation in the hunting behavior within this species. Lastly, observations of an exceptionally large community in Ngogo, Uganda, have complemented this view of a very flexible behavior in chimpanzees (Watts & Mitani 2000, Mitani et al. 2002a).

Hunting in chimpanzees is puzzling because while there are many similarities in hunting behavior between different populations, important differences have also been observed (Nishida et al. 1983, Goodall 1986, Boesch & Boesch 1989, Uehara et al. 1992, Boesch 1994, Boesch & Boesch-Achermann 2000, Mitani et al. 2002a). First, all chimpanzee populations have been observed to hunt mainly arboreal monkey species and, of those, red colobus monkeys (*Procolobus badius*) are generally the preferred species whenever they are present (Table 8.1). Second, since monkeys are the preferred prey, hunting occurs mostly in the trees. Finally, hunting success in chimpanzees is rather high, compared to many other predatory animal species (Table 8.1). For example, wolves are successful in 8% and 25% of their hunts of moose and deer, respectively (Mech 1970), which is two to five times lower than chimpanzee success rates. Similarly, lions in the Serengeti capture prey in 61% of purely opportunistic cases, but in only 19% of instances when they are stalking the prey, and in 8% of hunts that occur in the open plains (Schaller 1972). Hyenas in the Kalahari are successful in 32% of their hunting attempts (Mills 1990). One possible explanation for the chimpanzee's high rate of success may lie in the fact that, not relying on meat for survival, they tend to hunt only when the likelihood of a capture appears high, whereas social carnivores, being much more dependent on meat, hunt more depending on their level of hunger rather than according to the likelihood of success.

Table 8.1. Similarities in chimpanzee hunting behavior.

	Gombe[a]	Mahale[b]	Ngogo[c]	Taï[d]
Prey selection				
Red colobus	55%	53%	88%	81%
Hunting success	52%	61%	82%	52%

[a] Goodall (1986)
[b] Nishida et al. (1983), Uehara et al. (1992)
[c] Watts & Mitani (2002)
[d] Boesch & Boesch-Achermann (2000)

Differences in hunting behavior among populations of chimpanzees are quite striking (Table 8.2). However, consideration of the differences is complicated by the fact that in arboreal hunts with limited visibility, it is not always easy to distinguish hunters from non-hunters. This makes it difficult to compare observations from different chimpanzee populations. For example, if the hunt happens in a dense part of the forest or when a large number of individuals are present, it becomes very difficult to distinguish whether a given individual is actually actively trying to capture a prey or is simply looking on. This has led some not to distinguish between hunters and non-hunters (Ngogo: Mitani & Watts 1999, 2001, Watts & Mitani 2000; Gombe: Teleki 1973, Stanford 1995, 1998, Stanford et al. 1994a, 1994b), thereby making any discussion about the evolution of cooperation impossible because per definition hunters and cheaters are treated equally. Others have concentrated their analyses on hunts in which the distinction between hunters and non-hunters was based on the behavior of the individuals present (Gombe: Busse 1977, 1978, Goodall 1986; Taï: Boesch & Boesch 1989, Boesch 1994, Boesch & Boesch-Achermann 2000). In such cases, however, individuals may alternate between hunting for some time and just looking at hunting by others. In other instances, a pursued prey may fall to the ground, whereupon one of the individuals watching the hunt would capture it and by its action immediately become a hunter. This distinction between hunters and non-hunters should be made whenever possible, because it permits the proposal of scenarios concerning the evolution and stability of cooperation in hunting.

Despite the different approaches toward observation of chimpanzee hunting, it remains possible to see some clear differences between different chimpanzee populations. First, the tendency to hunt in groups is highly variable and is observed in only one-third of the hunts by Gombe chimpanzees, while group hunting is the rule in Taï and even more so in Ngogo (Table 8.2). Even more striking is the fact that the level of organization during hunts seems very different. Collaboration, in which different hunters perform different but complementary roles during a hunt to capture a prey (Boesch & Boesch 1989), has been regularly

Table 8.2. Differences in chimpanzee hunting behavior.

	Gombe[a]	Mahale[b]	Ngogo[c]	Taï[d]
Group hunt	36%	72%	100%	84%
Collaboration	19%	0%	rare	77%

[a] Busse 1977, 1978, Goodall 1986
[b] Nishida et al. 1983, Uehara et al. 1992
[c] Watts & Mitani 2002
[d] Boesch & Boesch-Achermann 2000

observed only in Taï chimpanzees (Table 8.2). In Ngogo, despite the fact that the chimpanzees hunt exclusively in groups, collaboration among hunters has been observed only rarely (David Watts & John Mitani pers. com.).

We now are going to discuss possible explanations of differences observed among chimpanzee populations in the level of cooperation in their hunting strategies. More specifically, is chimpanzee hunting better explained by kinship or mutualism? These explanations are not mutually exclusive, in the sense that if hunting is beneficial to all participants, it would also pay to hunt with kin and we would not be able to conclude whether kin selection or mutualism favors cooperation. If, however, kin do not hunt together, even though hunting is beneficial, we may argue that kin selection is not the prime factor explaining cooperation.

8.2.1 Kin selection hypothesis

The kin selection hypothesis predicts that related individuals will experience, through indirect benefits, greater paybacks from cooperating than would non-kin. There are exceptional circumstances in which kin selection might not be favored, such as when substantial levels of competition among kin exist (West et al. 2002) or when individuals vary in other ways in their suitability as cooperative partners (Chapais, this volume). However, substantial evidence exists for a role of kin selection in the social behavior of female-philopatric primates (Pope 2000a, Chapais et al. 1997; see Silk, this volume). For male-philopatric chimpanzees, we can make three testable predictions concerning the distribution of kin in social groups. First, males should be more related than females, since males are the primary hunters in chimpanzees. Second, to explain differences between Gombe and Taï, we would expect Taï males to be more related than Gombe males, as the first hunt so much more often in groups. Third, and even if the two first predictions are not supported, we should expect individual males within a group to choose to hunt more frequently with those individuals that are more related to them.

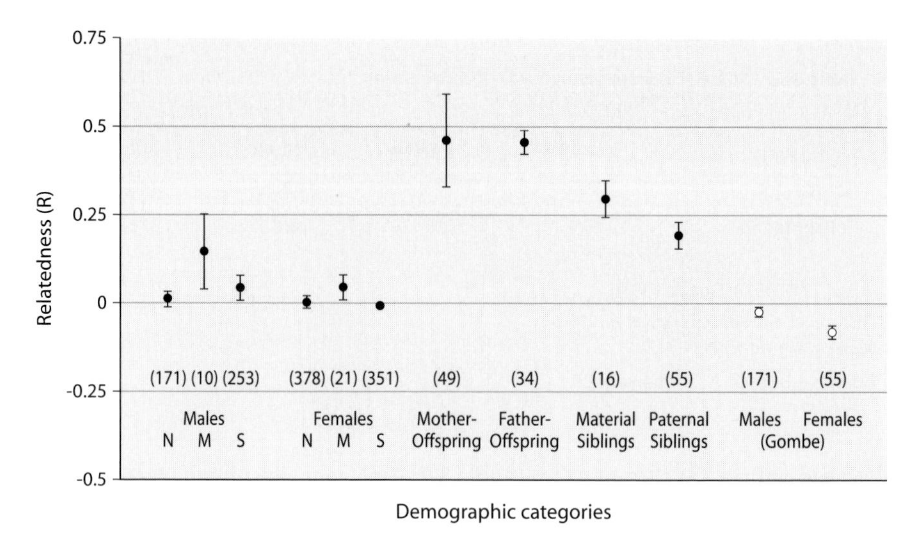

Fig. 8.1. Estimated relatedness (R) of the various demographic categories at Taï and Gombe.

Fig. 8.1 shows the average relatedness of the Taï chimpanzees as measured with microsatellite nuclear DNA markers (Vigilant et al. 2001). It shows very clearly that despite the fact that in Taï chimpanzees, males are philopatric and almost all females transfer between groups before reaching maturity, in each of the three groups on average the males are not significantly more related than the females. In addition, males in Taï chimpanzees and in Gombe chimpanzees have very similar, near zero average relatedness values. Therefore, the presently available data do not support the first two predictions that would originate from a kin selection explanation for male cooperative hunting.

However, the most interesting test is at the individual level, because males may selectively hunt with the individuals most related to them. The matrix in Table 8.3 shows the relevant data for joint hunting in Taï chimpanzees during two different time periods in 1987–1989 and in 1990–1994. Taï chimpanzees hunt mainly in groups containing an average of 3.5 individuals. To test for selective hunting according to relatedness, we established a second matrix with a pairwise comparison of genetic similarity for those males that we could sample (three individuals observed during the first period could not be sampled as they died before the initiation of the genetics project). Table 8.4 reveals that only one dyad out of 21 was judged significantly more likely to represent a half-sibling rather than an unrelated pair (using the program Kinship). The ability to determine a half-sibling from typical genotype information alone is limited (Blouin 2003), as is illustrated by the fact that a known maternal brother pair (Kendo-Fitz) was not significantly supported as a sibling pair when using Kinship. Nonetheless, all other dyads had low or negative (indicating lower relatedness than a random pair from the population) relatedness estimates, suggesting that no other pairs of relatives were present.

Table 8.3. Dyadic frequency of joint hunting in Taï males between 1987 and 1989 (upper half), and between 1990 and 1994 (lower half). Joint hunting was calculated as the number of hunts where two individuals were actively hunting divided by the number of hunts where both were present regardless of whether they were hunting or not.

	Bru	Dar	Ken	Mac	Rou	Sno	Uly	Ali	Fit
Bru	*	0.22	0.47	0.47	0.29	0.50	0.48	0.57	0.44
Dar	0.26	*	0.32	0.28	0.18	0.57	0.28	0.22	0.33
Ken	0.36	0.15	*	0.29	0.24	0.67	0.46	0.35	0.44
Mac	0.54	0.33	0.44	*	0.26	0.48	0.47	0.42	0.63
Rou					*	0.50	0.26	0.17	0.21
Sno						*	0.43	0.33	1.00
Uly							*	0.54	0.61
Ali	0.64	0.17	0.31	0.42				*	0.43
Fit	0.60	0.31	0.42	0.68				0.67	*
Mar	0.54	0.27	0.33	0.67				1.00	0.86

Table 8.4. Pairwise relatedness estimates for Taï male chimpanzees. In bold is the only dyad that was judged from genotype information to be significantly likely to be related at the half-sibling level, and underlined is the only dyad known to have the same mother.

	Bru	Dar	Ken	Mac	Ali	Fit	Mar
Bru	*	−0.186	−0.197	0.019	−0.080	0.002	−0.225
Dar		*	**0.218**	−0.166	0.038	−0.102	0.096
Ken			*	−0.021	−0.054	*0.184*	−0.065
Mac				*	−0.129	−0.033	0.016
Ali					*	−0.064	−0.092
Fit						*	−0.074
Mar							*

Table 8.5 presents the results of the correlation between these two matrices and clearly shows that relatedness had no significant relationship with joint hunting over both time periods. We checked for the importance of two additional factors, age and social rank, often proposed to be important in social interactions in chimpanzees. Of these, age had no effect but social rank played a significant role, in that males tended to hunt more frequently with males of

Table 8.5. Rowwise matrix correlations (Kr-test) of joint hunting with relatedness, age and rank in Taï male chimpanzees.

		Relatedness	Age	Rank
Joint hunt	Kr	1018	1480	1
1987-1989	p value	0.203	0.296	0.0002
Joint hunt	Kr	2308	232	1
1990-1994	p value	0.461	0.046	0.0002

similar social rank. Social rank was important in both time periods considered, despite the fact that some individual males died between the two periods and others occupied different social ranks (e.g. Macho decreased from the alpha position during the first period to the third position in the second, while Fitz was the seventh-ranking male in the first period but the alpha male in the second one).

Therefore, we can say that none of the three predictions were supported by the data and therefore kin selection generally appears not to be an important factor in explaining the hunting behavior of chimpanzees in the Taï Forest.

8.2.2 Mutualistic hypothesis

We now turn our attention to a mutualistic explanation, whereby we expect chimpanzees to hunt because they profit directly from taking part in this activity. Here again, we can make three different predictions. First, hunters should gain more when hunting in groups than when hunting alone. This would provide males with a strong incentive to wait for others to join or enlist others in the hunting activity. Second, for a given group size, we should expect hunters to gain more than non-hunters. If that were not true, we would not expect to see group hunting, as it would be better to cheat rather than to invest energy in a hunt. Third, we expect the first two predictions to apply more clearly to the situation in Taï chimpanzees than in Gombe chimpanzees and that this would explain the difference we have observed in the group hunting tendencies observed between these two populations (Table 8.2).

Fig. 8.2 shows the net benefit of three hunting strategies of male chimpanzees of the Taï Forest; namely hunter, bystander and latecomer. With regards to the first prediction, the success of hunters increases steadily and reaches a maximum when five individuals hunt together. This increase is significant and individual hunters gain significantly less when hunting in groups of three to five ($r_s = 0.78$, $n = 7$, $p < 0.05$; Boesch 1994). Similarly, Fig. 8.3 reveals that for hunting groups of three to five individuals, it pays an individual to be a hunter rather than a bystander or a latecomer (Wilcoxon signed-rank test: Hunter versus By-

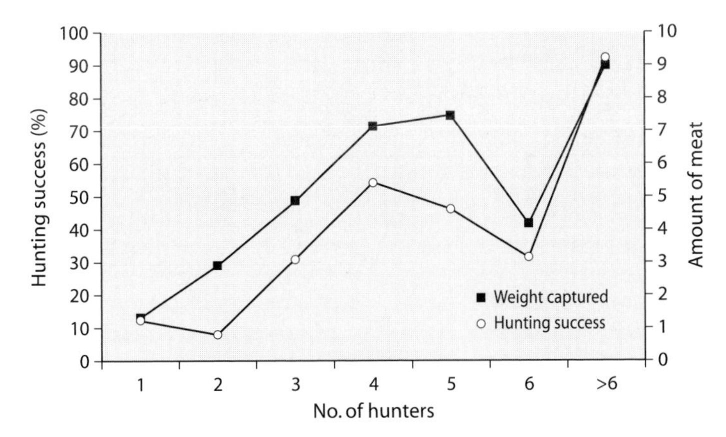

Fig. 8.2. Hunting success and weight captured as a function of the number of males participating in the hunt.

stander: for group size 3: $T^+ = 32$, $n = 8$, $p < 0.05$; for group size 4: $T^+ = 31$, $n = 8$, $p < 0.05$; Hunter versus Latecomer: for group sizes 3 to 5; $T^+ \geq 30$, $n = 8$, $p < 0.05$) (Boesch 1994). Therefore, there is not only an incentive for hunters to hunt in groups, but also for non-hunters to become hunters.

In Fig. 8.3, it is also possible to compare the relative success of both hunters and bystanders in Taï or in Gombe chimpanzees. There is a clear difference, for in Gombe bystander success does not differ from hunter success, except for groups of five hunters, in which case it is better to be a bystander (Wilcoxon signed-rank test: $T^+ = 30$, $n = 8$, $p < 0.05$) (Boesch 1994).

In conclusion, a mutualistic explanation, in which hunters receive direct benefits from cooperating appears to explain hunting in Taï chimpanzees. However, this does not appear to be the case in Gombe and this may explain why group hunting there is less frequent. An interesting correlate of this result is that meat-sharing among the males is done according to different rules in each population. In Taï chimpanzees, it is the behavior of the male that is the strongest predictor of the amount of meat he receives, with hunters receiving more than non-hunters and good hunters receiving the most (Boesch 1994, Boesch & Boesch-Achermann 2000). In Gombe, social dominance is the strongest predictor of meat access, with higher-ranking individuals receiving more, either through sharing or through stealing from the owner (Goodall 1986, Stanford et al. 1993, Boesch 1994). In Mahale chimpanzees, the dominant males secure the captured prey in the vast majority of the cases and favor allies when distributing meat (Nishida et al. 1992). Similarly, in Ngogo chimpanzees, meat-sharing occurs reciprocally with coalition partners of the meat owners (Mitani & Watts 2001). Thus, it seems that within each chimpanzee population, additional social factors are important and interact with the hunting behavior to result in different meat-sharing rules. It is notable that only in Taï chimpanzees do the sharing rules appear to support the stability of cooperation between individual active hunters.

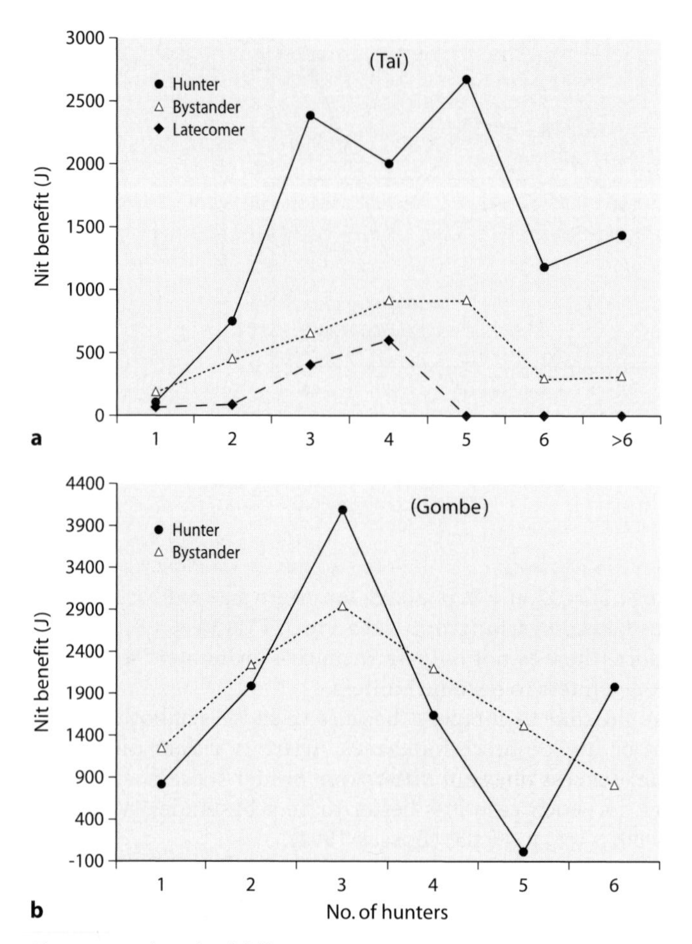

Fig. 8.3. Net benefit of different strategies used by chimpanzees when hunting red colobus monkeys in (**a**) Taï and (**b**) Gombe.

8.3 Discussion

Our analyses have revealed that cooperation in hunting among Taï chimpanzees is not readily explained by kin selection, but rather is the result of a mutualistic process in which all participants gain more than if they were acting alone. This result is consistent with investigations in two other chimpanzee populations of the importance of kin selection in explaining social behavior. A first study of the Kanyawara chimpanzees in the Kibale National Park of Uganda revealed that maternal genetic relatedness, estimated using mitochondrial DNA (mtDNA), did not predict social interactions such as association patterns or grooming interactions between individuals (Goldberg & Wrangham 1997). Similarly, in Ngogo chimpanzees of the Kibale National Park, Uganda, maternal genetic relatedness did not predict association patterns, grooming interactions, alliance between individuals, meat-sharing interactions or patrol participation among the males (Mitani et al. 2000).

The role of kinship in influencing behavior in social insects has been well documented, and there has been an expectation that kin selection would be a major factor for explaining social behavior in other animal taxa as well (Hamilton 1964, Wilson 1975, Maynard Smith & Szathmary 1995). In mammals, the role of kinship has been generally confirmed in social systems where dispersal patterns allow for related individuals to remain together, as in cooperatively breeding species in which helpers are often close kin or within female matrilines found in different primate species where social support may be given according to degree of relatedness (Chapais et al. 1997, Clutton-Brock 2002, West et al. 2002). However, in other studies, such as those on chimpanzees mentioned above, the influence of kin selection has not been found (Clutton-Brock et al. 2001b).

Two factors have been proposed to limit the generality of kin selection. First, groups of social animals may contain a high proportion of relatives and so competition between relatives may become a problem. It has been shown that competition between relatives can drastically limit the benefit of kin selection and that this might be a much more general phenomenon than previously estimated (West et al. 2002). Second, for some tasks it might be preferable to cooperate with particular individuals regardless of relatedness, as for example in a situation in which cooperation with an individual with particular skills is likely to lead to a better outcome (Chapais, this volume). In addition, motivation and predictability might be factors that explain why in Taï and possibly in Ngogo chimpanzees, it is similarity in rank and sometimes age for Ngogo (Table 8.5) rather than relatedness that are more important in predicting the distribution of social interactions and joint participation in hunting. Individuals of similar rank and age seem to share more similar social interests and are more likely to cooperate.

Mutualism has long been recognized as one of the mechanisms leading to the evolution of cooperation, but because of its obvious nature, 'if both gain more, then they should cooperate', theoreticians have neglected it and concentrated on less obvious mechanisms, such as delayed reciprocity and altruism (Maynard Smith 1982, Dugatkin 1997). However, a growing body of studies shows that mutualism has been underestimated because the costs of participating in the cooperative act have been overestimated and the direct benefits of cooperation have probably been underestimated (Clutton-Brock 2002). Hunting in Taï chimpanzees is best explained by a mutualistic process where each hunter gains more by hunting together with others. Cheaters that try to get access to meat without investing in the hunt have some success in obtaining meat, but clearly less than hunters, and that contributes to the stability of hunting in Taï (Boesch 1994).

The variability in chimpanzee hunting behavior seems to reflect an ecological difference, namely the difficulty of the habitat where the hunt is taking place. In the dense tropical forest of the Taï Forest, where monkey prey species have the possibility to escape in all directions, the hunting success of the chimpanzees is very low if they do not hunt in groups (Boesch 1994, Watts & Mitani 2002). In contrast, if they hunt monkeys in a disrupted forest, they can more easily corner them and hunting success seems much less affected by group size (Stanford et al. 1993, Boesch 1994, Watts & Mitani 2002). An additional demographic factor plays a role in which a larger number of hunters are able to disrupt prey defenses by overpowering numbers. This is best illustrated in the exceptionally large community at Ngogo where many individuals (up to 25 adult males) have been observed to hunt at the same time in a not very coordinated way but with a very high success rate (Watts & Mitani 2002). This has been observed in Gombe as well, where many individuals hunting at the same time in a non-coordinated way are able to make multiple captures (Goodall 1986, Boesch 1994). Similarly, in the Taï Forest, the number of potential hunters, for example adult males, in the community, has been found to influence both the frequency of hunts and the number of hunters (Boesch & Boesch-Achermann 2000).

Once they hunt, chimpanzees seem to have the ability to use the benefit of the hunt in a flexible manner. We suggest that when hunting is difficult, they are constrained to guarantee the stability of cooperation and therefore meat-sharing has to favor the hunters. However, when hunting is easier, cooperation is not necessary and the meat can be used to pursue other social goals. Thus, meat can become a currency used to pay for social services, allies in Mahale, social partners in Ngogo or simply to be taken away by dominant individuals in Gombe. This is somehow reminiscent of the situation that has been described for human hunter-gatherers. Among the Hadza, meat acquisition can be a quite solitary undertaking for men and meat-sharing then follows quite flexible social goals not directly related to hunting (Hawkes et al. 2001, Marlowe 2003), while for the net hunters like the Aka pygmies, meat-sharing follows very strict rules to ensure the cooperation of the different hunters (Bahuchet 1985).

In conclusion, cooperation in hunting among Tai chimpanzees does not seem to result from a kin selection process but rather from a mutualistic process. Mutualism not only explains some of the differences in the frequency of group hunts observed within the species but also the way meat is shared after a successful hunt. The importance of mutualism seen in chimpanzees has also been reported for other animal species and might lead to a revision of the relative importance of the different mechanisms that can lead to the evolution of cooperation in nature.

Acknowledgments

We thank the Ivorian authorities for supporting this study since it began in 1979, especially the Ministry of the Environment and Forests as well as the Ministry of Research, the Director of the Taï National Park, and the Swiss Research Centre in Abidjan. We thank Grégoire Kouhon Nohan and Honora Néné who helped to collect chimpanzee samples and behavioral data. We thank the organizers of the Freilandtage meeting in Goettingen in 2003 for inviting us to contribute to this volume. The Max Planck Society and the Swiss National Science Foundation supported this work financially.

Chapter 9

Toward a general model for male-male coalitions in primate groups

Carel P. van Schaik, Sagar A. Pandit, Erin R. Vogel

9.1 Introduction

Of the many forms cooperation can take in nonhuman primates, the formation of coalitions is perhaps the most spectacular. Coalitions can be defined as coordinated attacks by at least two individuals on one or more targets, often preceded by signaling between the attackers (side-directed communication: de Waal & van Hooff 1981, de Waal 1992a; cf. Smuts & Watanabe 1990). They may serve to protect against attacks by more powerful individuals, to defend or gain access to resources or to acquire the dominance rank of the target individual. Animals forming the coalitions often are friends, as defined by Silk (2002c), although this is not true in all cases (Noë & Sluijter 1995). Friendships that involve coalitions are commonly called alliances (Harcourt & de Waal 1992a).

Mothers of many mammalian species protect immature offspring, but coalitions among adults are remarkably limited in their taxonomic distribution to carnivores, cetaceans and primates (Harcourt 1992). This pattern suggests that there are obstacles to the evolution of adult coalitions. We will briefly discuss the two major ones. First, in a functional sense, coalitions may contain altruistic acts (Packer 1977, de Waal 1982a, Noë 1990, Chapais 1995, Dugatkin 1997). Theorists often consider the presence of delays between providing the agonistic support and reaping possible rewards an important differentiating factor between the non-nepotistic categories of cooperation. In coalitionary interactions, delays are common, even when the eventual outcome is mutualistic in that all coalition partners increase their rank, perhaps after several initially unrewarded coalitions. The delays can also be quite variable, depending on the rather unpredictable outcomes of coalitions and the unpredictable time until the next opportunity. Long and variable delays may make it hard for coalitions to become established.

A second obstacle to the evolution of coalitions is that most of them at least potentially involve a high risk of injury. This is especially clear where lower-ranking animals team up to attack a higher-ranking target (but even if high-ranking individuals team up to attack a single low-ranking target, there is always some chance that others, either powerful or simply numerous, will come to the victim's aid). In general, there is a serious risk of injury when a coalition partner defects in the middle of an escalated coalitionary fight. Thus, coalitions will tend to involve some cost to the participants, unless the partners can trust each other not to defect during the fight. We believe that this trust problem may be another serious obstacle to the evolution of coalitions.

The presence of these obstacles may explain the limited taxonomic distribution of coalitions (Trivers 1971). Coalitions are common in organisms that are long-lived, live in stable associations, and form friendships (Cheney et al. 1986). Living for years in a stable group provides frequent opportunities for reciprocation, and the trust needed to form coalitions is built and maintained in friendships in which a variety of social services are exchanged.

Coalitions are common among primates, but that does not mean that all species display them. The conditions giving rise to coalitions among female primates have received much theoretical attention (Wrangham 1980, van Schaik 1989, Isbell 1991, Sterck et al. 1997). The basic idea is that a high potential for contest competition (either within or between groups) leads related females to form nepotistic alliances and to be philopatric. It also produces predictable patterns of rank inheritance, the details of which may vary as a result of demography (Datta 1988, Chapais 1995) or the age trajectory of reproductive value (Chapais & Schulman 1980).

Coalitions among males, however, have not received similarly intensive scrutiny (but see Noë 1994), although their incidence is at least as spotty (van Schaik 1996) and within-taxon variation is also appreciable (Noë 1992, Pandit & van Schaik 2003). Although there is a trend for males to support relatives more than non-relatives (cf. Silk 1992a), nepotism does not explain much of the variation, if only because in many species the males in a group are not closely related. For instance, in male-philopatric chimpanzees, where close kin is at hand, allies are generally not close relatives (Goldberg & Wrangham 1997, Mitani et al. 2000, Vigilant et al. 2001). A different framework is therefore needed for males.

When we tally all the male-male coalitionary interactions observed in a group of primates over a given period, the emerging pattern may seem confusingly complex. Coalitions may occur in many configurations (Chapais 1995; see below for details), in various contexts (Noë 1992), in many different combinations, and not always consistently in support of the same partners or aimed at the same target (e.g. Silk 1993). They may occur spontaneously or as interventions in ongoing conflicts (de Waal et al. 1976), and some of them apparently beat their opponent(s) whereas others end in some kind of stalemate. It is not easy to identify the underlying strategic goals governing each male's decisions but it is a reasonable working hypothesis that their general goal is an increase in fitness. The most promising avenue for modeling, therefore, is to try to predict those outcomes that can be explicitly linked to fitness gains. If this approach can explain a considerable portion of the coalitions observed, we may then be able to identify the more intermediate tactical goals or as yet unknown strategic goals served by the remaining coalitions that are not immediately explained by the model. Here, we will adopt this approach, and will return to the coalitions not explained by the model in the discussion.

For the purpose of modeling, then, we are mainly concerned with outcomes. The aim of this chapter is to begin explaining the distribution of within-group male-male coalitions by developing and testing a general cost-benefit model for one class of male-male coalitions, offensive coalitions. We have so far produced

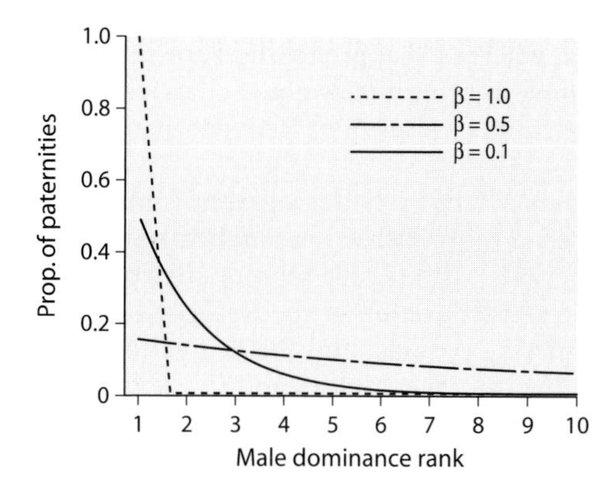

Fig. 9.1. The basic payoff curves with rank, for males competing over mating access to females: contest through priority of access in a constant-sum situation. Dotted line: $\beta = 1.0$; solid line: $\beta = 0.5$; broken line: $\beta = 0.1$

two technical papers on this subject (Pandit & van Schaik 2003, van Schaik et al. 2004). Our purpose here is to explain the basic logic underlying the models rather than the mathematical details, present the predictions and their tests, and discuss how the coalitions considered in this model fit into the overall scheme of coalitionary behavior in primates. To explain the basis for the model, we will begin with a review of the nature of male-male competition and coalitions in primates.

9.1.1 Male-male competition and coalitions in primates

Whether or not successful coalitions produce fitness benefits depends largely on the relationship between fitness and dominance rank. For males in a primate group, this curve is invariably and inevitably concave (at best approaching linearity), never convex (see Fig. 9.1). The reason for this is that males compete through contest for a set of constant-sum resources, fertilizations, which leads to a distribution of fertilizations determined by priority of access (Altmann 1962), also referred to as queuing (Alberts et al. 2003); the top-ranking male takes whatever he can monopolize of what is left, and so on down. In this model, a male's payoff is affected only by the number of males ranking above him, not by the number of males ranking below him or the total number of males in the group (as in Lomnicki's 1988 strict definition of contest). This process leads to negative exponential payoff curves with rank, except in the limiting case of complete scramble competition where males share all matings and therefore fertilizations equally.

Studies of primate males competing over females have generally confirmed the priority-of-access model (Altmann 1962, Bulger 1993, Weingrill et al. 2000, Alberts et al. 2003). We previously introduced the parameter β to describe the shape of the relationship or degree of despotism (Pandit & van Schaik 2003); β is the proportional reduction in payoff from one male to the next-lower ranking male. When β is near its upper limit of 1, virtually all fertilizations will be concentrated in the top-ranking males, whereas β approaching zero indicates pure scramble, in which all males will have approximately equal chances of fertilization regardless of their dominance rank, as a result of frequent matings by all and sperm competition.

The value of β is determined by a variety of factors: (i) demography (number of females in the group), (ii) ecology (degree of reproductive seasonality, producing more or less overlap in female mating periods), (iii) female reproductive physiology (the number of cycles per conception, the duration of the fertile window in each cycle, the presence of non-fertile mating periods, and the degree to which females actively synchronize or desynchronize their cycles), and (iv) female behavior (preferences for mating with dominant or subordinate males, or for polyandrous mating). We will postpone the discussion of how to estimate β until later.

For females in primate groups, on the other hand, the curve relating dominance to fitness will generally be near linear or even somewhat convex, but rarely concave. Females generally compete for access to food. Where food occurs in defensible patches, access to it can be contested. These patches can usually hold several females, who will all acquire approximately equal intakes. Even where the patches are smaller, however, high-ranking females will generally not systematically exclude others because that would deprive them from a major benefit of gregariousness; protection against predators (van Schaik 1983, Janson 1992). Moreover, the displaced individuals can usually find other food nearby or wait and subsequently gain access to the same food (albeit perhaps at higher risk). Thus, high-ranking females will show restraint to prevent the lower-rankers from avoiding them and forming a group on their own. Males generally have no such concerns since they tend to associate with groups of females.

We note that the resources over which the males in a group compete generally come in a fixed total amount (van Hooff & van Schaik 1992, van Schaik 1996). The number of fertilizations in a particular group during a particular period of time will be constant, and coalitions will not change this amount. As a result, the areas under each of the different curves drawn in Fig. 9.1 are all equal (an important factor in the modeling). Another implication of constant-sumness concerns leveling coalitions. Whereas these coalitions should always increase the number of matings with fertile females, they will not automatically bring increased fitness for both partners because they have to share this access. This property will therefore make it more difficult for them to gain a fitness benefit from forming coalitions; we assume approximately equal sharing of access to fertile females and hence paternity (see below; but see Noë 1990).

In sum, because of the sex difference in the shape of the curves and because nepotism is usually not a major factor among males (unlike the situation among females, we must develop an independent approach to modeling male-male coalitions within groups.

9.2 The Model

We always consider coalitions with a single target; this is also by far the most common pattern of within-group coalitions among primates. The basic approach is to identify the conditions in which coalitions are viable, i.e. expected to occur. We recognize two components of viability that have to be met simultaneously: profitability and feasibility. Coalitions are profitable when, for each coalition member, the direct benefits in terms of increase in fitness exceed the costs in terms of reduction in fitness (both relative to the situation without coalitions) through risk of injury or death and energy expenditure or stress. We will therefore employ a parameter, C, that denotes these costs, which we assume to be equal for all members (this assumption is especially reasonable for coalitions in which animals attack a higher-ranking target). Coalitions are feasible if they are strong enough to beat their target (and exceed their cost, again denoted by C), which requires that we have some way of adding up the fighting ability of the individual players.

Before proceeding to present the model, we need to insert a comment on terminology, because game theorists and behavioral biologists use coalition in a different sense. In animal behavior, the unit of analysis is the coalition, which is the actual interaction. In contrast, what in game theory (Kahan & Rapoport 1984) is called a coalition is what a behavioral biologist would call a successful alliance, i.e. the situation that arises when the alliance has achieved its goal. This difference is most acute for rank-changing coalitions, in which numerous coalitionary interactions may be needed before rank change is achieved and then may continue to be needed occasionally to maintain the new ranks. All of this is considered a single 'coalition' in the model. These different terms do not affect the model because costs and benefits are measured in the same units: fitness components per unit time. Incidentally, our model is not a game-theoretical model; it merely borrows useful concepts and terminology from Kahan & Rapoport (1984).

We consider three basic configurations of coalition members relative to their (single) target (Fig. 9.2): (i) all-down, (ii) bridging, and (iii) all-up. Chapais (1995) calls these conservative, bridging and revolutionary, respectively. It is obvious that all-down coalitions are always feasible; a combination of higher-rankers can always beat single lower-rankers. However, because there is no immediate fitness gain when the coalition members are already high-ranking and have priority of access to the limiting resources, they are not profitable in the sense used in the model (in the discussion, we will consider situations where they may bring indirect benefits to the participants). Bridging coalitions against a single target are always feasible as well because the highest-ranking coalition member acting alone can always beat the target. However, they are not profitable for the higher-ranking member of the coalition unless

[1] Alternatively, the supported individual is a non-relative providing some essential support to the high ranker in return, something we only expect in humans.

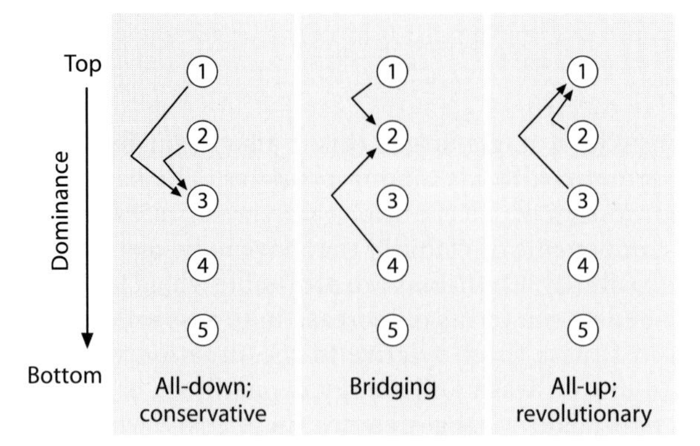

Fig. 9.2. Basic configurations of within-group coalitions (after Chapais 1995). Arrows indicate attacks.

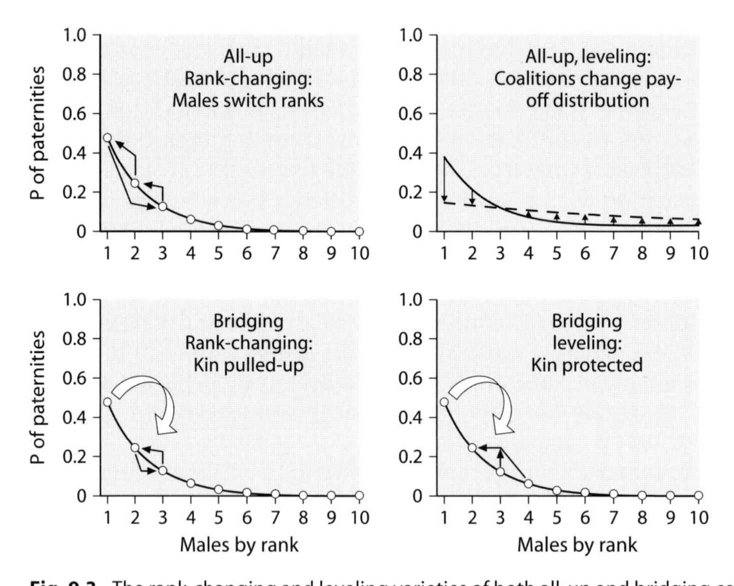

Fig. 9.3. The rank-changing and leveling varieties of both all-up and bridging coalitions.

the lower-ranking individual in the coalition is a relative, and the support can therefore improve the inclusive fitness of the supporter[1]. All-up coalitions may be feasible, when several low-rankers gang up to attack a higher-ranking target. When they succeed, they should also generally be profitable because of the improved ranks of the coalition partners, unless they are prohibitively costly due to high risk of injury or stress.

Both bridging and all-up coalitions can be profitable in two very different ways (Fig. 9.3). Coalitions can improve the ranks of the coalition members, but they can also improve the payoffs of their members by providing instantaneous (if temporary) access to the limiting resource, usually fertilizable females, without changing the ranks of the coalition members, a phenomenon we call leveling. In the case of bridging coalitions, this leveling takes the form of protection of lower-ranking relatives in contexts linked to competition. In the all-up configuration, such leveling is accomplished by coordinated attacks on a high-ranking male who is in consort with an estrous female. This context has been studied in detail in baboons (Packer 1977, Smuts 1985, Bercovitch 1988, Noë 1990, 1992), but in theory the same all-up leveling coalitions may also be found when no resource is directly at stake and then serve to induce the high-ranking male to show some restraint in mating competition. The model is really about these four kinds of situations: the rank changing and leveling varieties of bridging and all-up coalitions.

Feasibility needs to be estimated as well. To assess it, we have to solve the problem of determining the 'value' (i.e. strength) of the coalition. This problem has two components. First, we must decide how to estimate the strength of the coalition. We opted for the simplest assumption and simply added the strengths of the individuals in it, ignoring a possible effect of the number of participants. Second, we must determine which aspect of the individuals we need to sum. Here, again, we opted for the simplest solution and used the payoff before coalitions as the best estimate of a male's contribution to the alliance's strength (Pandit & van Schaik 2003, van Schaik et al. 2004). Ideally, we would like to use fighting ability, but that is very difficult to estimate, even if experiments are possible (cf. Noë 1990, 1994). Payoff, on the other hand, can be estimated by paternity analysis. Where β is modest, payoff and fighting ability are expected to show very similar functional forms with rank, allowing us to use payoff as our estimate of strength. At steep β, we expect the payoff differences to exceed those in strength, but fortunately, the model indicates no all-up coalitions (the only configuration for which feasibility is non-trivial) for $\beta > 0.5$, thus keeping the error modest.

We also incorporate a cost to feasibility because the allies need to coordinate their attacks with great precision and need to be prepared to do so at all times, and hence face some ecological and social cost. For simplicity, we have assumed that the cost C used to calculate profitability can be employed here as well. This assumption is probably not entirely correct, but the error is not likely to affect the predictions by much.

Although mathematically cumbersome, it is straightforward to calculate for each possible coalition (set of males attacking a particular target) whether it is profitable from the perspective of changing the ranks or payoffs of the allies, i.e. whether the formation of coalitions exceeds its costs, for each member. Feasibility can similarly be calculated. If for a given coalition, both conditions are satisfied for each member of the coalition, we predict the coalition to occur, although its frequency relative to other kinds of coalitions should of course depend on the net increase in payoff. We can then examine the features of these viable coalitions, such as the β range in which they are found, their sizes, the ranks of the participants, etc.

Table 9.1. Summary of predictions for within-group, male-male coalitions.

Type	Target	Members	Size	Despotism
All-up, rank-changing	Top or near-top	Just below top	Small (two or three)	Medium
All-up, leveling	Top or near-top	Mid- and low-rankers	Small-large	Low-medium
Bridging, rank-changing	Near-top	Top-ranker and relative not far below	Small (usually two)	More as despotism higher
Bridging, leveling	Anywhere	Variable	(variable)	Variable
All-down	Low-rankers threatening to form all-up coalitions	Top- and near-top-rankers	Probably small	(whenever all-up and bridging occur)

Here, we give an intuitive account of the predictions for the rank changing and leveling varieties of each of the two relevant configurations (all-up and bridging). Readers interested in the details should consult the technical papers (Pandit & van Schaik 2003[2], van Schaik et al. 2004). The predictions for each context are summarized in Table 9.1.

9.2.1 All-up, within-group coalitions: rank-changing

At constant *C*, the profitability of rank-changing coalitions increases as despotism increases, whereas their feasibility decreases. Hence, we expect them at intermediate values of β (and given our rule of calculating feasibility by comparing the sum of the coalition members' payoffs with that of the target, at $\beta < 0.5$). Because of the concave shape of the payoff curve, we expect these coalitions to be concentrated among the higher ranks; benefits are highest in that region. For the same reason, coalitions are expected to be small (larger coalitions will necessarily involve lower rankers for whom moving up in rank makes little difference in payoff). We therefore expect coalitions to concern the highest ranks, be fairly small, and involve mid- to high-ranking individuals.

[2] We have abandoned some terms and procedures used in that paper, which was our first exploration of the problem. In particular, we now let leveling refer strictly to the process of flattening the payoff curve (rather than to all-up as done in that paper). Moreover, we no longer assign a role to motivation in estimating the strength of the coalition, because the motivations (expected payoff differentials due to coalition formation) of the target and the coalition members may tend to cancel.

We can sharpen these predictions by including another consideration. High-ranking males can form effective counter-coalitions in an all-down configuration (which are always feasible) that prevent the occurrence of successful all-up coalitions. This is especially likely for the all-up, rank-changing configuration where the fitness loss of losing rank position is likely to outweigh the moderate cost of an all-down coalition (largely in the form of opportunity costs). Because the top ranker cannot form counter-coalitions with a male that ranks even higher than him, he would be the preferred target. Thus, the sharpened prediction is that all-up, rank-changing coalitions should generally be formed by a small number of males ranking immediately below the top-ranking male.

9.2.2 All-up, within-group coalitions: leveling

We use the same rule for estimating the feasibility of all-up leveling coalitions as for all-up, rank-changing coalitions. This may not be quite correct since the maximum risk level may be slightly lower, but as with other assumptions it should be close enough for this kind of strategic model. The profitability is obviously very different from the rank-changing variety; we are now asking whether the members gain enough in fitness from improved access to mates to outweigh the costs. To calculate this profitability, we have to resort to yet another simplification. For each particular coalition that is feasible, we check whether leveling the payoff curve leads to profitability for the top-ranking member of the coalition. Because he is the most likely not to gain from forming the coalition, if the coalition is profitable for him, it will be so for each member of the coalition. If so, we accept it as a viable coalition[3].

As can be intuited from comparison of the two curves in Fig. 9.3, the predictions are that all-up, leveling coalitions are expected to be relatively large and to be formed by mid- to low-ranking males targeting very high-ranking males. Mid- to low-ranking males stand to gain the most from these leveling coalitions in terms of improved fitness. They are expected to target very high-ranking males because those are the ones with access to the estrous females. Coalitions need to be relatively large in order to be able to beat the target (see Pandit & van Schaik 2003 for details). The curves in Fig. 9.3 also suggest that at higher values of *b*, individuals derive greater fitness benefits from rank-changing coalitions,

[3] We achieved this by introducing a new parameter α, which flattens the payoff curve, yielding a payoff function for the 1th male as a function of α and β (Pandit & van Schaik 2003). We calculate the optimum value of α, which is right where the highest-ranking member of the coalition starts losing payoff compared to not forming the coalition. If this optimum value of α is less than one, the payoff curve is flattened and the coalition is accepted as profitable. In practice, we therefore assume that the coalition will systematically attempt to intimidate all members ranking above the coalition, but because intimidating the top-ranking male will effectively intimidate all males ranking below him, we expect that most harassment will be aimed at the top-ranking male. This procedure introduces a small error in that it also changes the payoffs of those not participating in the coalition (ranking below them), but we considered this error acceptable relative to the complexity of alternative ways of modeling.

and thus would prefer to form those, so the leveling coalitions are especially likely at lower values of β.

All-down counter-coalitions are still feasible, but against all-up, leveling coalitions, we expect that they are less likely to be profitable. Perhaps the most important reason for this is that the payoff of other high-ranking males is not affected because they are not targeted and their ranks are not at risk. They may also face an opportunity cost to forming the all-down coalition if at the time of the coalition they are in consort with another female or with no female at all. If counter-coalitions occur, however, they should act to suppress leveling coalitions altogether.

9.2.3 Bridging, within-group: rank-changing

These kinds of coalitions are always feasible because they involve at least one member outranking the target. Assuming that they contain only two members, it is clear that they will never be directly profitable for the higher-ranking member. When the lower-ranking member is a close relative, however, kin selection may make it profitable for the higher-ranking male (we assume that they will always be profitable for the lower-ranker) if the rank increase of the lower-ranking member, corrected for the degree of relatedness with the high-ranker, outweighs the cost. Since these benefits increase as degree of despotism increases, we expect these coalitions especially at higher β values, and among males in the higher regions of the dominance hierarchy. Chapais (in press) has developed a similar argument for females.

9.2.4 Bridging, within-group: leveling

Higher-ranking males may always be available to protect lower-ranking relatives. However, to qualify as the equivalent among the bridging coalitions to the leveling among the all-up coalitions, this protection must increase the lower-ranking relative's access to receptive females. This requires that these coalitions happen in the context of males competing directly over access to receptive females. We expect them to be viable in a broad range of conditions, but they may be difficult to distinguish from the protection of relatives against attacks by others.

9.2.5 Estimating β

There are many different ways in which one could estimate β, the degree of despotism in payoffs. Payoffs can be estimated as fertilization rates (rather than, for instance, mating success), which we assume to correlate closely with fitness (perhaps best estimated as the number of offspring sired that survive to adulthood, but obviously rarely possible in naturalistic studies). Fertilization rates can be estimated through paternity analysis, or in some obvious cases behaviorally. Studies that estimate paternity through molecular techniques (e.g. micro-

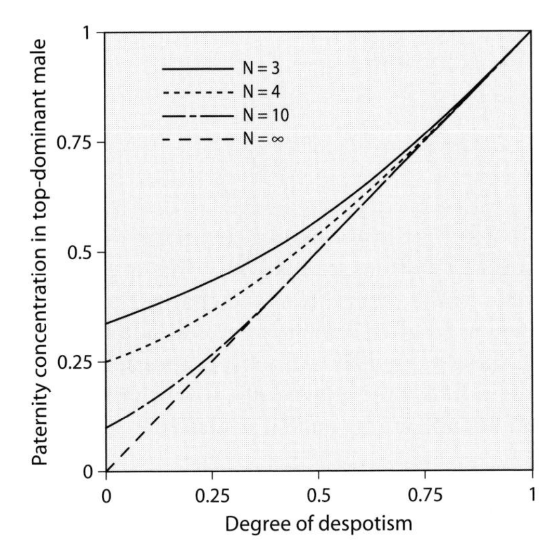

Fig. 9.4. The deviation between paternity concentration (payoff for the top-ranking male) and β. Note that the discrepancy increases as the number of males in the group (*N*) decreases.

satellites) are proliferating, and we rely on a recent compilation of these studies here (van Noordwijk & van Schaik 2004).

There are various ways of estimating β from a set of paternities across a range of male ranks: (i) the paternity of the top-ranking male (designated as paternity concentration in van Noordwijk & van Schaik 2004), (ii) the ratio of the paternity of the second-ranking male to that of the top-ranking male, and (iii) the slope of the regression of log (paternity) on rank (cf. Fig. 9.1). What is the best estimate depends on two main factors: sample size of fertilizations and the number of males (N). In smaller samples, the number of paternities going to individual lower-ranking males is likely to vary dramatically due to chance, and as a result we expect measure (iii), and even measure (ii), to be more dependent on sample size than the first one. This is indeed what we found in a simulation experiment where we had groups of 10 males compete for fertilizations and used sample sizes of 5, 10, 20 and 50 infants. Especially among smaller sample sizes, paternity concentration was both less biased and especially was more precise (i.e. it had far lower variance) than the other two measures. On the other hand, measure (i) is strictly speaking only valid for infinite N and should therefore be increasingly biased as the number of males decreases. Fortunately, however, the bias is largest at low β, where we expect N to be larger in any case (Fig. 9.4). We will, therefore, use paternity concentration (paternity of the top-ranking male) as our estimate of β.

9.3 Observations on primates

9.3.1 Model fit and discrepancies

The model attempts to explain only a subset of the many coalitions that can be seen among males in primate groups: all offensive coalitions that are strictly within a group with basically stable ranks, apart from the changes brought about by the coalitions. We have shown previously that the predictions of the model for these coalitions generally show a good fit with observations, despite the many simplifying assumptions (Pandit & van Schaik 2003, van Schaik et al., in press). Thus, the predictions listed in Table 9.1 by and large also represent the observations.

We found only four unambiguous cases in four species, in which a higher-ranking male supported a lower-ranking relative to the point that the latter moved up in rank to just below his supporter. Nonetheless, these nepotistic coalitions were observed in the very conditions where they were expected: despotic situations where the high-ranking relative is (near-) top-ranking. We did not encounter any reports of bridging, leveling coalitions in primates, in which higher-ranking males supported lower-ranking relatives in direct competition over access to females, although anecdotal accounts of support of presumed male relatives exist.

It is possible that bridging, rank-changing coalitions are so rare because close male relatives rarely find themselves together in the same group following dispersal and if they do, they may fail to recognize each other with sufficient reliability. Alternatively, the rarity of reported bridging coalitions may be an artifact. One would expect such situations to be rather common where the number of immigration targets for dispersing males is limited, if kin recognition rules are sufficiently reliable and β is high enough to produce high relatedness among male peers. New data on male relatedness will probably lead to a large increase in data on bridging coalitions and hence to further tests of the model. These data should at least show whether one form of nepotistic behavior (not modeled here) is quite common: males showing restraint to their lower-ranking relatives and thus smoothing their way to the top if the latter are pre-prime.

We identified seven cases in four species of all-up coalitions where males successfully challenged a high-ranker and switched rank, as envisioned in Fig. 9.3. The features of these coalitions were in close agreement with the model: challengers ranked near the top and challenged a top-ranker. We also identified six cases in six species of all-up, leveling coalitions. The review of empirical studies also found the predicted contrast between the rank changing and leveling varieties of the all-up coalitions, although the number of cases was small (see Table 9.1). While the targets of both tend to be top-rankers, the participants in leveling coalitions are mainly mid-rankers, whereas they tend to rank just below the top in rank-changing coalitions. However, we did not find the predicted larger mean size of leveling coalitions, although some of the leveling coalitions were indeed quite large. We will discuss leveling coalitions in more detail below.

Table 9.2. The number of reported cases for each of the three main kinds of coalitions examined here (see van Schaik et al., 2004 for details) in relation to the estimated values of β in groups of wild primates. Each β class covers a 0.25-section of the range from zero to one.

	low	medium	high	very high
All-up, leveling	4	2		
All-up, rank-changing		3	2	
Bridging, rank-changing		1	1	1

The various kinds of observed within-group coalitions also showed different ranges of β values, as expected under the model (Table 9.2; including free-ranging groups in nature only), although the observed range tended to be higher than the predicted one: rank-changing all-up coalitions were expected only at $\beta < 0.5$. We would caution, however, that estimates of β are not always from the same group or the same time period, and therefore the conclusion as to the effects of β is still preliminary. Moreover, there is a risk of circularity attached to this kind of *post hoc* testing, because successful alliances may actually produce reduced β in that males will be less likely to press their full advantage knowing that doing so may unleash all-up coalitions.

The good fit between model and observations suggests that the simplifying assumptions we have made (simple addition of the values of players to calculate the values of the coalition; use of payoff rather than some estimate of fighting ability to calculate this value; use of the α parameter in leveling coalitions) were not so far off from reality as to diminish the model's predictive value (see Pandit & van Schaik 2003 and van Schaik et al. 2004 for further details). However, the finding that coalitions are seen in a systematically higher range of β values than expected deserves comment. It indicates that simply adding up the participating players' strengths (payoff values) to arrive at the strength ('value') of their coalition is inadequate. This discrepancy may indicate some independent effect of the number of coalition members or of fighting abilities. What remains surprising, however, is not the presence of this effect but its rather modest influence. Still, there is an urgent need for field-based estimates of coalition strength based on natural variation (cf. Noë 1990, 1994).

9.3.2 More on leveling coalitions

Leveling coalitions show the only real discrepancy between model predictions and observations in that observed coalitions were smaller than expected. There may be several explanations for the model's failure. Our predictions assumed that all coalitions that moved the α parameter from one to a lower value would actually occur. In practice, many of these potential coalitions reduced α by tiny

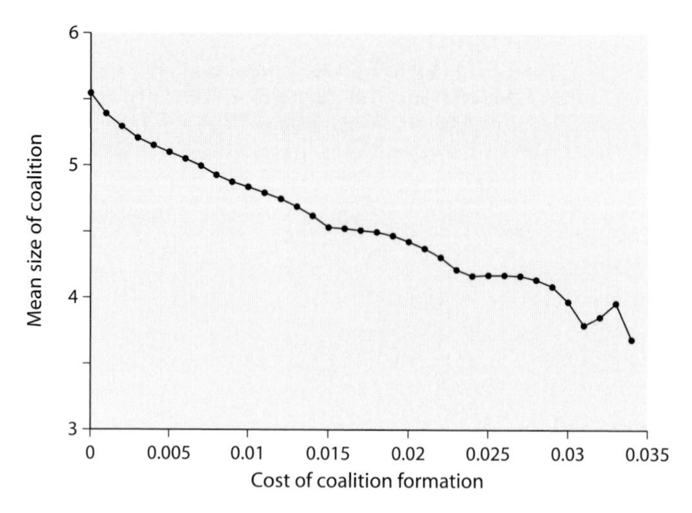

Fig. 9.5. The effect of introducing a fixed cost on the size of leveling coalitions through its effect on profitability.

amounts, and the presence of costs would almost certainly make them unprofitable in nature. Pandit & van Schaik (2003) attempted to deal with the issue of costs. Here, we re-examine this question: a fixed cost (corresponding to g_1 in our paper) does indeed reduce mean coalition size (Fig. 9.5). This cost will especially reduce the involvement of higher-ranking males.

A second problem is that we did not make N dependent on β. This issue will be discussed in more detail below, but in most cases we expect fewer males in groups with higher β. If we had made N dependent on β, the model would have generated fewer large coalitions, and hence fewer coalitions involving very low-ranking and rather high-ranking males.

Finally, we assumed that costs were constant for all males, but cost may be a function of rank distance between the coalition member and the target. If this modified assumption is used, the lowest-ranking males will be less likely to become coalition members, again reducing coalition size.

These three technical reasons contribute to explaining the discrepancy between model and predictions, but there may be biological reasons as well. In particular, it is possible that pre-prime males may avoid taking any risks that jeopardize their future rise to the top, or that males have trouble finding suitable partners for other reasons. More quantitative exploration of both model and empirical data is needed to fully resolve this problem.

Within savanna baboon groups (*Papio cynocephalus*), paternity monopolization by the top-dominant varies over time, due to demographic variation and the relative strength of the top-dominant male. Leveling coalitions vary widely in these groups. Consistent with the model, "dominance rank failed to predict mating success" (i.e. β approached zero) "when the number of adult males in the group was large, when males in the group differed greatly in age, and when the highest-ranking male maintained his rank for only short periods" (Alberts et al. 2003). It is therefore possible that, as suggested above, a demographic factor, namely the presence of many past-prime males, (even as the number of females remains constant) allows the formation of larger leveling coalitions, which the model predicts can then take place at higher β values (see Fig. 1 in Pandit & van Schaik 2003).

Because the observed paternity distribution includes the effects of effective leveling coalitions on mating access, there is some risk of circularity when testing the predictions (see Pandit & van Schaik 2003). Hence, either detailed comparisons or, if possible, experiments must be done. Pandit & van Schaik (2003) noted that the comparison between chacma (*Papio ursinus*) and savanna baboons supported the model because in chacma baboons, in which leveling coalitions are absent, top-dominant males show a strong tendency to monopolize paternity (Bulger 1993, Weingrill et al. 2000, Henzi & Barrett 2003), whereas that is usually not the case among savanna baboons (e.g. Alberts et al. 2003).

Similar coalitions involving various combinations of lower-ranking males and aimed at the top male can be seen in the absence of any direct competition over females. Such coalitions are often observed in chimpanzees (Goodall 1986). We believe they are best considered leveling coalitions as well, for the following reasons. One possible explanation for them is that they are attempts at unseating the top males. However, this interpretation lacks plausibility because they are formed in many different combinations and by (usually post-prime) males ranking well below the top rank, with little prospect of attaining that position. A second possibility is that their function is to reduce harassment by the top male, who frequently directs violent displays at subordinate males. The top-ranker may use harassment to reduce the likelihood of the formation of alliances that might later threaten to topple him. However, the involvement of past-prime males in these coalitions makes this possibility less plausible. The third interpretation is that these coalitions serve the same function as in baboons and macaques, i.e. to reduce the degree to which the top male will monopolize the matings in the community. Bettinger et al. (1993) mention that these all-up coalitions are more likely in the presence of swollen females. Thus, the mere possibility of leveling coalitions may intimidate the top male, who might therefore insist less on his priority of access.

9.4 Discussion

9.4.1 Further tests and extensions

One benefit of explicit modeling is that we can now also examine situations in which all-up, rank-changing or leveling coalitions are not expected because β is too high. We already noted the chacma baboons, but there may be other examples as well. For instance, Table 9.3 presents a summary of all male-male coalitions observed during 18 months of observation in one group of long-tailed macaques (*Macaca fascicularis*) containing between six and seven sexually mature

Table 9.3. Observed coalitions among males in group H of long-tailed macaques at Ketambe over an 18 month period during 1980-1981 (M. A. van Noordwijk & C. P. van Schaik, unpubl.)

Offensive within-group	**All-up (rank-changing or leveling)**	**0**
	Bridging, leveling	2
Defensive within-group	Challenger from within	21
	Challenging immigrant	16
	All-down, conservative	15
	Bridging, protective	17
Between-group		0
Other	(low-ranking male joins opportunistically)	3

non-natal males. Table 9.3 confirms that coalitions among males in this group occurred in a variety of contexts (see below), but not to achieve top rank (cf. van Noordwijk & van Schaik 2001). A high β value in this population has been confirmed (de Ruiter et al. 1994). More detailed work along these lines in populations with known β values will be useful in evaluating the model in greater depth.

The model also draws attention to puzzling exceptions. For example, male rhesus macaques (*Macaca mulatta*) have never been seen to form all-up coalitions, although low and medium β values are common. Dario Maestripieri (pers. com.) suggests that this absence may be due to the fact that female rhesus monkeys make good allies (see Chapais 1986, 1995), especially if they can easily recruit additional members of their matriline, thus diminishing the value of males to each other as allies. Alternative explanations might also be possible, but the important point is that the absence of male-male coalitions now becomes an issue to be examined.

9.4.2 The impacts of β

It is clear that the critical variable in the model is β. Fig. 9.6 presents the range of outcomes of male-male interactions over the full range of β values, as predicted by the model and supported by the preliminary tests conducted to date. The β values for the transitions between these outcomes are only approximately indicated because they depend on additional parameters. The appearance of clear dominance ranks depends on the cost of agonistic interactions relative to their benefit, which are a function of β. Above this β threshold, we expect all-up leveling coalitions; whether the ranks effectively disappear again as a result of the leveling coalitions is a function of both β and N, the number of males in the group (perhaps explaining why this is found only in the largest groups; Pandit &

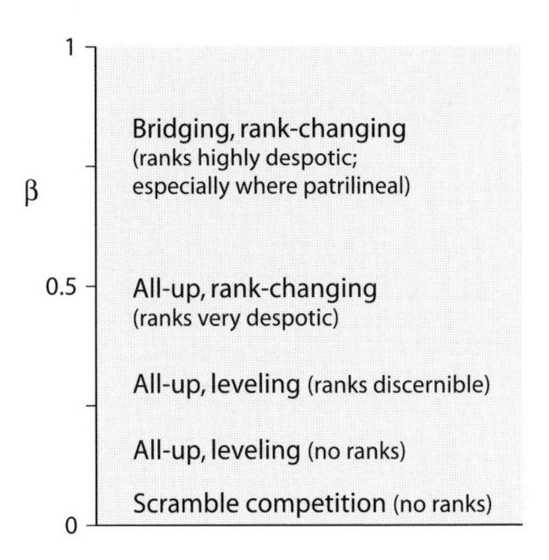

Fig. 9.6. A summary of the predicted offensive male-male coalitions in multi-male groups in relation to β. Switch points are approximate because their β value depends on additional parameters.

van Schaik 2003). The switch from leveling to rank-changing coalitions is determined by both β and the costs of coalition formation. Above $\beta = 0.5$ (or a somewhat higher value, depending on the detailed implementation of the feasibility rules), all-up coalitions should disappear, and the only coalitions expected are the bridging, rank-changing variety.

The influence of β reaches well beyond that of coalitions, however. Indeed, we expect to see major differences between low- and high- β situations, even within species. Van Noordwijk & van Schaik (2004) note that males in low- β situations tend to achieve top rank through a queuing or succession process rather than through active challenges, as at high β. In high- β situations, the top-ranking males are therefore males in their early prime, whereas as β decreases, the age of top-rankers will gradually rise, until in the very large groups, such as those of Japanese macaques, males rise to the top by default when the old top male dies or disappears. As a result, very old and visibly aged males can occupy top rank (e.g. Watanabe 2001). In high- β situations, we not only expect escalated fights over dominance to be much more common, but also for them to be concentrated among the top ranks (see Nishida & Hosaka 1996). In low- β situations, males tend to immigrate into groups with more favorable adult sex ratios, whereas males in high- β situations tend to move to groups in which the demographic situation is such that future prospects of achieving top rank are best, although older males understandably fall back on the low- β strategy. Male-female friendships may also differ predictably. Hence, the degree to which top males can achieve full priority of access to females is an important organizing variable for male socio-sexual strategies.

Another obvious impact of β is on N. Imagine the effect of imposing a variable cost to a male of living in a mixed-sex group, either due to the risk of injury because of attacks by other males or females (tangible even if males refrain from overtly participating in mating contests) or due to ecological costs imposed by differences in dietary preferences between males and females (van Schaik & van Noordwijk 1986). We assume that the males for whom ecological or social costs

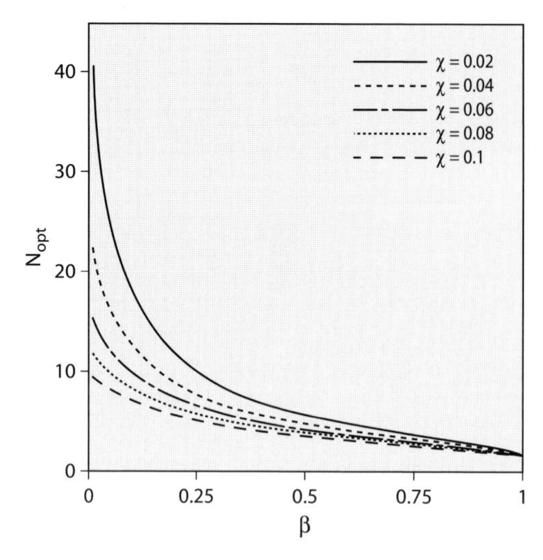

Fig. 9.7. The predicted relationship between degree of despotism (β) and the number of males in a group (*N*) of female primates when group membership entails a finite fitness cost (*C*).

outweigh the mating benefit can join all-male bands if there is also a cost to being solitary, or alternatively can join other groups in the population with lower β. Not surprisingly, such a small cost of group membership makes lower-ranking males more likely to leave the group as β increases, because lower-ranking males achieve increasingly reduced payoffs due to fertilizations. As a result, the relationship between β and N becomes concave, suggesting that the product $\beta \times N$ is approximately constant (Fig. 9.7). This expected concave shape of the relationship is actually consistent with observations (see figures in van Noordwijk & van Schaik 2004).

Within-group coalitions are only possible where there are at least three sexually active males in a group. It is therefore possible that an additional reason for the absence of coalitions at high β values is that there may not always be at least three males in the group (cf. Henzi et al. 1999)[4]. However, the most obvious relevance of the negative correlation is that it draws attention to the situation where N is unusually large despite fairly high β, which may produce leveling coalitions that would not otherwise occur. Two main conditions are expected to bring this about. First, the number of males may be largely independent of β; this could be due to male philopatry, as in chimpanzees, or the presence of (non-exclusive) pair bonds, as in humans. Second, females may reduce effective β if they benefit from the presence of additional males (e.g. van Schaik & Hörstermann 1994, Ostner & Kappeler 2004). Clearly, more work is needed to

[4] Obviously, the other predictions of the model still hold. The number of males does not explain the contrasting features (β-range, size, ranks of members and targets) of the three main types of male-male coalitions examined here.

establish the relationship between β and the number of sexually active males per group, both within and between populations, as well as its socio-ecological correlates.

9.4.3 Other kinds of coalitions

The model has explained some of the great variety of coalitionary interactions among male primates, but obviously it does not cover many others. We focused on one class, the offensive coalitions that bring explicit fitness benefits, either because one or more of the members of the coalitions rose in rank or because they managed to increase their payoffs by taking some of the resources away from high-rankers.

The presence of offensive coalitions may also have produced several other interesting social behaviors. First, the threat of all-up offensive coalitions may have led to separating interventions by high-ranking males (de Waal 1982a, 1992a, Perry 1998). The intervening male prevents affiliative contacts between possible coalition partners that may allow the latter to build up enough mutual trust to launch coalitionary all-up attacks. The presence of preventive all-down coalitions where leveling coalitions occur needs to be confirmed. Second, we see opportunistic all-down coalitions where males who normally form high-risk, all-up coalitions attack a weak target, who is unlikely to ever attack them. These coalitions have been suggested to test the partners' willingness to engage in more risky coalitionary interactions of the offensive type modeled here or the preventive types that follow from them (e.g. de Waal 1992a, Noë 1992). Alternatively, such opportunistic coalitions could be random acts to keep subordinates stressed and therefore less likely to mount challenges to higher-ranking individuals, as suggested by Silk (2002c) for dyadic aggression, although we should then see males of all ranks, and especially the higher-ranking ones, engage in them. Under either interpretation, their function is not linked to the outcome of the interaction, but rather to the maintenance of the alliance itself.

Many defensive coalitions directly follow from the existence of the offensive coalitions we modeled. Every successful all-up, rank-changing coalition will subsequently produce persistent all-down conservative coalitions in order to prevent a reversal to the original situation. Such all-down (conservative) coalitions that serve to maintain the *status quo* are well known for females (Chapais 1995, Preuschoft & van Schaik 2000). The threat of these preventive coalitions was also thought to affect the features of the all-up and bridging coalitions considered by the model. Thus, even though they are not part of the model, such defensive coalitions directly follow from it.

We believe that there are two classes of coalitions that require separate or additional modeling: (i) defensive coalitions against unranked targets and (ii) coalitions against coalitions. This second kind of defensive coalition does not directly follow from the model, which only considers situations in which male fighting abilities are stable and where males have explicit ranks (thus excluding both immigrants and disequilibria between ranks and fighting ability). First, resident males often form coalitions against (individual) immigrant males, where these cannot be ranked yet. Where these immigrants aim at achieving top rank, the highest-ranking residents form a defensive alliance (e.g. in long-tailed macaques; van Noordwijk & van Schaik 2001; see also Table 9.3). Second, very similar coalitions aimed at defending the participants' rank positions are seen against low-ranking individuals that are improving in fighting ability, usually due to maturation, to the point that they can soon pass several others to challenge for top rank. The benefit of these successful defensive coalitions is that all members maintain their rank positions, and thus the payoff rates associated with them, for a longer time than they would without having formed the coalition. Such coalitions need further modeling to assess the possible effect of β on their presence (perhaps they are most likely where β is high but not very high) and the possibility of rank changes within the coalitions.

The other class of coalitions not elaborated in the model is when the targets of coalitions are other coalitions. For within-group coalitions against coalitions, the model can easily be generalized. All-up, rank-changing coalitions will tend not to target multiple males because the steep β will make these attacking coalitions non-feasible and make the defensive all-down coalitions generally successful (as we saw above). For leveling coalitions, coalitions against coalitions are more likely to be feasible but require large numbers of males in the group; maybe they will be seen in very large groups.

When the coalitions reside in different groups (as in lions: Packer & Pusey 1982; howler monkeys: Pope 1990; or chimpanzees: Goodall 1986), the model no longer applies mainly because the competition is no longer over a constant amount of resources. In primates, coalitionary takeovers of groups, as in brown lemurs or capuchin monkeys (Jack & Fedigan 2004, Ostner & Kappeler 2004) also fall under this rubric. A separate model is probably needed to account for these between-group coalitions.

9.5 Summary and conclusions

In this chapter, we presented a model for within-group coalitions among primate males. Coalitions occur if they are both feasible, i.e. can beat the target, and profitable, i.e. lead to a fitness benefit for all coalition members. Based on simple logic, we predict the existence of different kinds of coalitions, whose main characteristics are the relative ranks of members and targets, and whether or not they change the dominance ranks of the participants. The key predictor for the different kinds of coalitions is the value of β, the degree to which dominant males can monopolize mating access to females.

The model fits what we know about primates rather well, but its main function is to draw renewed attention to male-male coalitions, which in turn should help the development of more encompassing models. Such an empirical cycle will not only produce a better understanding of the phenomenon of male coalitions (which in general are much more opportunistic than those found among females: de Waal 1982a, 1992a, Nishida 1983), but will also allow us to identify the decision rules used by males (cf. van Noordwijk & van Schaik 2001) and the flexibility in these rules when β varies. The latter task will help us to develop a far better appreciation of the cognitive complexity associated with coalitionary behavior.

Given the importance of β, it is not surprising to see the presence of behavioral tactical decisions of males that correlate with its value. Some of the variation in male socio-sexual strategies is observed intraspecifically, especially well documented among Japanese macaques (e.g. Sprague et al. 1998). Individual male tactics also change with age, especially with respect to dispersal decisions. Males also make opportunistic decisions. Groups with rank instability at the top attract more immigrating males, probably because monopolization by top-rankers is reduced at such times, and the additional males tend to disappear again after the ranks have stabilized (van Noordwijk & van Schaik 2001). Similar intraspecific variation is seen for coalitions (see van Schaik et al. 2004).

All of this suggests some flexibility in decision-making that is linked to the value of β, although the way(s) in which males derive their implicit estimate of β is completely unknown. Despite various attempts (Matsuzawa 2001, de Waal 2003), our ability to estimate the complexity of social behavior patterns used by nonhuman primates is limited. Hence, revealing the existence of these mechanisms may help us to estimate the cognitive demands of various social decisions.

Acknowledgments

We thank Peter Kappeler for the invitation to the fourth Freilandtage, Maria van Noordwijk for tallying the coalition data on the long-tailed macaques, and Bernard Chapais, Peter Kappeler, Dario Maestripieri and John Mitani for very helpful comments.

Chapter 10

Cooperative breeding in mammals

Tim H. Clutton-Brock

10.1 Introduction

Cooperative behavior between group members is common among mammals living in stable social groups (Dugatkin 1997) but cooperative care of young is less common and varies widely in development between species (Russell 2004). It is useful to distinguish four different types of cooperative breeding. In group breeders, multiple breeding females live and breed in the same social group. Group members may cooperate to defend resources against neighboring groups or to detect or deter predators, but direct alloparental care is limited or uncommon. Societies of this kind are common among macropods, bats and ungulates as well as in some families of primates and carnivores (Bradbury & Vehrencamp 1974a, 1974b, Jarman 1974, Clutton-Brock 1989b). In communal breeders, groups include multiple breeding females who share care of young born in the group. Not all females breed in each reproductive attempt and parents may be assisted by temporarily non-breeding females or by males. Well-studied examples include a number of social carnivores, including African lions, banded mongooses and spotted hyenas (Gittleman 1989, Lewis & Pusey 1997), and some bats also show communal care of offspring (Wilkinson 1987, 1992). In several cercopithecine primates, breeding females belonging to the same matriline cooperate to protect and support each other's offspring and these species, too, should, perhaps, be regarded as communal breeders (see Cheney 1977, 1990, Hrdy 1977, Chapais 1992).

In facultative cooperative breeders, neonates and juveniles are cared for by their parents and by non-breeding helpers of either or of both sexes, but the average number of helpers is usually low and parents are capable of breeding successfully without helpers and often do so. Well-studied examples include silver-backed jackals and European foxes (Moehlman 1979, 1989), marmosets and tamarins (Goldizen 1987b, French 1997), and a substantial number of social rodents (Solomon & Getz 1997). These species resemble avian societies where parents are often (but not always) assisted by helpers at the nest (Stacey & Koenig 1990).

Finally, in specialized cooperative breeders, breeding adults are seldom able to breed successfully without assistance from non-breeding helpers. Species of this kind are sometimes referred to as 'obligate' cooperative breeders but I use the term 'specialized' to avoid the objection that very occasionally parents may breed successfully without helpers. In most of these species, non-breed-

Fig. 10.1. Meerkat group. Meerkat groups consist of two to forty individuals with an approximately equal sex ratio. Typically, one dominant female and one dominant male are the parents of over 75% of pups born in the group, which are reared by all group members. Cooperative activities include allolactation, guarding the pups at the natal burrow, feeding pups of one to three months, clearing burrows and sentinel duty. Males normally leave their natal group voluntarily after they are 2-years-old while females may either inherit the breeding role or may be forced out of the group by the dominant female. Groups habituate very closely to observers, making it possible to weigh individuals repeatedly during the day and to estimate their foraging success in terms of weight gain per hour.

ing helpers out-number breeders and often contribute more heavily to some or all cooperative activities, so that young receive most of their care from helpers. Specialized cooperative breeders include African wild dogs (Courchamp et al. 2000a, 2000b, Creel & Creel 2002), Kalahari meerkats (Clutton-Brock et al. 2001b, Russell et al. 2003b) and naked mole-rats (Lacey & Sherman 1997). In at least two of these species, females that attain alpha status subsequently increase in size and weight (O'Riain et al. 2000b, Russell et al. 2005), presumably because this increases their reproductive potential (Clutton-Brock et al., in prep.) while helpers may also show physiological or behavioral adaptations (Clutton-Brock et al. 2004). A further distinction is sometimes made between obligate cooperative breeders and eusocial species (Sherman et al. 1995) but there is a disagreement over the criteria used to define eusociality (Crespi & Yanega 1995). For the purposes of this chapter, I include eusocial species within obligate cooperative breeders.

While the evolution of all forms of cooperative behavior poses problems to evolutionary biologists (Dugatkin 1997), the evolution of specialized cooperative breeding generates the most striking paradoxes since many helpers forego reproductive success for several years after reaching maturity and only a small proportion of individuals born ever attain dominant status. The central question of why helpers should forego reproduction and spend part or all of their time assisting in rearing the offspring of other group members is commonly broken down into three parts or sub-questions (Russell 2004). First, why instead of leaving immediately they reach sexual maturity should helpers delay dispersal from their group of origin? Second, why do they not attempt to breed themselves in their natal group? And, third, why do they assist in the reproductive attempts of dominants instead of remaining and conserving resources against a time when they will either disperse or attempt to displace the dominant breeder in their group? Below, I consider each of these three questions in turn, drawing extensively on our long-term study of Kalahari meerkats (Clutton-Brock et al. 1998a, 1998b, 1999a, 1999b, 2000, 2001c, Russell et al. 2002, 2003b)(see Fig. 10.1).

10.2 Why do helpers delay dispersal?

Students of cooperative birds have commonly argued that delayed dispersal occurs where available breeding habitat is saturated so that dispersers have little chance of locating suitable breeding habitat (Emlen 1984, Koenig et al. 1992, Ekman et al. 2004). Experiments with acorn woodpeckers have demonstrated that the creation of vacant territories close to a group's range increases rates of dispersal by helpers (Koenig & Mumme 1987, Koenig et al. 2000). Habitat saturation is also implicated as a cause of delayed dispersal in some communal or cooperative mammals (Doncaster & Woodroffe 1993, Creel & Macdonald 1995) but, in other cases, dispersal decisions appear to be independent of habitat availability (Cheeseman et al. 1993, Creel & Creel 2002). For example, in our study population of Kalahari meerkats, a prolonged period of drought led to the extinction of nearly two-thirds of all groups. Subsequent increases in rainfall led to the restoration of normal conditions and to successful breeding, but although the remaining groups increased in size, dispersal rates did not increase and many of the original territories remained vacant for several years (Clutton-Brock et al. 1999a). These results suggest that while habitat saturation may commonly depress dispersal rates, other factors are also likely to be involved.

One possible explanation of the failure of individuals to disperse when suitable breeding habitat is available is that the process of dispersal has high and inevitable dangers. Dispersal is associated with substantial increases in mortality rate in many mammals and especially in those living in closed social groups (Lucas et al. 1997, Creel & Creel 2002). However, where vacant habitat is available so that dispersers can settle rapidly, the costs of dispersal would need to be extremely high to offset the potential benefits of independent breeding and it seems unlikely that the risks associated with dispersal are as high as this.

An alternative possibility is that the potential benefits of remaining in an established breeding group to the individual's survival or subsequent chances of successful reproduction are large and exceed the benefits of dispersing (Clutton-Brock et al. 1999c). There is progressive evidence that these benefits can be large

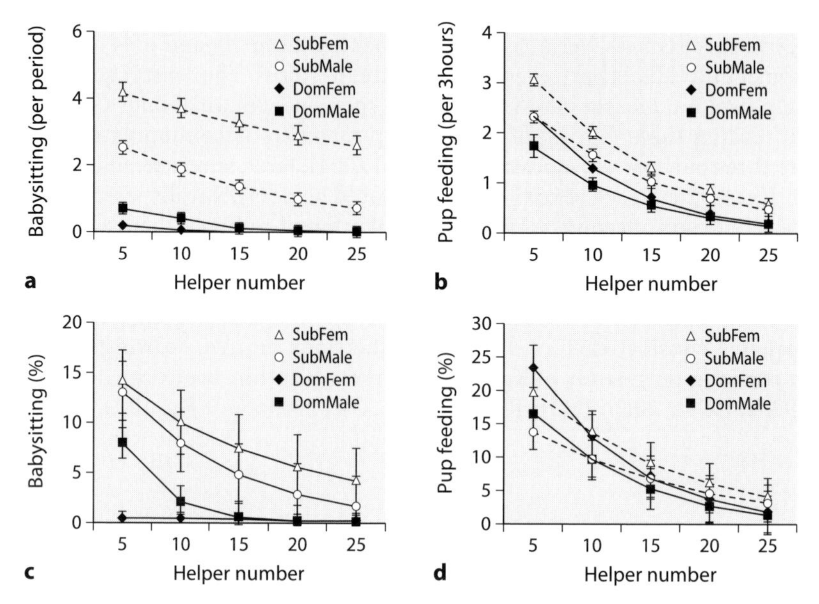

Fig. 10.2. Relation between helper number and contributions to (**a, b**) babysitting, (**c, d**) pupfeeding in Kalahari meerkats (from Clutton-Brock et al. 2004). For each activity, the first graph presented shows contributions as absolute measures and the second shows relative measures calculated as a percentage of total contributions by all group members.

in a wide variety of social mammals and can operate through several different ecological mechanisms. In many social mammals, the risk of predation and time spent on vigilance behavior decline in larger groups (Bertram 1978, Clutton-Brock 1989b). In other cases, hunting or foraging success increases while the costs of hunting decline. For example, large packs of African wild dogs are more successful in catching prey, spend less time running it down, hunt more frequently and catch larger prey than those in small packs and, as a result, the amount of meat available to each dog per day increases. In addition, where spotted hyenas are abundant, large wild-dog packs are better able to defend their prey and consume a larger proportion of it (Fanshawe & Fitzgibbon 1993, Creel & Creel 2002).

In other cases, large group size may play an important role in allowing groups to protect their ranges against neighbors. In many mammals, larger groups consistently displace smaller ones and individuals in smaller groups may suffer increased risk of being wounded or killed in fights with larger parties of neighbors, and parts (or, in some cases, all) of their ranges may be absorbed into those of neighboring groups (Wrangham 1980, van Schaik 1983, van Hooff & van Schaik 1992, Young 2003). One further consequence of these incursions for species that keep their dependent young in burrows or dens is that litters may be killed if neighboring groups locate the breeding burrow (Clutton-Brock et al. 1998b). Lastly, in cooperative breeders where helpers play an important role in raising young, the workload of all individuals is substantially lower in large, established groups than in small groups of founders, a phenomenon referred to as 'load lightening'. For example, in meerkats, per capita contributions to babysitting and feeding young decline with increasing group size (see Fig. 10.2). Where the benefits of remaining in a large, established breeding group are large, individuals may maximize their fitness by remaining in their natal group and queuing for breeding opportunities (Kokko & Johnstone 1999). This presumably explains why, in some mammals, members of one sex seldom or never leave their natal groups voluntarily, only dispersing if they are evicted by force (Clutton-Brock et al. 1998b, Russell et al. 2002, Ekman et al. 2004).

If the benefits of remaining in established breeding groups play an important role in discouraging dispersal, variation in the level of benefits should generate changes in the timing or frequency of dispersal. Though extensive data on dispersal rates is scarce in cooperative breeders, there is some evidence that this is the case (Russell 2004). For example, in marmots, delayed dispersal is more frequent when the thermodynamic benefits of communal hibernation are high and individuals that fail to reach adult body size in their first year are less likely to disperse (Armitage 1981, 1999, Blumstein & Armitage 1999).

10.3 Why don't helpers breed?

In many social mammals, females commonly remain and breed in their natal group (Pusey 1987, Clutton-Brock 1989b), so that failure to disperse need not prevent reproduction. In several cooperative mammals, subordinate females show lower levels of luteinising hormone (LH) or estrogen than dominant females, either throughout the breeding season or over the period of estrus (French 1997, O'Riain et al. 2000a, Creel & Creel 2002, Carlson, in prep.). In some cases, these differences disappear if subordinates are challenged with gonadotropin-releasing hormone (GnRH), indicating that suppression is temporary but, in others (including naked mole-rats), differences in LH levels between dominants and subordinates are not removed by GnRH challenge, and suppression is evidently more profound (Faulkes et al. 1990, Bennett et al. 1993, 1994, Faulkes & Abbott 1997).

At least three different processes probably contribute proximately to the failure of helpers to breed in cooperative mammals. First, in many mammals, females have a weight or age threshold below which conception does not occur (Albon et al. 1983, Creel & Creel 2002). In many cooperative mammals, some (but not all) helpers fall below this threshold and so are unlikely to breed. Presumably, the evolution of fecundity thresholds of this kind relate to the effects of age and weight on the costs of breeding or on the probability that individuals will raise offspring successfully. In meerkats, for example, individual foraging skills increase until animals are around 18-months-old and they rarely conceive before this age (Clutton-Brock et al. 2001b, 2001c).

Second, adult subordinates commonly lack access to unrelated breeding partners in the same group and (as in many social species) may delay reproduction until one or more unrelated males join the group (Greenwood 1980, Pusey

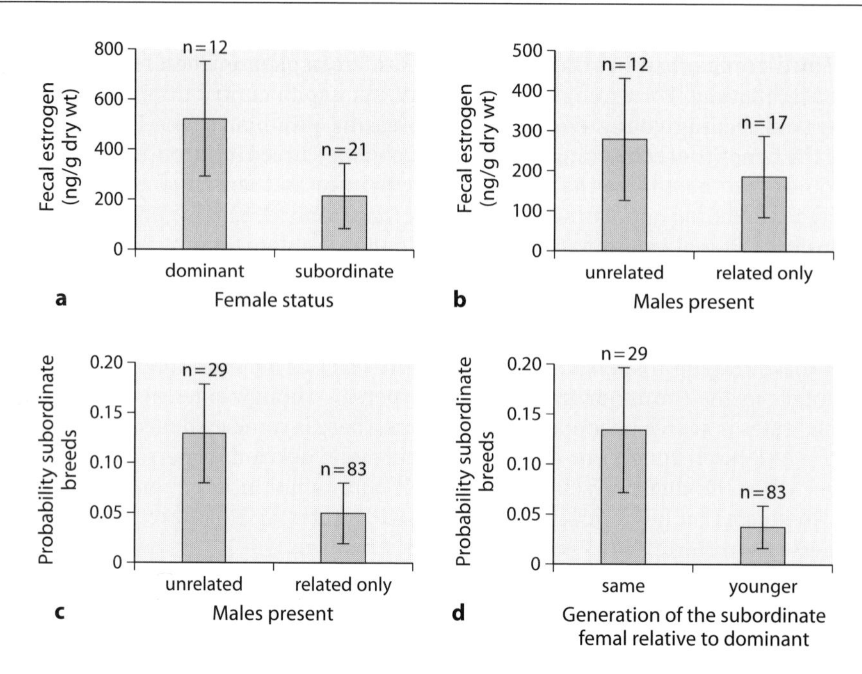

Fig. 10.3. Effects of absence of unrelated males on reproduction in subordinate female meerkats (from Clutton-Brock et al. 2001b). (**a**) Median levels of estrogen metabolites (±interquartile ranges) in fecal samples from non-pregnant dominant and subordinate adult females during the breeding season. (**b**) Median levels of estrogen metabolites in fecal samples from subordinate females of breeding age (>10 months) in groups with and without unrelated males. (**c**) Probability that subordinate females will breed in the presence and absence of unrelated males (with other factors affecting breeding frequency controlled). (**d**) Probability of breeding by subordinate females from the same generation (sibling or littermate) as the dominant versus a different generation (daughter or niece).

1987, Clutton-Brock 1989a). In several species, the absence of unrelated males is associated with reductions in LH and estrogen as well as in breeding in subordinate females (Russell 2004). For example, in subordinate female meerkats, levels of circulating estrogen and breeding frequency are both lower when there is no unrelated male in the group (see Fig. 10.3). Differences in estrogen level (but not breeding frequency) between subordinates and dominants disappear in groups where subordinate and dominant females both have access to unrelated males (Carlson et al., in prep.). Similarly, in captive groups of Damaraland mole-rats, experimental replacement of resident breeding males rapidly generates reproductive competition in previously quiescent subordinates (Cooney & Bennett 2000).

Third, the presence of dominant females may be associated with reduced levels of sex hormones or breeding frequency in subordinates. For example, in subordinate cotton-top tamarins, ovarian function is depressed by the presence of dominant females and is restored if they are removed (French et al. 1984, Widowski et al. 1990). Similarly, in meerkats, both the absence of unrelated males and the presence of dominant females appear to depress estrogen levels in subordinate females (Clutton-Brock et al. 2001c, Clutton-Brock, unpubl.). How dominants influence the fecundity of subordinates is unclear (Russell 2004). Though it was once thought that reproductive suppression in subordinates might be caused by high cortisol levels and social stress induced by dominants, recent evidence shows that, in several species (including wild dogs and meerkats), dominants commonly have cortisol levels that are higher than those of subordinates (Creel 2001, Creel & Creel 2002, Carlson et al. in prep.). However, it could still be that dominants inflict even higher stress levels on subordinates that challenge them, and that intermittent peak values rather than average cortisol levels prevent subordinates from breeding successfully. In several cooperative mammals, dominants commonly kill litters born to subordinate females (Clutton-Brock et al. 1998a) and, in these cases, it is possible that subordinate females increase their fitness by avoiding breeding when a dominant is present (Russell 2004).

While subordinate females seldom breed in cooperative societies, in most cases they do so occasionally. Two separate reasons why dominant females do not always prevent subordinates from breeding have been suggested. First, dominant females may, in some cases, be unable to control subordinate breeding, either because they are physically incapable of doing so or because the costs of suppression outweigh the benefits (Reeve et al. 1998). Under these conditions, the division of reproduction between dominants and subordinates may resemble a 'tug of war' whose outcome is affected by the relative asymmetry of their power as well as by a number of other variables, including their relatedness to each other (Johnstone et al. 1999, 2000). Alternatively, dominants may permit subordinates a share of reproduction if doing so reduces the risk that they will either challenge the dominant for her position or disperse (thus depriving the dominant of her subsequent contributions to raising young) (Keller & Reeve 1994, Johnstone 2000). Models of this kind are derived from studies of communal-nesting birds (Vehrencamp 1983a, 1983b) and are sometimes referred to as 'concession' or 'optimal skew' models since they assume that the dominant female has the ability to control subordinate reproduction but chooses not to do so in order to gain fitness benefits (Magrath et al. 2004). Two common predictions based on these models are that dominants should be less likely to allow subordinates to breed where the costs of dispersal are high (because they then have little option except to remain) or where the subordinate is closely related to them (because the subordinates then gain increased indirect benefits from rearing the dominant's pups) (Keller & Reeve 1994)

Satisfactory tests of these two predictions present problems and opinion is divided on the relative importance of concessions versus limited control in determining the frequency of subordinate breeding (see Clutton-Brock 1998a, 1998b, Emlen et al. 1998, Magrath et al. 2004). Some studies of cooperative mammals have produced evidence that supports the predictions of concession models. For example, in dwarf mongooses, older subordinate females are both more likely to disperse and are more likely to breed in their natal group than younger ones (Creel & Waser 1991). Similarly, in a number of cooperative animals, helpers that are closely related to dominant females are less likely to breed than more distant relatives or unrelated individuals (Reeve et al. 1998). However, these trends could also occur for other reasons; older subordinates may be more difficult for dominants to control while subordinates that are closely related to dominant females are commonly the offspring of the dominant pair and may consequently lack access to unrelated breeding partners (Clutton-Brock 1998b). As yet, few attempts have been made to control for effects of this kind and studies that have tried to do so have found little unequivocal evidence that concessions play an important role in affecting the distribution of subordinate breeding in cooperative societies (Clutton-Brock et al. 2001c, Haydock & Koenig 2002, 2003, Magrath et al. 2004).

In addition, there is still little firm evidence that reproduction modifies the behavior of subordinates to the dominant female's advantage, as concession models assume. For example, in meerkats, subordinates that have bred are no less likely to disperse or to contribute generously to rearing the dominant female's subsequent litters (Clutton-Brock et al. 2001c). Finally, several studies suggest that subordinates which breed are generally those that are likely to be most difficult for the dominant to control. For example, in meerkats, older, heavier subordinates are more likely to breed than younger, lighter ones; the dominant female's sisters are more likely to breed than her daughters; subordinate breeding is more frequent in years of high rainfall, when food is abundant and the animals are in good condition; and subordinates most commonly attempt to breed shortly after a dominant female has acquired alpha status, when her ability to control group members is weak (Clutton-Brock et al. 2001c).

10.4 Why do helpers help?

The question of why helpers should help lies at the heart of research on the evolution of cooperative societies since subordinates remain in their natal group and either breed ineffectively or fail to breed in a number of social species (Russell 2004). Since delayed dispersal and reproductive suppression in subordinates occur in a number of mammalian societies that do not breed cooperatively, the evolution of helping behavior presumably represents a separate development from failure to leave the natal group as well as from failure to breed. Explanations of the evolution of helping fall into four main groups. In some cases, apparently cooperative behavior may have immediate direct benefits to the helper's fitness and any effects on other group members may be coincidental (Bednekoff 1997, Dugatkin 1997). A second possibility is that group members are coerced into contributing to cooperative activities through harassment or the threat of punishment or eviction by dominant breeders (Gaston 1977, Clutton-Brock & Parker 1995). Theoretical models suggest that coercion is only capable of maintaining specialized cooperative behavior under rather restrictive conditions (Kokko et al. 2002). A third possibility is that cooperative behavior is usually directed principally at kin and is maintained through its effect on indirect components of inclusive fitness (Emlen 1984, 1991). Finally, helping may have evolved through mutualistic benefits shared by all members of large groups (Brown 1987, Kokko et al. 2001, Clutton-Brock 2002).

To assess the importance of different mechanisms in the evolution of helping, it is important to know what effects helpers have on other group members; whether or not helping has substantial costs; which individuals contribute most to cooperative activities; and how the level of contribution is controlled. The rest of this section considers each of these issues before assessing the relative importance of different evolutionary mechanisms.

10.4.1 Do helpers help?

In social birds, relationships between the presence or number of helpers and the reproductive success of breeders vary. In some cooperative species, the presence of non-breeding helpers apparently has little effect (Leonard et al. 1989); in others, the presence of a single helper increases the success of some breeders but the presence of multiple helpers either has no additional effect or depresses the success of breeders (Komdeur 1994b); and in some species, the success of breeders rises in an approximately linear fashion in relation to the number of helpers in the group (Heinsohn 1992, Ridley 2003). A similar range of effects occurs in communal and cooperative mammals. In European badgers, where several non-breeding females may share a set with breeders, their presence has no obvious effect on the reproductive success of breeders (Woodroffe & Macdonald 1995). Studies of several facultatively cooperative rodents, including pine voles and Mongolian gerbils, have also found no obvious relationship between the presence or number of helpers and litter size at weaning (French 1994), while in prairie voles, reproductive success increases with helper number in some populations but not in others (Solomon & Getz 1997). Positive relationships between helper number and breeding success have been found in several facultatively cooperative carnivores, including golden and silver-backed jackals as well as in several marmosets and tamarins (Goldizen 1987a, 1987b, Goldizen & Terborgh 1989), as well as in most of the specialized cooperative breeders that have been studied so far, including dwarf mongooses (Rood 1990), meerkats (Clutton-Brock et al. 1999b) and African wild dogs (Courchamp et al. 2000a, Creel & Creel 2002).

The presence of strong positive correlations between group size and breeding success in specialized cooperative breeders strongly suggests that helpers have a substantial influence on breeding success. For example, in meerkats, both the food intake of pups increases with helper number across groups (Fig. 10.4) and pups born in groups with multiple helpers are more likely to survive (Russell et al. 2002). However, it is still possible that these relationships could occur because some aspect of parental phenotype or territory quality leads to high juvenile survival in successive breeding attempts, generating positive correlations between the number of subadults in groups and the survival of juveniles. One way of testing whether the presence of helpers causes an increase in breeding success is to remove helpers from breeding groups (Brown et al. 1982). Experiments of this kind with colonies of prairie voles suggest that the effects on the development of young vary with environmental conditions; in colonies maintained at relatively low temperatures, the removal of alloparents reduced growth rates and delayed development in dependent young (Solomon 1991) while no effects of

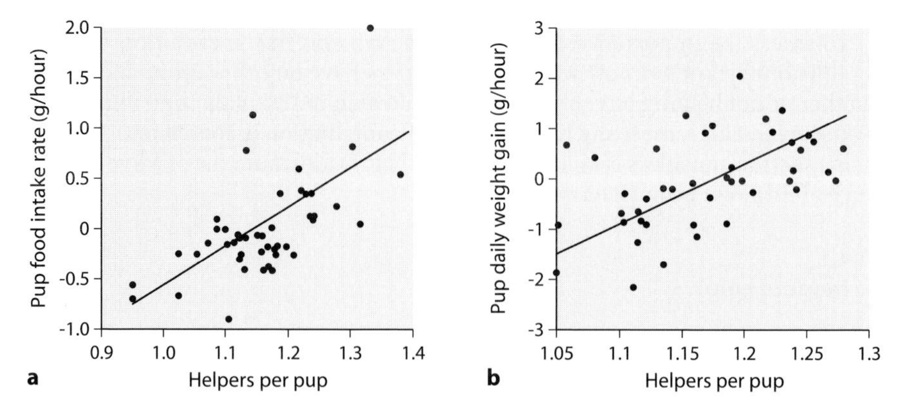

Fig. 10.4. Effects of helper number on (**a**) estimates of food intake and (**b**) daily weight gain of pups in Kalahari meerkats (Clutton-Brock et al. 2002).

the presence of helpers were found in colonies maintained at room temperature (Wang & Novak 1992). An alternative way of examining the effects of helpers is to manipulate the ratio of helpers to young. For example, in Kalahari meerkats, temporary reductions in pup number increased the weight gain of the remaining pups while increasing the number of pups (by adding pups of approximately similar age from another litter) reduced their weight gain (see Fig. 10.5a,b). In meerkats (as well as in prairie voles), early growth rates are related to weight at independence as well as in survival so these studies provide firm evidence of the effects of helpers on breeding success (Solomon 1991, Solomon & Getz 1997, Clutton-Brock et al. 2001a).

Recent studies suggest that the effects of helper number and group size are larger and more pervasive than has been supposed since early growth rates commonly have 'downstream' effects on survival and breeding success throughout the lifespan (Solomon & Getz 1997, Lindström 1999). In meerkats, the effects of helper number on growth and size at independence influences the subsequent survival of juveniles, their chance of breeding as subordinates and their chance of acquiring alpha status (Clutton-Brock et al. 2001a, 2002). In addition, larger breeding groups produce larger 'splinter' groups of dispersers which are able to feed more effectively after leaving their natal group and have a better chance of establishing themselves as new breeding groups than smaller splinters (Young 2003).

10.4.2 Is helping costly?

While the benefits of helpers in specialized cooperative societies may have been underestimated, the costs of helping have probably been overestimated. Where it is in the interests of subordinates to remain in their natal group even if they are prevented from breeding (see above), the costs of helping should not include the

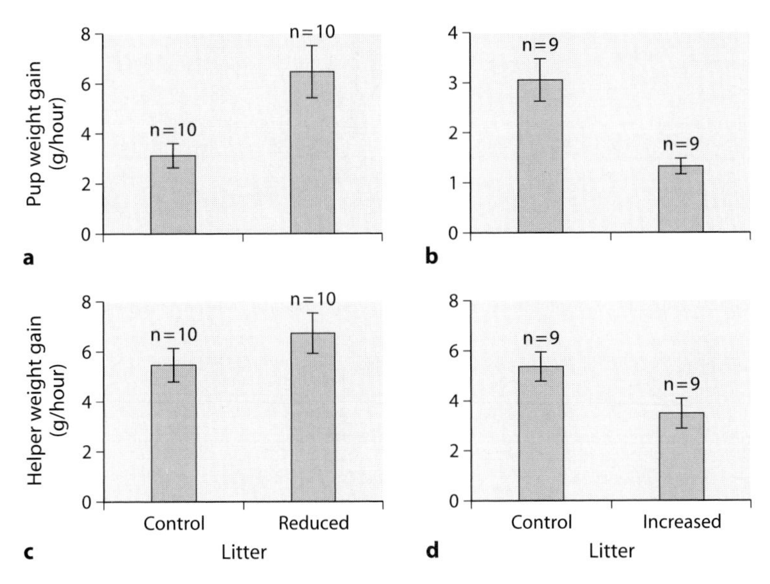

Fig. 10.5. Effects of manipulating helper:pup ratios on daily weight gain of pups and helpers in Kalahaari meerkats (from Clutton-Brock et al. 2001a). In this experiment, helper-pup ratios were either temporarily increased by removing 75% of pups or were reduced by increasing litter size by 75%. Control values were mean measures of daily weight gain in pups and helpers in the same group within two days of the experiment.

costs of delayed dispersal or the associated loss of reproductive opportunities and should be confined to the costs of helping *per se* (Russell 2004).

There is no doubt that helping can have energetic costs, both in cooperative mammals as well as in cooperative birds (Heinsohn & Legge 1999, Heinsohn 2004, Russell 2004). For example, in tamarins, helpers that carry infants spend less time foraging and suffer reduced calorie intake (Goldizen 1987a, 1987b, Price 1992, Tardif 1997). In meerkats, helpers that babysit pups at the burrow lose around 5% of their total body weight while those that forage with the group gain around 6% in the course of the day (Clutton-Brock et al. 1998b). In addition, helpers feeding pups give away around 40% of all the food items they locate and tend to give pups a high proportion of the larger, more valuable items (Brotherton et al. 2001). As would be expected, experimental reductions in pup:helper ratios tend to raise daily weight gain of helpers while increases in pup:helper ratios reduce helper weight gain (see Fig. 10.5c,d). When groups are babysitting or feeding pups, the average growth rate of younger group members is reduced to a greater extent in generous helpers than less generous ones (see Fig. 10.6).

Despite the obvious energetic costs of helping, firm evidence that helping incurs fitness costs is rare. For example, there is no evidence that generous meerkat helpers are less likely to survive, though they are slightly less likely to breed as subordinates (Russell et al. 2003b). One possible reason why the fitness costs of helping may be low despite the energetic costs is that contributions to most cooperative activities are conditional. For example, in meerkats, the level of con-

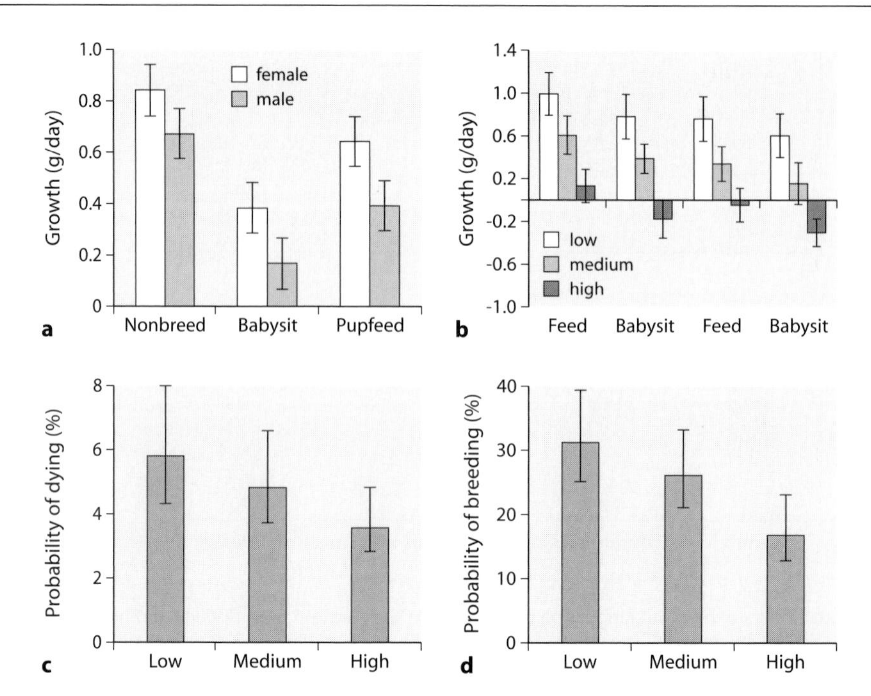

Fig. 10.6. Effects of stage of breeding cycle and relative contributions to cooperative activities on helper growth rates (g/day) in Kalahari meerkats (from Russell et al. 2003b). (**a**) Nonbreed: no breeding; Babysit: during periods of babysitting; Pupfeed: during periods of pupfeeding. (**b**) Low, medium and high refer to relative contributions to cooperative activities. (**c**) Association between cumulative contribution to cooperation and survival to two years in helpers. (**d**) Association between cumulative contribution to cooperation and probability of female helpers breeding as subordinates.

tributions increases with body weight and daily weight gain, and experimental feeding raises the contributions of female individual helpers (see Fig. 10.7). In addition, helpers alternate between being 'generous' and 'mean' in sequential breeding attempts so that although the weight of generous helpers is lower immediately after a breeding attempt, there is no effect on growth measured over the whole year (Russell et al. 2003b). Recent studies of other communal and cooperative mammals suggest that contributions to cooperative activities may usually be conditional (Gilchrist 2001, Hodge 2003, Ridley 2003) and, if so, the fitness costs of helping may be commonly lower than has generally been assumed.

10.4.3 Division of labor

To understand the evolution of helping behavior, it is also important to understand the extent to which different group members contribute to cooperative activities and the factors that affect their workload. In most cooperative societies,

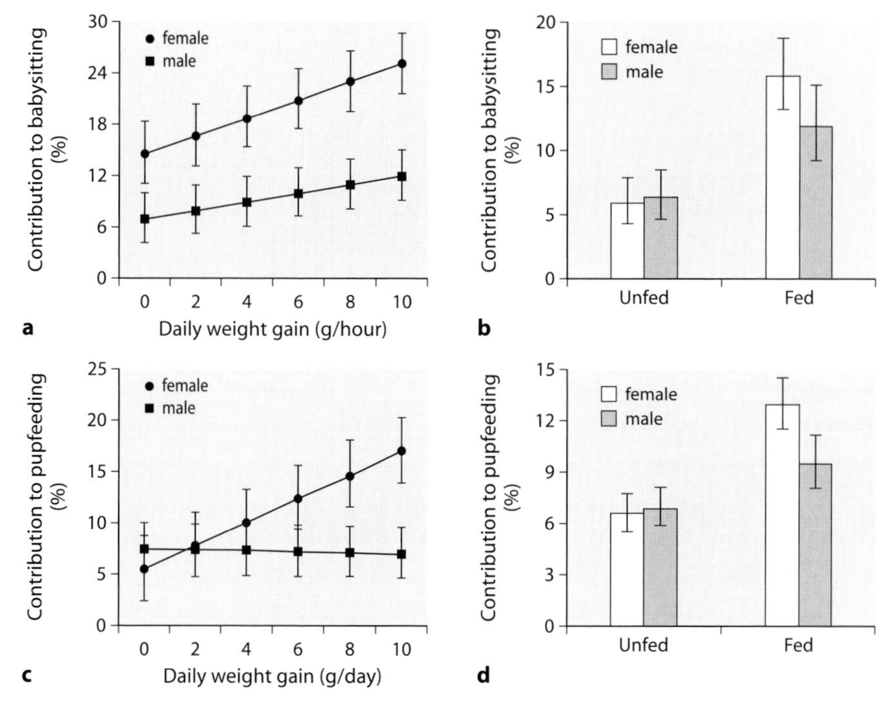

Fig. 10.7. Effects of daily weight gain (g/hour; an index of foraging success) on contributions by male and female meerkat helpers over 1-year-old to (**a**) babysitting and (**c**) pupfeeding (expressed as a percentage of total contributions to each activity) (from Clutton-Brock et al. 2002). Effects of experimental feeding of helpers on their contributions to (**b**) babysitting and (**d**) pupfeeding in the same breeding event.

the contributions of helpers initially increase with age. For example, in meerkats, juveniles only begin to make substantial contributions to babysitting and feeding subsequent litters of pups after they are 6-months-old (see Fig. 10.8). Subsequently, their contributions increase before declining as dispersal approaches (Clutton-Brock et al. 2002). Individual differences in average contributions are large and are consistently related to body weight and foraging success (see above).

The sex of helpers also affects their level of contributions. In cooperative societies where helpers include males as well as females, there are commonly differences in the overall level of contributions between the sexes (Stacey & Koenig 1990, Cockburn 1998). In many cooperative societies, members of the 'philopatric' sex (the sex that tends to remain and breed in the group) generally contribute more to rearing young than members of the other sex (Clutton-Brock 2002). For example, in meerkats (see Fig. 10.8) and brown hyenas, females may remain and breed in their natal group, and females helpers typically contribute more to rearing young than males whereas in African wild dogs (where males may remain and breed in their natal group), males generally contribute more than females

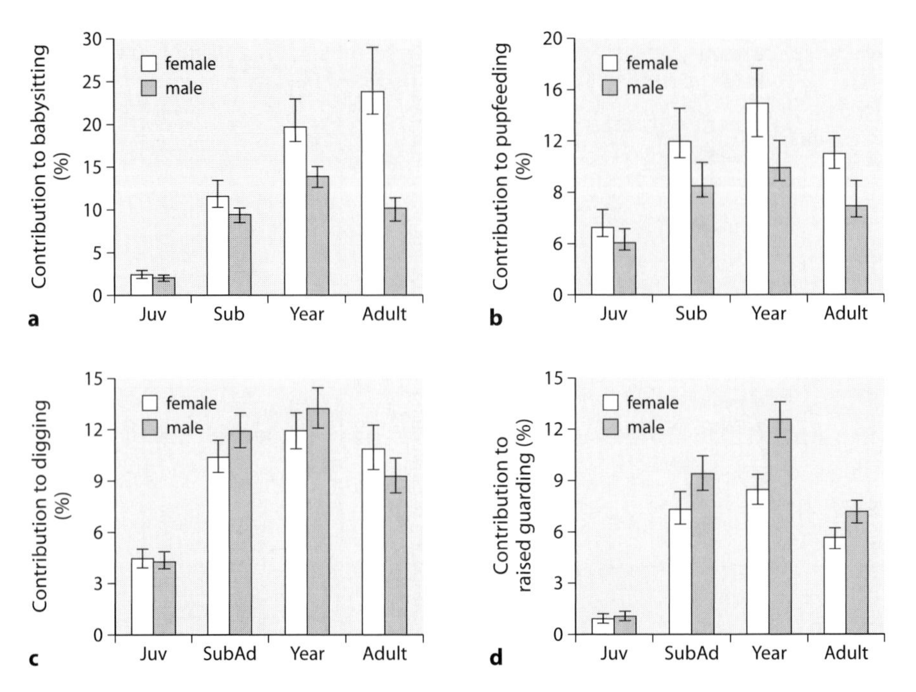

Fig. 10.8. Age-related changes in percentage contributions by male and female meerkat helpers to (**a**) babysitting, (**b**) pupfeeding, (**c**) social digging and (**d**) raised guarding (from Clutton-Brock et al. 2002). Juveniles: 3-6 months; Subadults: 6-12 months; Yearlings: 12-24 months; Adults: >24 months.

(Malcolm & Marten 1982, Owens & Owens 1984, Rood 1986, 1990). One additional line of evidence supports an association between workload and philopatry. In naked mole-rats, a small number of subordinate males adopt a contrasting growth to most other individuals, laying down more body fat, contributing little to cooperative activities and showing evidence of sexual behavior at an earlier stage. O'Riain, who described these morphs, categorizes them as "fat, lazy and promiscuous" and has shown that, unlike other individuals of both sexes, they commonly disperse from their natal colony (O'Riain et al. 1996). Male and female helpers also differ in the extent to which they contribute to particular cooperative activities. For example, in meerkats, females contribute more than males of the same age to babysitting and to feeding pups while males contribute more than females to sentinel duty (Clutton-Brock et al. 2002). Sex differences in contributions to different cooperative activities are most pronounced in large groups, where the demands on individuals are relatively low as well as in well-fed individuals.

In contrast to social insects, there is little evidence that individual helpers specialize in particular activities. Early studies of naked mole-rats described smaller helpers that were principally involved in the maintenance of tunnels and food collection, and larger ones that played an increased role in colony defense;

it was speculated that the latter animals might be a soldier caste (Jarvis 1981). However, subsequent studies have suggested that smaller helpers are younger animals and that the contrast in contributions to different activities represents an age-related polyethism (Lacey & Sherman 1991, 1997). In meerkats, the youngest helpers contribute relatively little to babysitting (which involves a full day without food) as well as to sentinel duty (Clutton-Brock et al. 2002) but these differences are temporary and change with age. When age effects are allowed for, there are still large differences in the extent to which individuals contribute to particular activities but the level of contribution is positively correlated across activities so that some individuals are consistently 'generous' and others are consistently 'mean' (Clutton-Brock et al. 2003).

The existence of large individual differences in the 'generosity' of helpers has stimulated attempts to investigate whether or not these are correlated with kinship. As yet, there is little evidence that helpers adjust their workload in relation to their degree of kinship to breeders or juveniles (Clutton-Brock 2002). For example, in meerkats, individual contributions to cooperative rearing are not consistently correlated with kinship when the effects of age and body weight are controlled (Clutton-Brock et al. 2000, 2001b). Although the great majority of helpers are related to the young they are rearing, some male helpers are unrelated immigrants (Clutton-Brock et al. 2000, 2001b). These animals seldom breed with the dominant female but contribute as much as other group members to guarding and raising her pups (Clutton-Brock et al. 2000, 2001b). The scarcity of convincing evidence that non-breeding helpers adjust their workload in relation to their kinship to the young they are rearing contrasts with a recent meta-analysis of helping behavior in cooperative vertebrates which suggests that help is most closely focused on relatives in societies where it has the greatest impact on offspring fitness (Griffin & West 2003). However, this analysis is not confined to resident helpers and includes a proportion of species where failed breeders subsequently rejoin and assist relatives. There is good evidence from several birds that failed breeders selectively join and assist relatives (Emlen & Wrege 1988, Russell & Hatchwell 2001), and this tendency is probably responsible for the relationships demonstrated by this meta-analysis.

10.4.4 The control of cooperative behavior

In bi-parental birds, parents adjust the rate at which they feed chicks to the frequency and intensity with which chicks beg for food, which, in turn, is related to their hunger level (Horn & Leonard 2002, Kilner 2002). A similar process of negotiation (Godfray 1995a, 1995b, Johnstone 2004) between juveniles and helpers probably controls the rate at which juveniles are fed in cooperative mammals. For example, dependent meerkat pups beg continuously for food during periods when the group is foraging (Manser 1998). The rate at which pups give begging calls is related to their hunger level and can be manipulated by feeding or food-depriving pups (White 2001). Helpers respond to increases in call rate by increasing the proportion of food items that they find which they give to pups; experimental playback of pup begging calls increase the rate at which helpers in

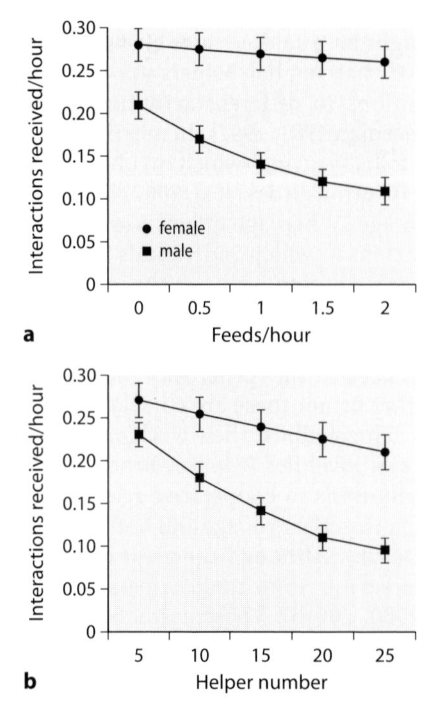

Fig. 10.9. Cooperation and rates of received aggression in meerkats (from Clutton-Brock et al., in press). Plots show the rate of aggressive interactions received by subordinate males and females over 12-months-old that fed pups at different rates. Absolute levels of aggression received are exaggerated in these plots since only individuals that were observed to receive at least one aggressive interaction were included.

groups with dependent young bring food to pups and can cause helpers that have ceased to feed pups to start again. As pups develop and spend more time foraging independently, the rate at which they beg declines and helpers gradually cease to bring food to them (Manser 1998).

In some cooperative societies, lazy or recalcitrant helpers may be punished by breeders and coercion may play a role in maintaining helping behavior (Mulder & Langmore 1993, Clutton-Brock & Parker 1995, Balshine-Earn et al. 1998), though there is no firm evidence that lazy helpers are more likely to be ejected from breeding groups than generous helpers in cooperative mammals (see Young 2003). In several cooperative mammals, breeders 'stimulate' inactive helpers and direct more frequent aggression at lazy helpers than at generous ones. For example, in naked mole-rats, breeding females shove inactive helpers and their removal causes a reaction in the level of helper activity, especially among less-related individuals (Reeve & Sherman 1991), though effects of this kind are not found in all colonies (Jacobs 2000). In meerkats, coercive tactics may be used to stimulate cooperative activities in males; individuals that feed pups relatively infrequently receive more aggression (mostly from breeders) than those that feed pups frequently (see Fig. 10.9). No similar relationships between workload and received aggression were found in females.

While it seems unlikely that breeders control the cooperative behavior of helpers directly, they may be able to influence the level of helper contributions through their effect on the hormonal status of helpers. The hormonal basis of cooperative behavior is still poorly understood but several studies of cooperative birds have found that individual differences in workload among helpers are positively correlated with prolactin levels (see Vleck et al. 1991, Schoech et al. 1996, 2004) and there is also evidence of an association between prolactin levels and alloparental care in common marmosets (Roberts et al. 2001). In contrast, contributions to pupfeeding among meerkat helpers appear to be more closely correlated with cortisol levels than with prolactin (Carlson, in prep.). This raises the possibility that breeders could manipulate the contributions of helpers through the imposition of social stress but an alternative possibility is that high cortisol levels are a consequence rather than a cause of elevated contributions to cooperative activities. In line with this suggestion, playback of pup begging calls to helpers also raises cortisol levels in helpers as well as the frequency with which they feed pups (see Carlson, in prep.).

10.5 The evolution of helping behavior

Detailed studies of cooperative behavior in helpers help to discriminate between alternative explanations of its evolution (Cockburn 1998, Clutton-Brock 2002, Koenig & Dickinson 2004). Both the energetic costs of helping behavior, the extent to which contributions vary between individuals and the adaptation of work levels to the need of juveniles suggest that, at least in specialized cooperative societies, it is unlikely that cooperative behavior is an unselected consequence of parental care (Jamieson 1989) or a byproduct of some activity which has an immediate benefit to the helper's survival (Bednekoff 1997). Coercion appears to occur occasionally in most cooperative societies (see above) but lazy helpers are rarely, if ever, ejected from breeding groups (see above) and in most cases, helpers appear to determine their own level of contributions and adjust these to the needs of dependent young. Moreover, models suggest that coercion is likely to maintain specialized cooperative behavior only under rather restrictive conditions (Kokko et al. 2002).

Kin selection (Hamilton 1964) is widely thought to play a dominant role in the evolution of specialized cooperative societies, where breeding individuals rely on the assistance of non-breeding helpers to raise their young (Brown 1987, Emlen 1991, Dugatkin 1997, Cockburn 1998). In all specialized cooperative birds and mammals, most helpers are relatives of dominant breeders that have not yet left their group. However, the view that kin selection provides a satisfactory general explanation of cooperative societies now appears less compelling than it did twenty years ago (Cockburn 1998, Clutton-Brock 2002) for several reasons. First, most permanent groups of social animals consist of relatives and, if the haplodiploid Hymenoptera are excluded, it is not clear that the degree of relatedness is consistently higher in cooperative species than in other species that live in stable groups but do not breed cooperatively (though see Griffin & West 2003). In many societies of vertebrates as well as invertebrates, differences in contributions to rearing young do not appear to vary with the relatedness of helpers (Strassmann et al. 1997, Cockburn 1998, Queller & Strassmann 1998, Clutton-Brock et al. 2000), and several studies have shown that helpers can be unrelated to the young they are raising and that unrelated helpers invest as heavily as close relatives (Dunn et al. 1995, Magrath & Whittingham 1997, Clutton-Brock et al. 2000). In addition, the relative importance of 'indirect' fitness benefits acquired by helping collateral kin may have been overestimated. Estimates of indirect benefits have sometimes incorporated the effects of helping on direct descendants (offspring and grand-offspring) as well on collateral kin. Benefits received by helpers from their kin and those they confer on kin have often both been included, leading to double accounting of kin-selected benefits (Creel 1990). Also, costs arising from competition between relatives for resources or mates have seldom been set against the indirect benefits of cooperation (West et al. 2001).

In contrast, evidence that helpers may gain substantial direct benefits from cooperative behavior has increased. Although reciprocal altruism (Trivers 1970) does not provide a satisfactory framework for the evolution of cooperative breeding since helpers and dependent young rarely exchange benefits, it is now clear that helpers can gain a variety of direct benefits from their actions and that shared benefits or 'generalized reciprocity' is common (Clutton-Brock 2002). In particular, positive correlation between group size, survival and breeding success appear to be usual in specialized cooperative societies (see above) and 'group augmentation' models show that, where group members automatically share benefits derived from increased group size, cooperative behavior can be maintained in small groups by mutualistic benefits even if their members are unrelated (Kokko et al. 2001). Where the benefits of increasing group size are not automatic and depend on the contributions of group members to cooperative activities, cooperation can still be maintained by group augmentation, but its initial evolution requires some previous tendency to help which could be provided by byproduct mutualism (Jamieson 1989, Bednekoff 1997) or by kin selection. The behavior of group members in specialized cooperative societies is consistent with the view that direct benefits play an important role in maintaining cooperative behavior. For example, members of whichever sex is more likely to remain and breed in the natal group generally contribute more to rearing young, and unrelated individuals commonly contribute as much as close relatives (Clutton-Brock 2002) while unrelated individuals commonly contribute as much as close relatives (Cockburn 1998, Clutton-Brock 2002). Moreover, the benefits of increasing group size may explain why group members sometimes kidnap unrelated juveniles from neighboring groups (Heinsohn 1991) and why groups commonly kill litters born to their neighbors if they discover them (Clutton-Brock et al. 1998a).

While both kin selection and mutualism could, in theory, maintain cooperative breeding on their own, the empirical evidence that helpers are generally close relatives of breeders (Emlen 1984, 1991) emphasizes the artificiality of attempting to distinguish between these explanations. The current need is to assess the relative importance of direct and indirect benefits in maintaining cooperative behavior, and the extent to which variation in benefits of these two kinds predicts the distribution of cooperative behavior within and between species.

Chapter 11

Non-offspring nursing in mammals: general implications from a case study on house mice

Barbara König

11.1 Introduction

Reproduction in female mammals is associated with lactation, which involves relatively high energetic costs and influences a mother's future reproduction (Fuchs 1981, Bronson 1989, Clutton-Brock 1991). Because of these high costs, we do not expect females to provide milk to non-offspring. Hence, if they engage in such potentially altruistic or mutualistic behavior, careful study of its evolutionary causes and mechanisms is warranted.

Non-offspring nursing (also communal nursing or allonursing) is known from both breeding and non-breeding individuals, most probably exclusively done by females (to my knowledge, there has been only one rather anecdotal documentation of lactating males in free living Dayak fruit bats, *Dyacopterus spadiceus*; Francis et al. 1994).

Among species with some kind of communal care of young, singular breeders (i.e. typically one breeding female per social unit) form the majority in most mammalian taxa, as they do in social birds and insects. Singular breeders are species with high reproductive skew, and frequently with helpers-at-the-nest (non-breeding individuals that help caring for the dominant's offspring). Sometimes, a subordinate female can also produce pups, as in suricates, dwarf mongooses, callitrichids or wild dogs. For recent reviews, see Stacey (1990), Emlen (1991), Jennions (1994), Creel (1997) and Solomon (1997).

Plural breeders, instead, are species with several breeding females per group and more egalitarian reproduction among females, as in lions, house mice, most bats, most primates and most ungulates. Females in some of these species cooperate in some kind of communal care, as for example babysitting, social thermoregulation, communal defence of young, provisioning of food to pups, or non-offspring nursing (Packer et al. 1992, Jennions & Macdonald 1994, König 1997, Solomon & French 1997).

Nevertheless, there are principal differences between singular breeders with helpers-at-the-nest and cooperating plural breeders. As Lewis & Pusey (1997) have emphasized, non-breeding helpers sacrifice their direct reproductive effort in the short term, whereas cooperation among breeders does not necessarily imply a loss of current direct fitness. In singular breeders, the focus of interest is primarily on the following questions addressing non-breeding subordinates: (i) Why not disperse? (ii) Why delay breeding? (ii) Why help? In plural breeders, instead, questions regarding the value of breeding in groups rather than alone are most important: (i) Why live and breed in groups? (ii) Why help or nurse non-offspring?

Cooperative care of young has mainly been studied in singular breeders, and relatively fewer studies analyze species with shared reproduction among breeding group members. Here, I will focus on communal nursing as an example of a specific cooperative behavior, and I will discuss for a species with plural breeders why lactating females nurse non-offspring. I will summarize our understanding of the ultimate causation of non-offspring nursing, and will present experiments analyzing its proximate mechanisms. Furthermore, I will suggest a novel hypothesis for why it occurs, and speculate on its distribution among mammals.

11.2 Non-offspring nursing in mammals

Non-offspring nursing has been described for approximately 70 species in 12 orders. Field observations indicate that it is more common in pigs than in other Artiodactyls, and that it is more common among rodents and carnivores than in primates and bats. In carnivores, non-offspring nursing is ubiquitous in canids, but also occurs in felids like lions and domestic cats. Furthermore, it has been observed in otters, coatis and some populations of Eurasian badgers. In primates, it has been documented in the field among marmosets (*Callithrix*), in *Alouatta*, *Cebus*, *Erythrocebus*, *Homo sapiens*, *Lemur*, *Microcebus*, *Miopithecus*, *Presbytis*, *Varecia*, and maybe in Goeldi's monkeys (*Callimico*). In rodents, non-offspring nursing has been documented for members of the Cricetidae, Gliridae, Muridae, Sciuridae, Cavidae and Hydrochoeridae. However, in only 10% of all species in which non-offspring nursing is recorded were non-offspring nursed as much as one's own young (for recent reviews, see Packer et al. 1992, König 1997, Lewis & Pusey 1997, Solomon & French 1997, Hayes 2000).

11.2.1 Why do females nurse non-offspring?

Several hypotheses have been suggested to explain the phenomenon of non-offspring nursing, and there is some controversy as to whether communal nursing confers a reproductive advantage or not.

11.2.1.1 Non-adaptive hypotheses

Two non-adaptive hypotheses have been proposed. First, some authors consider non-offspring nursing to be milk theft by other females' young, making it obviously non-adaptive for the donor (McCracken 1984, Boness 1990). The second hypothesis is that it represents a byproduct of providing parental care in a group-living context. Jamieson & Craig (Jamieson 1989, Jamieson & Craig 1987) suggested that alloparental behavior occurs simply because the social structure of those species in which it is found provides an opportunity for parent-like behavior. A similar explanation was offered by Pusey & Packer (1994) for non-offspring nursing in lions. Female lions live in groups and raise their young in crèches because of the advantages of defense against infanticidal males. Non-offspring nursing then occurs as an inevitable consequence of group rearing, with the costs of rejecting non-offspring being higher than the costs of allowing some nursing by non-offspring.

11.2.1.2 Adaptive hypotheses

Adaptive explanations, on the other hand, include kin selection, or direct benefits. Such direct benefits can accrue due to improved survival, growth, or future reproduction of own offspring, or due to improved breeding success of mothers in the presence of lactating peers. I will discuss later which mechanisms can result in direct benefits for either mothers or pups.

11.3 Non-offspring nursing in house mice: a case study

In order to assess which of these hypotheses best account for non-offspring nursing, we study this phenomenon in house mice. We are interested in non-offspring nursing at both the ultimate and proximate level, to complement evolutionary approaches with mechanistic ones.

House mice (*Mus domesticus*) are short-lived rodents with a high reproductive output. They have a rather flexible social structure, but most typically they live in small groups that consist of a dominant male, one or several adult females with their litters and several subordinate animals (DeLong 1967, Lidicker 1976, Bronson 1979, Berry 1981a, Singleton 1983, Gray et al. 2000). Litter size of wild house mice increases from the first to the second lactation, and decreases again after the fifth lactation (Pelikán 1981, König & Markl 1987). Fifty years ago, Southwick (1955) published for the first time that females of the same reproductive group can pool their litters in communal nests. Since then, this behavior has been documented both in the field and in captivity (Sayler & Salmon 1969, Wilkinson & Baker 1988, König 1993, Manning et al. 1995).

To analyze whether non-offspring nursing in house mice is adaptive, we quantified the fitness consequences of communal rearing of young under laboratory conditions. Experimental animals were first- to third-generation wild-caught house mice, born and reared in the lab. Under otherwise standardized conditions, we simulated different social structures that are known to occur in feral or commensal house mice, and measured the females' lifetime reproductive success. We defined lifetime reproductive success as the number of offspring weaned during an experimental lifespan of six months (for a detailed description of the methods used see König 1993, 1994b). Although average life expectancy of newborn house mice is only 100-150 days, an experimental lifespan of six months is realistic for females that survived at least until maturity (Berry 1971, Berry 1981b, Pennycuik et al. 1986).

In all experiments, females always reared litters in a communal nest as soon as more than one female in a group gave birth to pups. Moreover, nursing of pups within a communal nest was indiscriminate (König 1989, 1993).

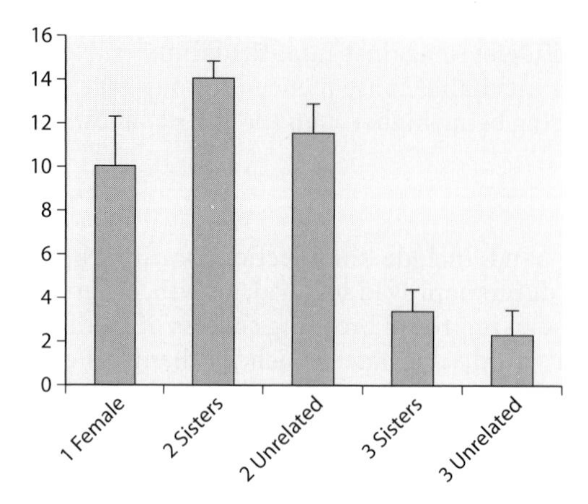

Fig. 11.1. Number of offspring weaned during an experimental lifespan of six months (median ± SE) of female house mice as a function of group size (number of females per group ranged between one and three) and of genetic relatedness. Sisters were familiar full-sibs that grew up together; unrelated females were previously unfamiliar and genetically unrelated females. An unrelated adult male was always present. Independent sample sizes (number of groups per treatment): 1 Female: n = 21; 2 Sisters: n = 21; 2 Unrelated: n = 24; 3 Sisters: n = 10; 3 Unrelated: n = 10. Data modified from König (1994a).

We manipulated group size (the number of adult females per group) and relatedness among females. At the age of 7-8 weeks, females were mated with an adult, unrelated male and during the following four months lived either monogamously (one female plus one male) or in polygynous groups (for further details see König 1993, 1994a, 1994b). In polygynous groups, females were either two or three genetically full-sibs, reared together (simulating the situation of sisters staying together), or two or three genetically unrelated and previously unfamiliar females (simulating the situation of females immigrating into a group).

Lifetime reproductive success of individual females differed significantly as a function of both group size and relatedness among the females, and reached a peak for females living with one sister (Fig. 11.1). In a group of three females, however, individual lifetime reproductive success was lower than in a monogamous situation, irrespective of the females' relatedness. Offspring weight at weaning did not differ significantly among the groups (König 1993, 1994b).

The reason why females differed in individual reproductive success as a function of group size and relatedness is that females varied in the probability of reproduction and of successfully weaning young within the experimental lifespan. Not all females weaned young, due to competition over reproduction despite communal nesting. The extent of this competition is illustrated by the index of reproductive skew for the females involved (Fig. 11.2; index of reproductive skew according to Reeve & Keller 1995; data on house mice from König 1994a).

This index varies between zero and one. When a single individual produces all the offspring, the skew is one, reflecting a despotic society; when reproduction is perfectly equitable among all group members, the skew is zero, indicating egalitarian reproduction among females.

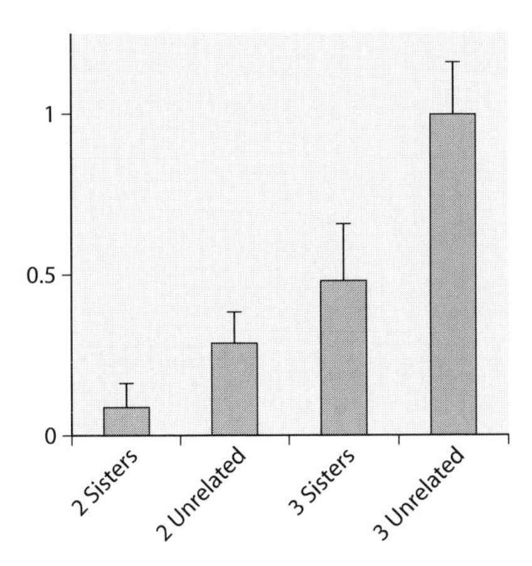

Fig. 11.2. Index of reproductive skew (median ± SE) among female house mice as a function of group size (two or three females) and relatedness. The index was calculated according to Reeve & Keller (1995), for all groups in which at least one female produced a litter and in which individual reproductive success was known (groups in which females gave birth to litters on the same day were excluded, because of lack of information about maternity). Independent sample sizes: 2 Sisters: n = 20; 2 Unrelated: n = 20; 3 Sisters: n = 6; 3 Unrelated: n = 7. Data on individual lifetime reproductive success is from König (1994a).

The lowest index was found for pairs of sisters. In such units, females are not only egalitarian in terms of the probability of reproduction but also in terms of the number of offspring weaned. The median degree of reproductive skew increased significantly towards despotic relationships with decreasing relatedness among the females within a group, and with increasing group size.

These findings permit two conclusions. First, non-offspring nursing is an integral part of the reproductive behavior of female house mice in egalitarian groups. Thus, the non-adaptive hypothesis that it is milk theft by young can be rejected in this case. The milk theft hypothesis should result in more variable occurrences of non-offspring nursing, with an increase with increasing age and mobility of young. Furthermore, female house mice have the option to breed solitarily even when another female reproduces within their territory (Weidt & König, unpublished observations from a population of wild house mice in a barn), which allows us to exclude the hypotheses of misdirected maternal care, and of a byproduct of group living.

Second, where a female established an egalitarian reproductive relationship, communal nursing increased her individual lifetime reproductive success, irrespective of the degree of relatedness or familiarity to the female partner (König 1994c). However, the probability for such mutualistic cooperation was highest when a female shared a nest with a familiar sister to form a low-skew society. As a consequence, non-offspring nursing of female house mice in pairs with egalitarian reproduction proved to be adaptive, and involved mutualistic, direct fitness benefits for both partners. The fact that communal nursing was most efficient among familiar relatives may indicate that kin selection played a role during the evolution of communal nursing. However, because neither familiarity during juvenile development nor high relatedness are necessary pre-requisites, direct benefits of cooperation seem to stabilize non-offspring nursing among female house mice.

11.4 Direct benefits of allomaternal care

The experiment demonstrated that female mice gained a direct mutualistic benefit from forming a communal unit characterized by allonursing. Several hypotheses have been suggested to explain why female mammals or their offspring gain direct benefits when mothers exhibit allomaternal care (licking, huddling over, or carrying non-offspring) or nursing non-offspring (for previous reviews see Packer et al. 1992, Lewis & Pusey 1997, Hayes 2000, Hayes & Solomon 2004).

11.4.1 Improved survival of pups

Communal nursing can reduce pup predation either by the dilution effect (Hoogland 1989; in analogy to communal care of eggs in birds as in ostriches, Bertram 1992), or due to improved protection against infanticide committed by non-group members, as suggested by Manning et al. (1992). When females alternate nursing pups in a communal nest, offspring are left alone less often and thus have a lower probability to be killed by unfamiliar conspecifics compared with pups reared by a single female.

11.4.2 Improved future reproduction of pups

Packer and co-workers further raised the idea of improved cooperation, based on their long-term observations of lions. Group size is critical for reproductive success in both male and female lions (Packer et al. 1990, 1991, Pusey & Packer 1997). Thus, communal care would result in their own young having more potential allies later in life, even if no full-sib survived.

Nevertheless, both hypotheses mentioned so far cannot explain why communally nursing female house mice weaned more offspring within their lifespan in our experiments under rather luxurious environmental conditions, with unlimited food, in a favorable climate, and in the absence of predators or cannibalistic non-group members.

11.4.3 Improved growth of pups

According to Caraco & Brown (1986), allomaternal (pluriparental) care may reduce starvation of young if at least one of the participating parents finds sufficient food to allow for lactation. When there is a cost of starvation, cooperative provisioning of young might evolve through reciprocity given that breeders feed the young asynchronously. The authors further suggest that even when food is plentiful, offspring may benefit because of reduced time between meals. In house mice, communally nesting females do not nurse simultaneously so that pups are almost always cared for by one lactating female (unpubl. pers. obs.). Litters that grew up in a communal nest have a relatively high weaning weight compared to same-sized litters from solitarily nursed mothers (König 1993; see also Table 11.1). It is not known, however, whether shorter time intervals between meals cause this effect, or whether other energetic benefits of communal nesting are involved, as suggested by the following hypotheses.

11.4.4 Immunological benefits for pups

As an alternative hypothesis to explain intra- and inter-specific variation in allosuckling frequency, Roulin & Heeb (1999) suggested immunological benefits (the immunological function of the allosuckling hypothesis). We modified the authors' hypothesis and tested the prediction that house mouse pups gain a more variable immunocompetence through milk provided by several females (Ramsauer & König, submitted).

Newborn mammals do not yet have a functioning immune system and are dependent on immune factors received through maternal milk. During the first two weeks of lactation in house mice, immunoglobulin and lymphocytes reach the pups' intestines through the milk, and then are passed on into the blood (Janeway & Travers 1997). Due to indiscriminate nursing of own and alien young in communal nests, pups might benefit by acquiring a broader immunocompetence when reared communally in comparison to pups raised by just one female. The major histocompatibility complex (MHC) is crucial for the production of immunocomponents and plays an important role in pathogen recognition. Receiving variable MHC products through maternal milk supplied by both the mother and another lactating female might thus allow for a better defense of pups against pathogens and be of importance for the growth and viability of offspring.

Immunocompetence typically is a matter of genetics and experience. The social behavior of female house mice, however, might offer a non-genetic tool to influence offspring immunocompetence through cooperative nursing. We therefore predicted improved growth and/or earlier weaning of pups reared by females of different MHC, and differences in the immunocompetence of subadult house mice that have been nursed by two mothers compared to those receiving milk from one mother. We tested these predictions by cross-fostering newborn house mouse pups from our population of wild-caught animals to a communal nest of two lactating foster mothers either of the same or of different MHC types (Ramsauer & König, submitted).

Foster mothers were from two congenic strains differing in the MHC ('A' = B10BR/OlaHsd and 'B' = C57BL/10ScSnOlaHsd). Each replicate consisted of three newborn full-sisters: one reared by 'AA'-foster mothers, one by 'BB'-females, and the third by one 'A'- and one 'B'-female. Litter size of communal nests was always standardized and consisted of 13 congenic offspring, with a sex ratio of seven males and six females, plus one wild female pup; independent sample size was 12.

Growth and weaning weight did not differ significantly for females reared by two foster mothers of either the same or different MHC. In collaboration with Andrew MacPherson from the Institute of Immunology at Zürich University, we measured immunocomponents in the pups' blood. Our treatment did not significantly influence immunoglobulin concentrations (IgA, IgM and IgG) of young at day 15 (before the immune system of pups is fully functional). Lymphocyte concentrations (B220 representing B-cells, and CD4 representing T-cells), however, differed significantly at day 28, with intermediate values in females raised by 'AB'-foster mothers (at the age of four weeks, subadult house mice are already immunocompetent). 'A'-females had rather high concentrations of CD4 lymphocytes in their milk which is reflected in high concentrations in pups that had been nursed by 'AA'-foster mothers; 'B'-females, on the other hand, had rather high concentrations of B220 lymphocytes resulting in similarly high values in their offspring (Ramsauer & König, submitted).

Immunological components that are transferred via milk influence the immunocompetence of wild-type house mouse pups irrespective of their own genotype. Such influence on immunocompetence, however, did not result in energetic benefits of young as reflected in improved growth or earlier weaning under our experimental conditions. Nevertheless, a female house mouse that chooses a partner for communal nursing according to MHC characteristics might be able to influence her offspring's future survival and reproduction. Under more natural conditions, when offspring encounter a variety of pathogens, we therefore may expect that MHC characteristics contribute to structuring among females within social groups in house mice. Even if the influence of maternal milk on offspring immunocompetence cannot explain our observation of improved reproductive success of communally nursing females, it might influence a female's choice of a social partner, which remains to be tested.

11.4.5
Physiological benefits for the mother

Wilkinson (1992) suggested that female evening bats, *Nycticeius humeralis*, nurse non-offspring to dump excess milk prior to the next feeding trip, thereby obtaining immediate energetic benefits and maintaining maximum milk production. House mouse pups, however, are limited in their growth by the milk available from the mother (König et al. 1988), and especially in communal nests with many pups, it is not plausible that females have to face the problem of getting rid of excess milk before they leave for a foraging trip.

For relatively small mammals such as rodents, communal care might involve direct energetic or metabolic benefits as improved thermoregulation or improved milk production, and thus allow for a higher weaning success of females that nurse non-offspring (Sayler & Salmon 1969, Boyce & Boyce 1988, Hayes & Solomon 2004).

To test whether females are more efficient in converting solid food into offspring body mass during cooperative care of young, we measured the energy costs of lactation of females rearing litters either solitarily or communally with a familiar sister. Litter size of experienced females (rearing at least the second litter) was standardized to 6–7 pups directly after birth, and litters of communally nursing sisters did not differ by more than six days in age. The animals had *ad libitum* access to food and water, but were kept in a climatic chamber at an ambient temperature of 15°C. This should reflect rather natural conditions for house mice and avoid missing an effect due to climate conditions that are too luxurious (Barnett 1965, DeLong 1967, Berry 1981a; for detailed methods, see Diedrichsen 1993).

Daily food consumption of females was measured from day 2 until day 13 of lactation with the help of an automatic feeding device (Neuhäusser-Wespy & König 2000). This device allows measuring individual food consumption of group-living animals without any disturbance. At day 14, we milked females with a milking device (König et al. 1988) and measured the amount of milk produced (after four hours of separation from the litters), and its energy content from lipids and total solids.

Neither litter weight at birth and weaning, nor the individual female's food consumption or milk production differed significantly for solitarily or communally nursing females (Table 11.1).

To quantify the females' allocation of energy into lactation versus maintenance, we calculated Calow's index of reproduction (Calow 1979). This index (I) was analyzed for day 14 of lactation, by using the following equation:

$$I = 1 - \frac{\text{(Energy consumed)} - \text{(Energy invested)}}{\text{(Energy consumed when non-reproducing)}}$$

Energy consumed = energy equivalents of maternal food consumption at day 14 of lactation; Energy invested = total energy of milk produced at day 14; Energy consumed when non-reproducing = energy equivalents of daily amount of food eaten (averaged over five consecutive days) when the adult females were non-pregnant and non-lactating. Energy equivalents of food pellets (Altromin rat and mouse) were 12.5 kJ/g (information according to the producer).

The index, I, relates a female's energy investment during lactation to her maintenance metabolism. A value equal or less than zero indicates that females compensate the energetic demand of reproduction (or lactation) through increased food consumption. For a value larger than zero, females meet the energetic costs of lactation at the expense of their maintenance metabolism, or by using lipid stores or other reserves that they accumulated before reproduction.

Energy allocation during reproduction did not differ significantly in both social groups (Table 11.1). Females did not allocate more energy to milk production, and did not lactate more efficiently, when nursing communally compared to mothers nursing solitarily.

Table 11.1. Energy allocation during lactation of female house mice rearing litters solitarily or communally with a familiar sister. Litter size was standardized at birth to six pups.

	Females rearing litters			
	Solitarily (n = 7)	Communally (n = 11 pairs)	z (U-test)	
Female weight day 1 (g)	30.1 ± 5.5	31.3 ± 3.0	−0.498	ns
Female weight day 23 (g)	30.9 ± 5.6	31.3 ± 3.0	−0.045	ns
Offspring weight day 1 (g)	1.6 ± 0.1	1.5 ± 0.1	−0.126	ns
Offspring weight day 23 (g)	8.5 ± 1.4	9.2 ± 1.4	−1.907	$p < 0.10$
Maternal food consumption (days 2-13; kJ)	1971 ± 98.9	2002 ± 113.8	−0.226	ns
Milk production at day 14 (g)	1.0 ± 0.3	1.0 ± 0.3	−0.317	ns
Energy provided through milk (at day 14; kJ/day)	63.7 ± 29.1	67.1 ± 24.9	−0.402	ns
Calow's index I (see text for explanation)	−0.7 ± 0.3	−1.1±0.6	−1.407	ns

11.4.6 Metabolic peak load reduction

In the experiment described before, both solitarily and communally nursing females met the energy need for lactation through increased food consumption from days 1–4 until days 13–16 (see also König et al. 1988). Lactating house mice were able to rear a growing litter both by increasing the amount of milk produced and by improving the quality through an increase in total solid and fat concentrations until day 16 of lactation. At the age of 17 days, offspring shift to solid food and are fully weaned when they are 23-days-old (König & Markl 1987). As a consequence, females go through a period of peak energy demand during lactation that is reflected in a drastic increase in daily food consumption, by over 200% in comparison to the non-reproducing state. This energetic demand can be further increased by simultaneous pregnancy during lactation. Conception during the postpartum estrus results in the birth of one litter every 28 days, on average. Nevertheless, female house mice are limited in their maximal (or peak) sustainable metabolism especially when nursing a large litter (Hammond & Diamond 1992). This effect is called 'metabolic ceiling'.

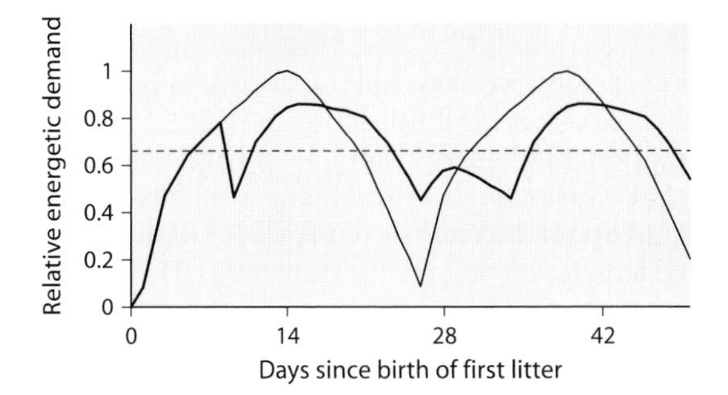

Fig. 11.3. Model of the relative energetic demand during lactation of a female house mouse over two consecutive litters. Thin line: female rearing litters solitarily. Bold line: female rearing litters communally with a female conspecific, the two litters differing in age by eight days. Dotted line: average demand in both situations. Maximum demand is set to one. The curves were derived from data of laboratory mice rearing a medium-sized litter (König et al. 1988), calculated as daily amount of milk produced times the proportion of dry weight. For the communal situation, we assumed equal contributions of both females to both litters.

In our population of house mice kept in polygynous groups over an extended period, litters in communal nests showed an average age difference of eight days (König 1994b). Due to indiscriminate care of young, females in such a situation are nursing more or less continuously. Based on these observations, we assumed that the energy budget of communally nursing females remains at a rather constant and medium level because both litters do not simultaneously reach the period of highest energy need (Fig. 11.3). We therefore formulated the hypothesis of benefits due to peak load reduction (Müller & König, submitted). By nursing litters communally, lactating females avoid peak energy demand. Because peak energy demands at the metabolic ceiling are especially costly, females that avoid such peaks will benefit by improved reproductive success.

To test this hypothesis, we manipulated the energy demand of females rearing litters alone, so that other possible benefits of communal breeding were absent. We analyzed the energy output of two groups of lactating females in which the total amount of energy spent on rearing a litter was the same, but energy allocation was timed differently.

In the manipulated group, we simulated a constant, medium-level energy output for lactating females by cross-fostering two older pups against younger ones every 2–3 days, beginning at day 8 of the first lactation, and continuing during the females' second lactation. As a control, we used females in which handling was done in the same way, but without cross-fostering. Manipulated and control females reared similar-sized litters (litter size was standardized at day 1 of lactation: six pups for the first litters, and seven for the second ones). Energetic demand of the manipulated females during lactation was assumed to have the same mean but lower variance as that of the control females, without a prominent peak two weeks after giving birth (for further details see Müller & König, submitted).

To quantify energy output, we measured the females' food consumption and resting metabolic rate. As fitness correlates, we analyzed the interbirth intervals and the size and weight of the females' second and third litters.

In accordance with our assumption, total energy output was similar for manipulated and control females that consumed similar amounts of food when rearing both the first and the second litters (there was a tendency for higher total food consumption in manipulated females during the first lactation). Daily food consumption of control females increased significantly from day 9 until day 15 of lactation and significantly decreased afterwards. No such variation was observed in manipulated females, with significantly lower food consumption at peak lactation, and higher food consumption during day 28 than in control animals, both during then first and the second litter.

Resting metabolic rates (RMR) of manipulated and control females were measured twice during each lactation period (at day 14 and at day 28), which allows further examination of the assumption of a rather constant energetic burden throughout lactation in manipulated females. RMR of control females decreased significantly, as expected, from peak lactation to weaning in both lactation periods. The RMR of manipulated females, however, did not change significantly.

Given that the assumptions of our model were fulfilled, we tested the prediction that females not experiencing peak loads had lower reproductive costs than control females, reflected in shorter interbirth intervals and/or larger litter sizes of manipulated females on the next reproductive occasion.

Neither the number of young at birth of the second and third litters, nor the proportion of females that mated *post partum* differed significantly between manipulated and control females (Müller & König, submitted). These observations support data from a former study on house mice by Fuchs (1981, 1982), who found an effect of the burden of lactation on the interval to the following litter, but not on its size.

As predicted, intervals between the first and second litters were shorter in manipulated than in control females. This effect, however, was only significant for those females where standardization of litter size directly after birth resulted in an experimentally increased litter (manipulated females gave birth to the second litter two days earlier than control ones, on average). Females, whose litters had been decreased in size at day 1 of lactation, might not have been confronted with an energy demand at their metabolic ceiling.

These data suggest that peak load reduction results in lower future reproductive costs at least for females that suffer an energetic burden near or at their maximum metabolic capacity.

Intervals between the second and the third litters, however, did not differ significantly between the two groups (Müller & König, submitted). Recent work by Johnson et al. (2001) has shown that the metabolic ceiling does not remain constant throughout the life of a house mouse but that it increases from the first to the second litter. Presumably, with a litter size of seven young, females were not forced to invest at their metabolic ceiling during the second lactation, and peak load reduction therefore did not result in lowered reproductive costs.

Nevertheless, the experiment suggests that communal nursing can modify a female's energy output, and can reduce peak energy demand of lactating females if litters differ in age by several days. Peak load reduction may thus affect fitness parameters of lactating house mice, and we further suggest that this effect is most pronounced if the peak forces them to approach their metabolic ceiling. However, it remains to be shown that this is not only the case in the rather artificial setting used during the experiments, but also under conditions of communal nursing.

11.5 Can peak load reduction explain non-offspring nursing in mammals?

Metabolic benefits due to peak load reduction are a prime candidate for explaining the observed higher reproductive output of communally versus solitarily nursing females. It is therefore tempting to speculate that such energetic benefits can also underlie other cases of non-offspring nursing, including those that have been interpreted as non-adaptive.

Packer et al.'s (1992) investigation of the effects of a variety of factors on the frequency of non-offspring nursing (excluding data from captive studies!) revealed three significant findings. First, non-offspring nursing increases with litter size across taxa. Second, non-offspring nursing is more common and better tolerated in polytocous (average litter size larger than one) than in monotocous species. In species that typically nurse only one pup, non-offspring nursing is more likely to be classified as milk theft (as for example in Mexican free-tailed bats, or Northern elephant seals). In contrast, in polytocous species, non-offspring nursing is less likely to be classified as milk theft, and also occurs in species where females can discriminate their own from foreign young, as in African lions (Pusey & Packer 1994). In both situations, non-offspring nursing correlates with increased energetic costs of lactation that peak shortly before weaning (Oftedal 1984, Oftedal & Gittleman 1989), increasing the probability that females invest at their metabolic ceiling. These two findings are consistent with the peak load reduction hypothesis; females are not expected to carry such a heavy energetic burden when litters are small and their life history is not as fast-paced.

Their third finding was that non-offspring nursing is most common in polytocous species when group size is small, and decreases significantly as group size increases. This observation is also in accordance with the hypothesis of peak load reduction if the probability of avoiding simultaneous peaks during lactation decreases with increasing number of breeding females in a group. Furthermore, the risk of exploitation by non-mutualistic individuals increases with group size and thus will hinder the evolution of stable cooperation.

The hypothesis of peak load reduction requires that females increase energy allocation during lactation up to their metabolic ceiling. Furthermore, females within a group have to be synchronized in reproduction, so that there is considerable overlap in lactation (perfect synchrony, however, that is giving birth on the same day, should not occur). Such constraints might explain why non-offspring nursing, although not very rare, nevertheless is limited to rather few taxa.

11.6 Concluding remarks

During communal nursing, female house mice do not discriminate between own and non-offspring, and gain direct, mutualistic benefits. Non-offspring nursing therefore is a cooperative behavior that allows females to improve weaning success of pups in a reciprocal manner, even among unrelated partners, once they have established an egalitarian relationship.

Such cooperation, however, may run the risk of being exploited. The most extreme case would be a highly pregnant female that drops her litter into another lactating female's nest and deserts. The benefits of such free-riding behavior are high. A non-lactating female will give birth to her next litter on average six days earlier than a female simultaneously being pregnant and lactating (König & Markl 1987). The deserted female, on the other hand, has to invest into non-offspring because she cannot tell them apart from own young (König 1989, 1993). It is not known whether such brood parasitism exists in house mice, but some aspects of the females' social behavior suggest protection against exploitation by non-cooperative partners.

First, female aggression is rather rare within groups and among relatives. Females, however, are very aggressive towards foreign females, not belonging to the same group, and especially so, when they are lactating (Crowcroft & Rowe 1963, Haug 1978, Kareem & Barnard 1982). Second, females preferentially share nests with a familiar relative (Manning et al. 1992, Dobson et al. 2000). Interestingly, familiarity during juvenile development is of paramount importance for improved reproductive success of females in egalitarian pairs, and overrides the effect of genetic relatedness (König 1994c), despite the fact that house mice of both sexes use genetic cues to discriminate against unfamiliar kin during mate choice (for a recent review, see Penn 2002). The importance of familiarity may suggest either that a physiological mechanism is involved which requires some period of adaptation to or synchronization with a partner, or that information about the partner is of significance for successful cooperation. The rather simple rules of thumb to communally nurse with a familiar group member and to aggressively keep away strangers might prevent females from being exploited by the opportunistic free-riding of other females.

Nevertheless, even during communal nursing females might benefit when reducing their investment, given that the partner will do more. In rodents, lactation performance is influenced by litter size *in utero*, which determines *pre partum* mammary growth (Jameson 1998), but more so by the number of sucking pups (Mann et al. 1983). Due to indiscriminate nursing, we assume that lactating females do not adjust milk production according to their own litter size but that energetic investment is shared equally among the members of a communal nest (we are currently testing this assumption). Such equalized investment therefore might be a prerequisite for stable cooperation among female house mice.

Acknowledgments

For stimulating discussions and helpful comments, I cordially thank Carel van Schaik. I further thank Gabi Stichel for excellent animal care and help with some experiments. Financial support of the Swiss National Science Foundation (31-50740.97/1 and 31-59609.99/1) is gratefully acknowledged.

Part V
Biological Markets

Chapter 12

Monkeys, markets and minds: biological markets and primate sociality

Louise Barrett, S. Peter Henzi

12.1 Introduction

The sight of a monkey group, huddled together in pairs, each individual taking turns to comb diligently through the other's fur inevitably brings to mind the old cliché, "you scratch my back and I'll scratch yours" and makes it obvious why primate grooming behavior is often seen as the quintessential act of cooperation and reciprocity. Among the monkeys and apes, grooming is also seen as the defining act of sociality; the fact that individuals put considerable effort into their grooming relationships, groom some group members more than others, and work to sustain time for grooming in the face of opposing pressures (Sade 1972, Dunbar & Sharman 1984), suggests that grooming helps to serve an individual's social goals, as well as enabling animals to stay clean and healthy. Understanding how primates cooperate and perform successfully in the social world means, to a great extent, understanding the dynamics of grooming.

Traditionally, it has been assumed that dominance and competition are the factors that explain the intensity with which female primates, in particular, engage in grooming (e.g. Seyfarth 1977, Harcourt 1988). Competition among female primates arises as a consequence of group living. Living together in a cohesive social group can itself be seen as a cooperative act; joining together with others enables animals to receive benefits, like reduced predation risk or decreased vulnerability to infanticidal males (van Schaik & Kappeler 1997, Henzi & Barrett 2003), that are unavailable to solitary animals. However, living in a group is not cost-free; the unavoidable corollary of living in close proximity to others is conflict over access to scarce local resources, such as food or predator-risk reducing spatial positions (van Schaik 1989, van Schaik & Kappeler 1997). Although these effects may be ameliorated by the fact that females often reside in kin-based groups, they nonetheless remain trapped by the need to remain safe, on the one hand, and the need to secure sufficient resources for themselves and their offspring, on the other. This dilemma, seen in both proximate and ultimate perspective, generates the subtle and complex patterns of cooperative interaction that are associated with female-bonded primate social systems.

Among the most important of these cooperative interactions, and the ones most frequently linked to grooming in a causal manner, are the coalitions that females form during aggressive encounters, whereby one individual comes to the aid of another to help fight off an attacking individual (Silk 1987). The consensus view is that, among the primates, females form long-term mutually-beneficial alliances with specific individuals in order to buffer themselves against the negative effects of competition within their groups. This buffering is thought to take the form of coalitionary support during agonistic interactions combined with the use of grooming to build trust and alleviate stress. Grooming, thus, has two (non-exclusive) functions within this scenario; it builds bond strength and thereby establishes the trust on which coalitionary relationships may be built (Dunbar 1984) and/or it acts as a currency that can be exchanged in anticipation of future coalitionary support (Seyfarth & Cheney 1984).

12.2 The problem with chacma baboons

This idea that grooming is a means of servicing coalitionary relationships is neat, coherent and fits well with notions that monkeys and apes are highly 'political' animals (see e.g. Byrne & Whiten 1988). However, there is a problem with the above scenario: chacma baboons, (*Papio hamadryas ursinus*), our chosen study animal, very rarely form coalitions, despite the fact that females sustain grooming relationships and compete over access to resources (Ron et al. 1994, 1996, Silk et al. 1999, Barrett & Henzi 2002). At De Hoop, our current study site, for example, we have seen only two female-female coalitions in approximately 30000 observer-hours. Moreover, recent work on yellow baboons in Amboseli, Kenya, reveals that females form coalitions against other adult females at extremely low rates in this population as well (1-4 interventions per 100 disputes) (Silk et al. 2004). Low rates of coalition formation may therefore be characteristic of all baboons, and not just the southern African sub-species. Silk et al. (2004) suggest that coalitions confer significant individual benefits on the females that participate in their formation but, as suggested by Henzi & Barrett (1999), their overall rarity makes it unlikely that they are the organizing principle of female social strategies.

Another pertinent fact is the finding that chacma females from the Drakensberg Mountains continue to form grooming relationships and adjust their time budgets to conserve grooming time despite the fact that the distribution of resources in their environment means that they experience almost no competition for food and consequently show little aggression (Henzi et al. 1997). Specifically, once the size of the female cohort of a group exceeds a critical number, Drakensberg females cut back the number of different individuals with whom they engage in grooming. This allows them to increase the length of individual grooming bouts with their chosen partners and, more importantly it seems, keeps levels of grooming reciprocation high (Henzi et al. 1997); female clique size is reduced at precisely the point at which reciprocal grooming with all other female group members can no longer be sustained.

The significance of these findings is further highlighted by other data from Amboseli revealing that grooming has significant fitness benefits for baboons, even though grooming is not causally related to coalition formation in this population. Females that are highly social and who groom frequently have significantly higher offspring survival rates than less sociable females (Silk et al. 2003).

Among baboon females, then, grooming remains significant and has positive fitness effects even in the virtual absence of coalition formation. Consequently, the notion that the function of grooming is to cement coalitionary alliances cannot be taken as a general explanation for the prevalence of grooming across the primates as a whole. On a more personal and immediate level, the rarity of coalition formation among chacma females meant that, in our own studies, we had to start thinking about grooming and its dynamics differently, resulting in a simple and very obvious insight into the problem; namely, that a comprehensive explanation of the role of grooming in primate societies should encompass its utilitarian benefits, as well as its social ones (Barrett et al. 1999, Henzi & Barrett 1999, Barrett & Henzi 2001).

It is clear that grooming retains its original hygienic value, despite its social importance, since animals spend time grooming themselves as well as directing it to others, and because the grooming they receive from others is generally directed at areas they cannot easily reach (Barton 1985). The targets of this grooming are ectoparasites such as fleas and lice (Tanaka & Takafushi 1993). A greater parasite load means a greater loss of blood, greater irritation, and increases the probability of infection (Johnson et al. 2004), so keeping levels low is of clear benefit to animals. In addition, it is possible that grooming has thermoregulatory benefits by maintaining the loft of the fur, thus aiding heat retention and dissipation. It is also clear that the physical contact that grooming involves is highly pleasurable for the recipient and is, in fact, associated with the production of β-endorphins (Keverne et al. 1989). This latter feature cannot be viewed in the same utilitarian light as the removal of parasites, as it is presumably a derived feature that proximately reinforces grooming behavior. Nevertheless, it can result in grooming being exchanged for its own hedonic benefits, rather than for 'political' reasons.

Grooming is also costly for its participants. Not only are there opportunity costs associated with grooming another animal (an individual could be engaged in other activities like foraging, or indeed being groomed themselves) but there is also a risk of acquiring parasites from grooming partners if the parasites are able to move from one animal to another during the course of a grooming bout (see Johnson et al. 2004 for a theoretical approach based on the ideal free distribution). Removing another animal's parasites therefore comes with the simultaneous cost of acquiring a few of them oneself; a problem that will be exacerbated in larger groups because these tend to have higher average parasite loads than small groups (Johnson et al. 2004). Grooming is therefore a cooperative act since animals cannot obtain all the grooming they require to be parasite-free by their own actions, and the benefits of grooming another must be traded off against the costs of doing so.

12.3 The social market place

While this idea is obviously not new, the focus on the social function of grooming has pushed aside the rather more mundane role that grooming serves. By highlighting the broader utility of grooming, it becomes easier to appreciate that grooming is a valuable commodity in itself. The fact that an animal must trade with other individuals in order to reduce its parasite load means that, regardless of its ability to facilitate other social interactions, grooming is a valuable service that one animal can supply for another. Johnson et al. (2004) go further than this, however, by emphasizing that levels of parasite infestation can produce patterns of behavior (e.g. group fission) that have traditionally been attributed solely to complex social processes. Social dynamics may actually be linked to parasite loads in a fundamental way, making it impossible to divorce the hygienic from the social function of grooming in the way that some authors have suggested (e.g. Dunbar 1988).

A more utilitarian perspective on grooming also frees us from the assumption that coalition formation is inevitably tied to grooming; although females potentially are able to trade grooming for this service, there is no necessary link between these two behaviors from our perspective. Other conceptual approaches to the issue require that they are connected (e.g. Dunbar 1988, van Schaik 1989), even though the evidence to support such a mandatory link is equivocal at best (Henzi & Barrett 1999).

This notion of grooming as a tradable commodity thus leads neatly to the adoption of biological market (BM) theory as an explanatory framework. BM theory, as put forward by Ronald Noë and Peter Hammerstein (Noë et al. 1991, Noë & Hammerstein 1994, 1995), holds that (i) where individuals control resources or can provide a service to others, these constitute commodities that can be exchanged but not taken by force (they are 'inalienable'); (ii) trading partners are chosen from a range of alternatives, via a mechanism of outbidding competition, in such a way that profit is maximized. This, in turn, means that the prevention of defection is not a driving force in a BM framework, in contrast to models based on the iterated Prisoner's Dilemma (Axelrod 1986). (iii) Supply and demand determine the bartering value of commodities exchanged; thus, within primate groups, animals may trade grooming with each other on a mutualistic basis in order to reap the benefits that grooming itself offers (reciprocal traders), or they can exchange grooming for other commodities that are, in some sense, value equivalent (interchange traders) (Hemelrijk & Ek 1991). It should therefore be possible to distinguish 'trader classes' of females that exchange grooming in different ways. Possible sources of interchange commodities are tolerance around feeding or drinking sites, mating opportunities, tolerance and access to infants (for further details see Barrett & Henzi 2001). Coalitionary support is also a potential interchange commodity for species other than baboons, although there are reasons to suspect that this is less likely to occur than other exchanges (see below).

In the absence of coalitions and alliances, adult female baboons generally gain access to resources on the strength of their own power (females can gener-

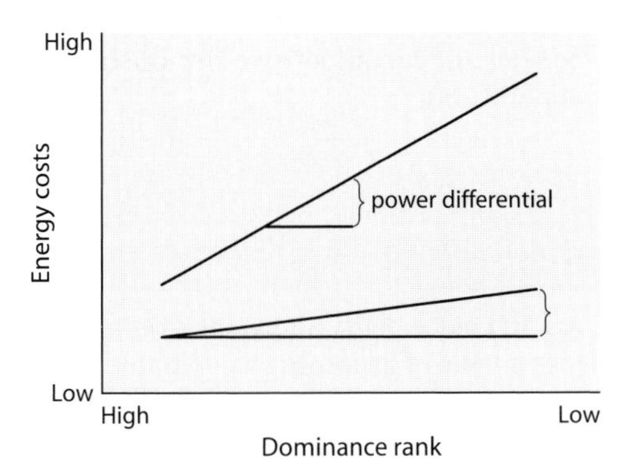

Fig. 12.1. A notional graph to illustrate the concept of a power differential. As dominance rank gets lower, the energy costs associated with receiving aggression from others increase. High-ranking females can therefore exert a much greater effect on another female's fitness than low-ranking females. The two slopes represent groups with different dominance gradients. When gradients are steep (upper line), the negative effect exerted by one female on another can be achieved at lower rank distances than in a group where the gradient is shallower. Adapted from Barrett & Henzi (2001) with permission.

ally be placed in a linear, transitive dominance hierarchy, which describes priority of access to resources). Consequently, we predict that interchange trading should occur only when the 'power differential' between two participants is great enough that access to the commodity cannot be achieved by the less powerful individual alone. Power differentials among adult females can be expressed in terms of the distance between the animals' respective dominance ranks, and the gradient (or 'steepness') of the hierarchy (Barrett & Henzi 2001). As Fig. 12.1 shows, the same power differential may represent a difference of only one rank position in troops where the dominance gradient is steep (upper line) but can encompass the whole dominance hierarchy in troops where the gradient is shallow. Thus, the power differential between the highest- and lowest-ranked females in the latter group is equivalent to the differential between two adjacently-ranked females in the former.

Gradients are expected to be shallow when competition is low and resources are non-monpolizable. In such cases, reciprocal traders should dominate the market place since females cannot exert sufficient power over others to induce interchange trading. As competition increases and resources become monopolizable, gradients are expected to become steeper. Rank distance will therefore exert a stronger influence over females' ability to obtain access to resources, and they will be in a position to trade grooming for access to commodities; interchange traders should therefore come to make up a significant proportion of the market. Importantly, this implies that reciprocal grooming should make up a significant proportion of grooming under all circumstances in all groups, regardless of dominance gradients. This is because all females will always have the ability to exchange this commodity with each other; a notion similar to Chapais's notion of low-competence cooperation (Chapais, this volume). Interchange traders, on the other hand, should only be seen under competitive circumstances (cf. Chapais's competence-dependent cooperation).

12.4
Why markets?

So far, so good. But does the notion of a market place really add value over a standard optimality analysis of behavior? Does a view of grooming as valuable in its own right necessarily require buying into a whole new theoretical framework? Are terms like 'commodities' or 'trader classes' really essential to understanding how females use grooming for their own particular ends? Not surprisingly, perhaps, we would argue that the answer to these questions is 'yes' and that a BM framework is more than just a new bottle for some rather old wine. The value of BM from our perspective is three-fold. First, unlike other models of cooperation, BM explicitly focuses on partner choice as a factor influencing the kinds and levels of cooperation that one sees; it thus recognizes the inherent dynamism represented by primate social groups. This means that, as well as giving some insight into the ultimate function of cooperative behavior, a BM framework also places great emphasis on the proximate mechanisms by which these cooperative outcomes are negotiated.

This focus on the process by which individuals choose partners in relation to the state of the market means that 'noisy' relationships can be transformed into highly informative ones: the variance around a mean level of interaction between two individuals does not have to be viewed as potential error, but can be investigated as a contingent response to fluctuations in the supply and demand of the commodities on offer. In addition to the relative balance between reciprocal and interchange traders in the market place, potential partners can also vary in value depending on their health, reproductive state, seasonal changes in the competitive regime, and on the presence or absence (through migration or death) of other individuals. A BM approach can deal with this kind of dynamic change within groups in a way that simply cannot be matched by analyses based on a static assessment of the costs and benefits of interacting with others.

Second, a market-based analysis does not treat primate groups as monoliths in which all females are assumed to show the same response to a given competitive regime (Barton et al 1996); rather, it takes a more individual-based approach in which traders are predicted to behave differently depending on what they are trading and with whom (see also Silk et al. 2004, who make a similar argument for individual benefits in the context of coalition formation). This is in contrast to more standard socioecological models that characterize groups as 'despotic' or 'egalitarian' and implicitly assume that all females will follow the same set of behavioral 'rules' (see e.g. Sterck et al. 1997). In a BM formulation, a female can be both egalitarian and despotic in her interactions at the same time; she may trade in a reciprocal (egalitarian) manner with one female, but interchange (despotically) with another. Again, this emphasis on individual dynamics over time is a more realistic approach to understanding primate social interactions. It exploits potentially informative variability within and between females, rather than attempting to smooth out all the bumps and wrinkles in order to force them into a specific category of social interaction.

Third, and on a more practical level, perhaps, BM offers a way out of the kin selection-reciprocal altruism impasse. While BM was originally envisaged as a way of explaining how unrelated individuals (even those of different species) could achieve cooperation, there is actually no reason why kin should not trade commodities with each other if trade is necessary for each of them to achieve their goal. As Silk (this volume) and Chapais (this volume) both point out, one should not mistake kin-biased behavior for kin selection, nor should we expect individuals to always favor kin for cooperative tasks. A BM approach does not force an immediate distinction between kin selection and reciprocal altruism as explanations for cooperation, but allows one to remain agnostic on this thorny issue, while including factors like relatedness as variables likely to influence partner choice and commodity exchange rates.

In addition to these benefits concerning the analysis of cooperation within groups, we also feel that the BM approach has the potential to add to other areas of research beyond understanding cooperation. As detailed below, we believe that a view of primate groups as market places, with trade as central to group social dynamics, can help to shed light on other areas of evolutionary significance, such as the cognitive differences that have evolved within the primate order.

12.5 When is a primate group also a market?

There is one more point about a BM framework that needs emphasizing before we go on; any attempt to investigate whether market forces structure the grooming dynamics of primate groups requires a focus on the dynamic part of the equation, and not just the grooming. The BM approach is concerned with the manner in which individual behavior reflects changes in the market place and the supply and demand of valuable commodities. In other words, it is concerned with responses to variation in local circumstance. If circumstances do not vary, then it becomes impossible to test whether supply and demand for commodities structure the market, since by definition both of these variables will remain constant. Valid tests of a BM framework therefore require fluctuation in the market; if this is not the case, then one may rightly conclude that market forces do not explain behavior, but for the wrong reason; if markets do not vary, then market forces will not be apparent, but this does not mean that they do not exist at all. As is recommended in all cases where a particular theoretical framework is applied, *a priori* reasons for why dynamic market effects are expected should be generated, rather than merely assuming that they are present.

The other reason for emphasizing this dynamic element is that BM has been seen by some as an alternative to Seyfarth's (1977) model of primate grooming when, as Noë & Hammerstein (1995) originally pointed out, the latter is actually a form of market model. As in a standard BM formulation, the key components of Seyfarth's model are partner choice and competition for partners that differ in value. However, the crucial difference between the BM approach and Seyfarth's

(1977) model is that the latter is entirely static; it assumes that the value of high-ranking females remains constant over time. Seyfarth's model also deviates more significantly from a BM approach in that the competition between females does not take the form of 'outbidding' competition, whereby females who supply a better quality product (or ask a lower price) do better within the market. Instead, competition applies on a 'first come, first served' basis where females are able to prevent others from grooming merely by virtue of their rank; high-ranking females get to choose first and they remove females from the 'grooming pool'. In doing so, they prevent other females from entering into a 'bidding war' because the excluded females never get an opportunity to engage with such partners and make them a better offer. This again results in a static, as opposed to a dynamic, market place, where partner choice precludes outbidding rather than promoting it. In this sense then, Seyfarth (1977) does not present us with a true market-based model.

Finally, not all instances of grooming need represent 'market trading': in some instances, individuals will groom for purposes related to tension-reduction, bonding with offspring and the like, in a manner that is not dictated by market forces. Again, this means that it is important to give *a priori* reasons as to why market effects should be in operation, and to test this assumption, rather than merely proceed under it.

12.6 Testing the framework: market forces and grooming reciprocity

So, how well does market theory do when put to the test? As a first step in exploring the applicability of a market-based approach, we tested whether grooming reciprocity between females was influenced by dominance gradients and power differentials, using data from two contrasting populations of South African chacma baboons. Data from two troops living in the Drakensberg Mountains of Kwa-Zulu Natal were compared with two troops at De Hoop, an area of coastal fynbos (Mediterranean scrub vegetation) in the Western Cape (see Barrett & Henzi 2002 for an overview of this site), matched for female cohort size.

Differences in the level of food competition experienced by females in the two populations were substantial. In the Drakensberg Mountains, the sparse and relatively even distribution of food (Henzi et al. 1992) meant that agonistic events between females occurred at a rate of only one in every 500 hours of observation, whereas at De Hoop, individual females were engaged in aggression at least once per hour on average (Barrett et al. 1999, Barrett et al. 2002). Consequently, females at De Hoop could be ranked in a strong linear dominance hierarchy, whereas this was not possible for the two mountain troops. We inferred from this that interchange trading would be possible at De Hoop since power differentials were likely to be high, whereas this was unlikely to be the case in the Drakensberg. In the latter population, we predicted that females would be limited to reciprocal exchange, able only to trade grooming for its own intrinsic value.

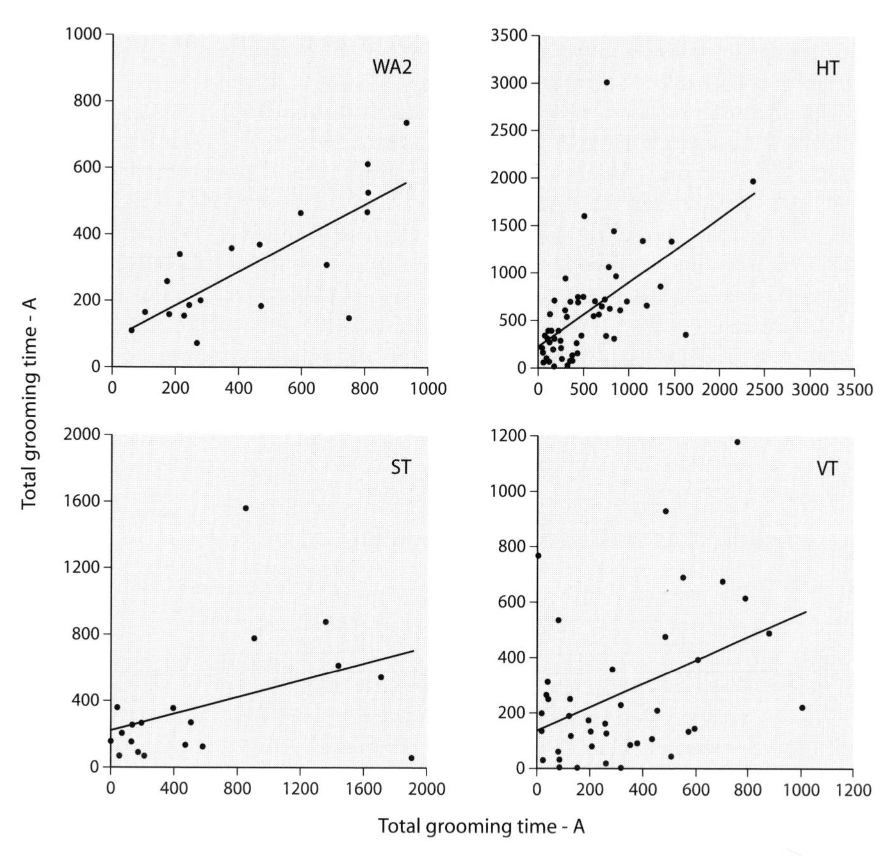

Fig. 12.2. Time-matching across four chacma baboon troops from two different South African populations. WA2 and HT are from the Drakensberg Mountains, Kwa-Zulu Natal, and ST and VT are from De Hoop, Western Cape. Troops were matched for female cohort size (WA2 and ST: n = 7; HT and VT: n = 12). In each case, there is a significant relationship between the amount of time spent grooming by the initiator of bout (groomer 1) and its reciprocating partner (groomer 2). However, time-matching is more precise for the Drakensberg populations in terms of both explained variance and a slope that more closely approximates a 1:1 fit (WA2: $r^2 = 0.588$, $b = 0.50$; HT: $r^2 = 0.588$, $b = 0.67$; ST: $r^2 = 0.163$, $b = 0.25$; VT: $r^2 = 0.168$, $b = 0.42$). See Barrett et al. (1999). Reprinted from Proceedings of the Royal Society, Series B, London.

In line with our prediction that reciprocal traders should make up a significant share of the market, regardless of the potential for interchange trading, we found that females in both populations showed significant levels of 'time-matching'; that is, there was a significant positive correlation between the grooming contributions of individuals to a grooming bout (Barrett et a. 1999) (Fig. 12.2). The fact that females showed a significant tendency to match their partners' grooming contribution (they 'give as good as they get'; Barrett et al. 2000) suggested that being a good value partner required 'fair trade'. Experimental studies of capuchin monkeys support this notion that individuals are capable of generating expectancies about what they can expect to receive based on what their partner gets (Brosnan & de Waal 2003).

It is also interesting to note that individuals did not supply their grooming partners with a single lengthy bout of grooming, which was then reciprocated in kind. Rather, individuals divided their grooming into a number of short bursts that were traded back and forth over the course of the bout (see Barrett et al. 2000). This 'parceling' of grooming fits with Connor's (1995) theoretical demonstration that such behavior increases the costs of finding an alternative partner by ensuring that one always remains a valuable partner, thus removing any temptation to defect. The need to be a good value partner in a market place where many other individuals can supply the commodity in question, plus the parceling of bouts into short grooming bursts, appears to keep females honest and well out of the clutches of the Prisoner's Dilemma.

12.7 Partner control as well as partner choice?

In the Drakensberg, females time-matched more precisely than those at De Hoop; the relationship between individual partners' contributions was much stronger for the Drakensberg troops in terms of both the amount of explained variance and a slope coefficient that was closer to a one-to-one fit (see Fig. 12.2). Originally, we suggested that this reflected the limited potential for interchange trading in the Drakensberg (Barrett et al. 1999), with the result that the market place contained only reciprocal traders. The poorer fit at De Hoop was attributed to the fact that dominance effects, and hence the potential for interchange, introduced more noise into the relationship found, thus resulting in poorer time-matching.

However, it is also possible that high power differentials at De Hoop provide dominant animals with more 'leverage' (*sensu* Lewis 2002) to extract a higher amount of grooming from subordinates during reciprocal bouts (so that, for example, one unit of grooming from a dominant requires two in return from a subordinate). This in turn could be due to a market effect created by coercion, punishment or other forms of partner control (Barrett & Henzi 2001, Bshary & Nöe 2003). Support for this interpretation is provided by a significant relationship between rank distance and time-matching of bouts in the De Hoop population; subordinate individuals tended to groom for much longer than dominant individuals within bouts. Overall, for each unit increase in rank, there was a 28 second discrepancy in the amount of grooming provided by the subordinate animals compared to the dominant. This suggests that dominant individuals were indeed able to use their increased power to extract a higher price in grooming from their subordinate counterparts.

The introduction of partner control into the mix represents a departure from the original BM formulation, which dealt only with trade and outbidding in the absence of physical force or coercion. However, as Noë (2001) points out, multiple sources of power are needed to understand cooperative interactions completely. Thus, while it is true that the original BM formulation dealt only with inalienable resources, this does not imply that market forces cannot coexist with the use of coercion or other forms of leverage (Ronald Noë, pers. com.). In line with Bshary & Nöe's (2003) views on cleaner fish, partner control seems to be crucial to an understanding of the baboon market.

Moreover, Bowles & Hammerstein (2003) note that human economic models which take account of power have long existed, and that market theory in economics now "takes as its foundational assumptions the incomplete nature of contracts (biologically speaking, the possibility of cheating, exploitation etc.)" (p. 157). Another factor that may also be relevant is that of asymmetric price transmission, whereby variation in supply and demand is not passed on to consumers and producers equally. This can cause prices to 'stick' at artificially high or low levels depending on whether transmission fails to consumers or producers, respectively (e.g. Azzam 1999, Goodwin & Holt 1999, Bunte & Peerlings 2003). Thus, we should view partner control as itself determined by a market situation in which partner choice options are exercised.

Overall then, these initial results showed how differences in ecology, and hence competitive regime, have the effect of setting up differential market forces that influence the strength of grooming reciprocation seen between partners. This, in turn, leads to an asymmetry in the payoff for grooming bouts between distantly-ranked animals compared to closely-ranked animals (see Barrett et al. 1999 and Barrett & Henzi 2001 for a more detailed discussion).

12.8 Time-matching in other primate species

Time-matching and rank effects have also been investigated in samango monkeys (Payne et al. 2003), capuchins and captive bonnet macaques (Manson et al. 2004). Among samango monkeys, time-matching occurred at approximately the same level as the Drakensberg baboons and overall levels of reciprocated grooming were similar (50% and 40%, respectively). This is a point worth noting because samango monkeys have been characterized as an archetypal 'egalitarian' species (Rowell et al. 1991), with the reciprocal nature of female grooming held up as a key characteristic of egalitarian societies in general. The fact that 'despotic' baboons show a pattern of grooming similar to that of the 'egalitarian' samango illustrates our point that a BM approach cuts across static categorical designations, and emphasizes that individual females and populations will show patterns that reflect their individual circumstances.

Similarly, capuchins and bonnet macaques also time-matched significantly (Manson et al. 2004). This study demonstrated, too, that the length of time that an individual spent grooming was a significant predictor of whether its partner would reciprocate at all. However, when the two species were analyzed separately, time-matching remained significant only for the capuchins and the relationship was much weaker than in either the baboons or samangos. Interestingly, immediately-reciprocated bouts accounted for only 5–7% of the total grooming observed among the two bonnet macaque groups and only 12–27% among the capuchin monkeys. Moreover, among the macaques, grooming was significantly unbalanced over longer time spans (although this needs to be interpreted cautiously since it is difficult *a priori* to determine the timeframe over which data should be analyzed; Barrett et al. 1999). This figure is much lower than for baboons and samangos suggesting that reciprocation is of less importance to capuchins and macaques and that the nature of the market place thus differs. One major source of difference is likely to be the fact that, unlike baboons and samangos, female capuchins (O'Brien 1993, Di Bitteti 1997, Parr et al. 1997) and bonnet macaques (Sinha 1997) are known to direct grooming down the dominance hierarchy from high-ranking to low-ranking animals. It is therefore possible that a higher proportion of capuchin and bonnet macaque grooming represents appeasement of subordinates by more dominant animals; it is a signal of 'benign intent' towards subordinates (Silk 1996), rather than an example of market-based trade for hygienic/hedonic benefits.

Despite these differences in reciprocity, rank effects were nevertheless apparent in both bonnet macaques and capuchins, with distantly-ranked dyads showing greater grooming discrepancy than closely-ranked dyads. However, for each unit difference in rank distance, a 5.8 second discrepancy in grooming was predicted for the capuchins and a 2.25 second discrepancy was predicted for the bonnet macaques; values far lower than the 28 second discrepancy predicted for the baboons. As Manson et al. (2004) suggest, rank may therefore provide a relatively poor measure of a partner's market value in these species. Alternatively, grooming simply may be a more valuable commodity for baboons compared to capuchins and bonnet macaques. As wild terrestrial animals, baboons are more likely to have higher ectoparasite loads than arboreal animals, like capuchins, or captive animals, like the bonnet macaques in Manson et al.'s (2004) study. The value of grooming is likely to be greater among baboons in much the same way that a glass of water is worth more if one is dying of thirst in the Sahara desert than if one is sitting in the middle of a lake.

12.9 Shifting power relations and the balance of trade

In addition to these cross-population and cross-species effects, a market-based approach can also help explain behavioral differences within populations of the same species over time. By monitoring temporal ecological variability, it is possible to test whether females are able to track the value of commodities and adjust their behavior accordingly. As such, it entails a more dynamic and individual-based approach to issues of power and dominance among females. Therefore, the ability to test for such effects requires that ecological conditions vary sufficiently to have an impact on the competitive regime. Fortuitously, this was possible at one of our study sites, De Hoop, where the ecological regime of one of our study troops, VT, changed markedly over a short period of time (see Barrett et al. 2003 for details). This involved the loss of an entire habitat type, a dry lake bed, through natural flooding. The net result of this was a significant reduction in food competition as the troop was forced to range and feed in areas where resources were more uniformly distributed and less monopolizable.

As a consequence of this change in habitat availability, aggressive interactions dropped from over two agonistic interactions per female per hour to less than one interaction per female per hour (Barrett et al. 2003). This was because, during the low competition post-flood period, there was little benefit to be gained from using dominance to exclude females from food resources. As a result, the dominance gradient became shallower and power differentials were significantly reduced (Fig. 12.3a). Conditions at De Hoop therefore became much more like those in the Drakensberg. Related to this, we also found that aggression was targeted much more towards females of adjacent rank during the low competition period, so that there was a negative relationship between aggression rates and rank distance (Fig. 12.3b). No such significant relationship had existed during the period of high competition, indicating that females were equally likely to direct aggression to distantly-ranked, as well as closely-ranked, opponents (Fig. 12.3b). Thus, the changes in the competitive regime produced by the flood resulted in dominant females losing some of their leverage over low-ranking females; acting aggressively no longer imposed such severe costs on subordinate females or achieved high benefits for dominant females.

Given this loss of leverage by high-ranking females, and the more relaxed competitive regime reminiscent of the Drakensberg, patterns of grooming were predicted to show increased levels of time-matching, reflecting both the loss of opportunity for interchange trading for feeding tolerance and the reduced leverage of dominant females to secure themselves a better rate of exchange through the threat of potential force. In line with this prediction, time matching was more precise during the period of low competition than during the period of high competition and much more like that of HT, the Drakensberg group of equivalent size (De Hoop VT: $r^2 = 0.298$, $b = 0.558l$; Drakenberg HT: $r^2 = 0.331$; $b = 0.67$) (Fig. 12.4; Barrett et al. 2003). In addition, the relationship between rank distance and grooming time discrepancy found prior to the flood was no longer present during the subsequent period when competition was low (Barrett et al. 2003). Thus, our notion that, during the high competition period, the exchange rate for reciprocal bouts was determined by the capacity for interchange plus the increased leverage of dominant females was supported by these data.

Our most recent analyses (Henzi et al. 2003) have tackled long-term patterns of grooming in relation to ecological variability and show the same patterns as these within-bout analyses. During the period of high resource competition, we found that female grooming clique size (the number of other individuals that a given female grooms) and partner diversity were higher than during the post-flood period of low competition. This is because the steeper power gradient meant that more females were in a position to exchange tolerance for grooming when competition was high. In the absence of strong competition during the post-flood period, females needed to exchange grooming only for itself, which they were able to do with a smaller set of closely-ranked partners (Henzi et al. 2003).

These results are particularly interesting because, when making our predictions regarding changes in social dynamics, we also took Seyfarth's model and determined the predictions this would make if it contained a dynamic element. According to this model, when resource competition is high, competition among

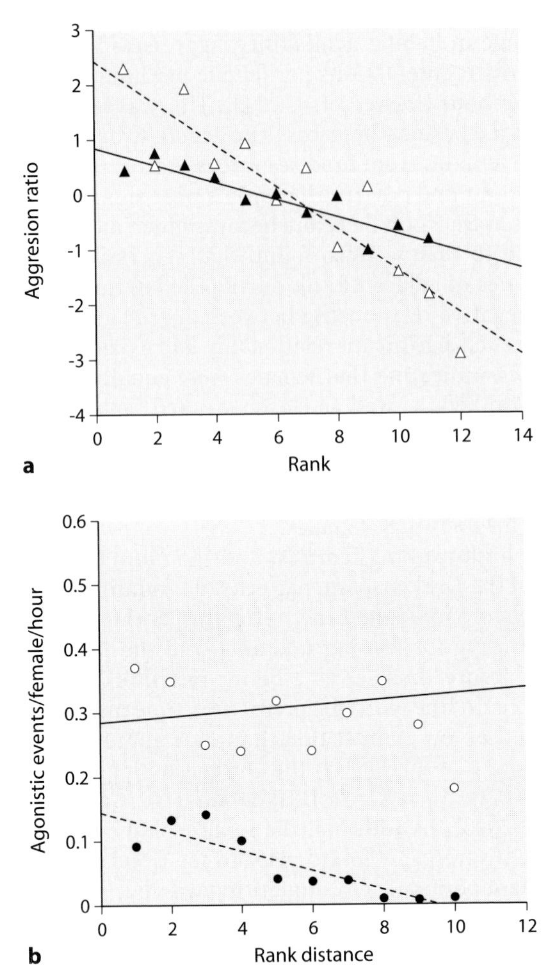

Fig. 12.3. (**a**) Relationship between aggression ratio (aggression given by a female – aggression received by a female) and rank for De Hoop females. The aggression ratio represents the dominance gradient of the group (Barrett et al. 2003). During the post-flood low competition period (closed triangles, solid line), the relationship between aggression ratio and rank has a significantly shallower slope than during the pre-flood, high competition period (open triangles, dotted line), indicating that the dominance gradient was reduced during the post-flood period (low-competition period: $b = 0.16$; high competition period: $b = 0.39$; $t_{19} = 3.5$, $p < 0.005$). (**b**) Relationship between overall rates of aggression and rank distance between females at De Hoop. During the low competition period (closed triangles, solid line), aggression rates decline significantly as rank distance increases ($r^2 = 0.77$, $p = 0.001$), indicating that aggression is mainly directed at females of adjacent rank. During the high competition period (open triangles, dotted line), there is no significant relationship between the two ($r^2 = 0.025$, $p = 0.645$), indicating that aggression is directed to females of all ranks (see Barrett et al. 2003). Reprinted with permission from Elsevier.

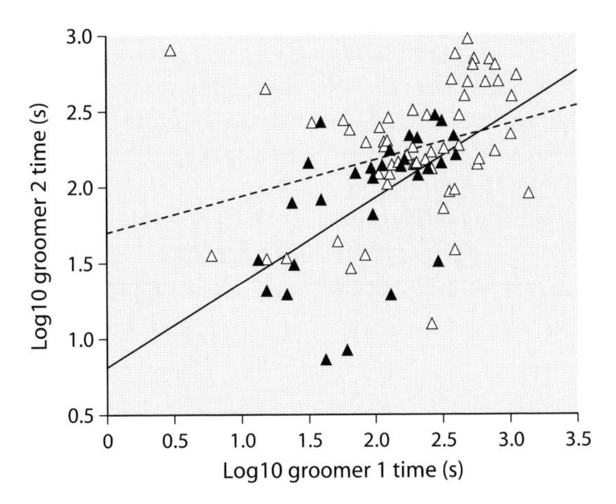

Fig. 12.4. Time-matching between De Hoop females during the periods of high competition and low competition. Time-matching is more precise during the low competition period (closed triangles, solid line), when dominant females' leverage is reduced, than during the high competition period (open triangles, dotted line) as indicated by greater explained variance and a slope that more closely approximates a 1:1 fit (low competition period: $r^2 = 0.298$, $b = 0.558$; high competition period: $r^2 = 0.099$, $b = 0.237$) (see Barrett et al. 2003). Reprinted with permission from Elsevier.

low-ranking females to gain access to high-ranking females should result in a grooming distribution where females spend most of their time grooming those of adjacent rank. Consequently, under conditions when competition is reduced and the pressure to seek out high-ranking females is relaxed, a wider grooming distribution is predicted. As should be apparent, our results are directly opposed to this prediction, demonstrating that even when a dynamic element is brought into Seyfarth's (1977) model, it still does not function as a true market-based model. This probably stems from the model's assumptions about why females seek particular partners. The fact that baboon females increase the rank diversity of their partners at times of high competition suggests that, as mentioned above, partner choice and dynamic outbidding competition structure the market, and not exclusion by dominants; when there is a greater need to interchange grooming for tolerance, high-ranking females are more often sought out as partners by all females and they are able to gain sufficient access to achieve these goals. Alternatively, Payne (in prep.) suggests that high-ranking females may use their increased leverage during high competition periods to 'extort' grooming from a wider variety of females, forcing lower-ranked females to give them more grooming. According to this argument, females do not groom to gain tolerance, but to avoid increased intolerance from dominant females. Either way, these findings support Chapais's (this volume) ideas regarding partner choice in relation to kinship. Baboon females apparently choose their partners in relation to their competence at providing a particular service, rather than directing all behavior preferentially to kin because of presumed inclusive fitness benefits.

Lazaro-Perea et al. (2004), in a study of wild marmosets, also found evidence for competence-dependent trade. In this study, the breeding female in a marmoset group tended to groom non-breeding females in an asymmetrical manner, giving much more than she received. This was interpreted as 'payment' for the services that non-breeding subordinate females had to offer; subordinate females are known to carry and share food with infants, are more active in territorial defense, and participate in alarm calling and mobbing behavior (Lazaro-Perea et al. 2004). These findings are particularly gratifying because they come from a species of non-female bonded New World monkey, showing that a market-based approach applies more broadly than just female-bonded societies in general, and Old World monkeys in particular.

12.10
The baby market: supply, demand and leverage

Although the above findings are consistent with a BM interpretation, and imply that interchange trading occurs, they do not actually show that this is the case. In order to provide full support for the BM framework, we need to show that the behavioral interaction of two trader classes is determined by fluctuations in the supply and demand of a commodity that can be exchanged for grooming.

To demonstrate interchange grooming in the De Hoop population, we exploited the fact that new-born infants are a source of great attraction for female baboons. Adult females frequently attempt to interact with both infants and their mothers in the first few months *post partum*, despite the fact that mothers are very reluctant to expose their young infants to the attentions of other group members. This set-up allowed us to measure the impact of grooming on an individual's ability to interact with new infants. If grooming increased tolerance around infants, then females could potentially 'buy' access to these commodities by grooming the mother (Henzi & Barrett 2002; see also Muroyama 1994 who initially made this suggestion with reference to allomothering in patas monkeys). More specifically, the length of the grooming bout associated with infant handling should vary according to the supply of infants so that the 'price' (in terms of grooming bout length) should be higher when fewer infants were available. In order to test for this, we partitioned our data set into cases where the mother was lower ranking than the female handling the infant ('handlers' hereafter) and cases where the mother was higher ranking. This was both to control for the effects of dominance on interchange indicated in our previous work and to test whether dominance-related differences in leverage influenced exchange rates between mothers and handlers.

As predicted, grooming bout lengths were significantly influenced by the number of infants present in the group for cases where the handler outranked the mother and there was a strong trend in cases where the mother ranked above the handler (Henzi & Barrett 2002). Specifically, an increase in the supply of infants led to a reduction in the grooming bout length needed to gain tolerance, representing a classic market effect within the group. The influence of partner control within the market place was also apparent in these analyses, with higher-ranking mothers apparently able to gain more grooming than lower-ranking mothers for a given supply of infants. Plotting the relationship between the

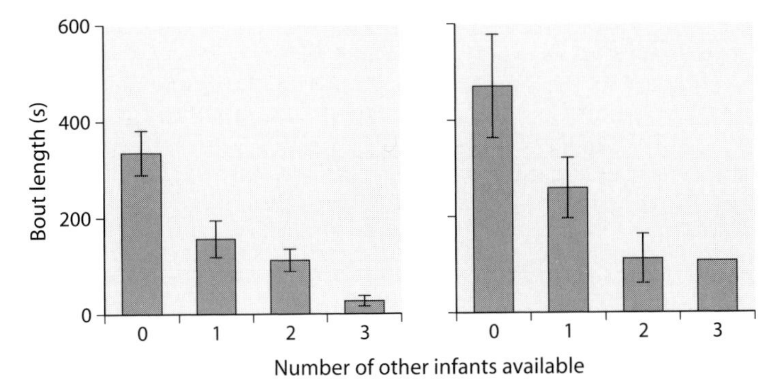

Fig. 12.5. The baby market. As the number of other infants present, in addition to the focal infant in the group, increases, so the amount of grooming given to its mother to obtain tolerance decreases. The value of the commodity (infants) is thus dictated by the supply of infants relative to the demand for handling.

rank distance of handlers and mothers against grooming time revealed a significant negative correlation; higher-ranking mothers could demand a higher price for access to their infants (Henzi & Barrett 2002). This was interpreted as a form of asymmetric price transmission; for dominant mothers, an increase in the supply of infants was not transmitted to handlers in form of reduced price. Instead, their price seemed able to stick at a higher level compared to the situation when handlers outranked mothers. However, re-analysis of these data with an enlarged data set reveals that, while a much stronger market effect is present across all females (Fig. 12.5a,b: two-way ANOVA, number of other infants available: $F_{3,32} = 3.276$, $p = 0.034$), there is no significant main effect of maternal rank ($F_{1,32} = 0.929$, $p = 0.342$) nor any interaction between maternal rank and infant number ($F_{3,32} = 0.881$, $p = 0.461$), and the correlation between rank distance and grooming bout length is no longer significant ($r_s = -0.279$, $n = 40$, $p = 0.08$, two-tailed; Fig. 12.6). However, a trend is still apparent in the data, at least for instances where there are one or fewer other infants available (Fig. 12.5a,b), and it is possible that partner control and asymmetric price transmission can only be exercised by the very highest-ranking females. The inclusion of more middle-ranking females into the dataset suggests that, overall, market forces prevail; the supply of infants is the main factor that determines the exchange value of grooming in the baby market.

Lazaro-Perera et al. (2004) also looked for interchange trading in their study of marmosets. Contrary to their predictions, breeding females did not groom other females more in times of greater need; for example, when there were more dependent infants in the group or following inter-group encounters. Payne et al. (2003) obtained similar results from samango monkeys, suggesting either that services are not, in fact, interchanged or that the exchange of services is not immediate, a point we return to below.

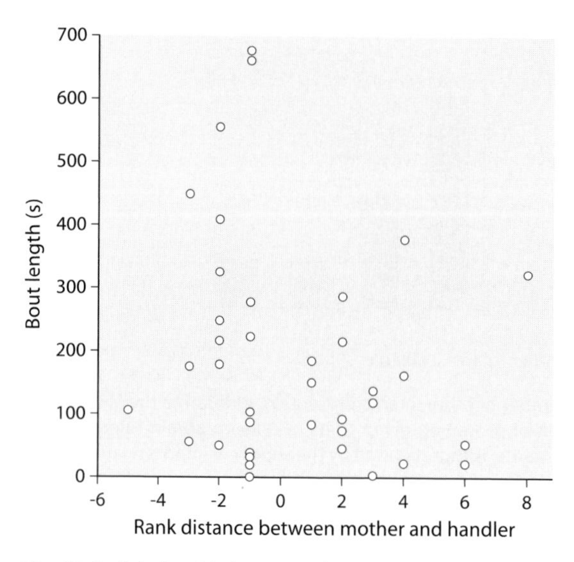

Fig. 12.6. Relationship between the rank distance of mothers relative to handlers and the grooming bout length given to mothers. There is a non-significant trend for higher-ranking mothers to receive relatively more grooming than lower-ranking mothers in exchange for infant-handling.

12.11 A market for brain power

The value of a market-based approach to understanding patterns of primate sociality seems clear and, on the strength of these results, it seems appropriate to extend work on primate markets to other species and to other arenas where commodity exchange is to be expected, such as access to resources, mating opportunities and coalitionary support. The manner in which market-based trading influences other aspects of primate social behavior, such as reconciliation and other forms of conflict-management, is also worth considering (see Aureli & Schaffner, this volume). In addition, we feel that a BM approach may also pay great dividends in studies of primate cognition and tests of the 'Machiavellian intelligence' (Byrne & Whiten 1988) or 'social brain' hypotheses (Dunbar 1998). In particular, we suggest that differences in market structure may help explain differences in monkey and ape cognitive capacities, which seem to exist, yet remain poorly characterized (Barrett et al. 2003).

In primate market places, individuals track the price of commodities and respond flexibly to changes in supply and demand as we have shown. This requires cognitive and behavioral flexibility; an ability to learn rapidly and to update one's view of the world swiftly in the light of new information. A market-based approach to primate cognition therefore agrees with the 'Machiavellian intelligence' hypothesis (Byrne & Whiten 1988) that sociality has driven brain evolution. It differs, however, by discarding the assumption that animals have been selected to cope with increasingly elaborate strategies and counter-strategies, the goals of which are to 'outwit' the competition. We argue that brain size and structure have, instead, been driven by a need to track short-term fluctuations in commodity value.

Monitoring the market place is intrinsically complex; the value of a particular partner is contingent on the value of others. Each of these values may shift with changes in reproductive state, health, dominance and ongoing social behavior. Those who are good value today may not be so tomorrow. This constant state of flux means that keeping tabs on the social market is very different from the other kinds of contingent monitoring that primates must do, such as tracking fluctuations in fruit availability (Milton 1988). Fruits, unlike conspecifics, do not make decisions in response to primate behavior (except in an evolutionary sense). This inherent contingency in primate market places thus requires the ability to track the contingencies between one's own behavior in relation to others. More importantly perhaps, it also requires the ability to track the behavior of third parties in relation to each other and the behavioral consequences that this may have for one's own behavior. This has clearly selected for what Call (2001) refers to as a 'knowledge-based' understanding of others, as evidence from monkeys and, to an even greater degree, apes has shown. Nevertheless, there remains a cognitive difference between monkeys and apes that, although not precisely identified, is apparent when comparing their performance on psychological tests (Tomasello & Call 1997, Hare et al. 2001, 2003).

Our suggestion is that these differences arise as a consequence of both increased spatial and temporal dispersion in ape market places compared to those of monkeys. While monkeys are all highly gregarious and live in cohesive groups in which individuals encounter every member of their group every day, the apes (in particular, the chimpanzee and orangutan) live in more fragmented societies due to the impact of food competition, which forces females to forage in small parties or on their own. The apparent exceptions to this distinction, group-living gorillas and fission-fusion spider monkeys, are less problematic than they appear. Among gorillas, group living may be a relatively recent adaptation in response to infanticide by adult males (Harcourt & Greenberg 2001), and fission-fusion is likely to have been the ancestral ape state. The nature of spider monkey fission-fusion is not well studied and may differ from apes in important ways. If, however, their behavior is truly ape-like, then we have an ideal test case; we would predict that they manifest ape-like cognitive abilities.

In fission-fusion societies, individuals see each other only at infrequent intervals, often weeks apart, yet each recognizes and remembers the members of its community and is capable of maintaining long-standing relationships. In such systems, there will be greater pressures on individuals to mentally represent those animals that are not currently present and to retain and manipulate information about them for substantially longer periods of time than is common in spatially- and temporally-stable monkey groups, where animals are only out of view for hours at the most. This is not to say that monkeys are incapable of representation; their high performance on delayed response tasks shows that they are able to represent objects in their absence (see Tomasello & Call 1997 for a review). Rather, the issue at hand is the length of time over which this information must be retained and manipulated. Thus, while the studies we have reviewed here reveal that baboons are highly competent market-traders, they also highlight the fact that most of the social decisions made by these animals occur over a very short timeframe; females respond to current need (access to an infant, high resource competition) when making their grooming decisions, with little indication that they plan strategically for the future by grooming in anticipation of future need (see also Barrett & Henzi 2002).

In a fission-fusion society, however, monitoring the state of the social market place requires the ability to track and update any changes observed in the interactions of others after coming into contact after a period of absence, and animals must use individuals' absence, as well as presence, to predict reliably the occurrence of certain behaviors in others. Thus, while all anthropoid primates are capable of tracking 'third-party relations' (relationships between two other animals without reference to self) (Tomasello & Call 1997), the ability to generate a causal understanding of such behavioral interactions in the absence of certain individuals would seem to be much more demanding cognitively. Thus, the key to social survival in dispersed systems is the ability to work with a social world that is partially virtual, rather than purely physically instantiated. The fact that chimpanzees are apparently able to represent the relative spatial locations of crude stone tools (hammer stones and anvils) and to use this information in a flexible manner (Boesch & Boesch 1984) supports the notion for a similar capacity in the social domain.

Recent work by Boroditsky (2000) arguing that, in humans, the sense of time emerges via a metaphorical analogy from a sense of space provides us with a means of extending our argument beyond the spatial domain. A sense of space could, with sufficient additional cognitive control, be used to develop an extended sense of time. This would then enable animals to predict future states of the market place, as well as track current changes, in a very effective way and to be able to project key aspects of social interaction and relationships onto an uncertain future. An animal with this predictive capacity would have a clear advantage over one that could only track current states and respond after the fact. Evolutionarily, once animals had a well-developed ability to understand a virtual spatial world of trading partners, this could have scaffolded the development of an understanding of temporally-dispersed trading partners as well, enabling animals to sequence social events into causal chains. This ability would enable animals to start predicting the likely consequences of behavior beyond the immediate present, enabling them to plan ahead effectively and to inhibit responses that could have negative repercussions (Barrett et al. 2003; see also Tulving 1983, Suddendorf & Corballis 1997).

One important point to note here is that we are not arguing that a dispersed social system *per se* selected for these higher cognitive abilities. After all, there are many species of lemurs and other prosimians that have dispersed social systems (see e.g. Eberle & Kappeler 2002), but that apparently have brains somewhat smaller than their testes (Peter Kappeler, pers. com.). Rather, it is the specific historical contingency of evolutionary events in the anthropoid line that produced this state of affairs. The shift to a diurnal lifestyle and group living that arose with the evolution of the anthropoids created the selection pressures for social market places like those described above. The skills needed to trade grooming for other goods, to respond to fluctuations in the value of goods, and to play individuals off against each other were honed in the context of these stable diurnal social groups, which is why both monkeys and apes have relatively large brains, relative to prosimians and other mammals.

The evolution of the great apes as ripe fruit specialists then gave rise to dispersed social systems in which the group-based social skills of these animals were placed under the new selection pressures we outline above. A further point to emphasize is that only the social changes within the market place created the pressure to expand brain size; the ephemeral and dispersed nature of fruit supplies served to create a more fluid social system, but did not have any impact on brain size *per se* according to our hypothesis (cf. Potts 2004). Thus, ours is not a general explanation of the consequences of dispersion on brain size and intelligence, since we assume that most of the skills needed for dealing with a market-based system were already in place by the time such systems arose. Instead, it is a historically-based hypothesis dealing with the particular evolutionary pathway taken by the anthropoid apes (see also Potts 2004).

12.12 (Neuro)biological markets

In humans, the ability to plan ahead, to contemplate the future and reflect on the past, are all faculties associated with the pre-frontal cortex (PFC) (see Fuster 1989). Miller & Cohen (2001) have suggested that the actions of the PFC enable a high level of 'cognitive control' of exactly the kind that we suggest is required in a dispersed market place; namely, the ability to take charge of one's actions and direct them towards future, unseen goals. Put simply, they suggest that the role of the PFC is to guide activity flow along the neural pathways needed to solve the task, ensuring that these pathways are activated even when there is strong competition from more frequently used, but inappropriate, pathways (Miller & Cohen 2001). The impressive expansion of the PFC across the primate order suggests that monkeys, apes and humans will differ in their ability to achieve cognitive control. Both the frontal lobes (Semendeferi et al. 1997) and the PFC (Fuster 1989, Passingham 1993) of monkeys are significantly smaller than those of humans and apes (in the latter case, 11% of total cortical volume, compared to 17% and 36% for chimpanzees and humans, respectively; Fuster 1989). Neurobiological evidence thus backs up our argument that monkeys should be more limited than apes and that, by the same token, apes should be more limited than humans in their ability to plan ahead effectively over more than a few hours, or to inhibit behavior in order to achieve long-term goals.

The PFC is not the only element crucial for producing cognitive control, however. The allocation of such control is thought to be dependent on the anterior cingulate cortex (ACC), an area variously associated with error detection, response selection and, most relevant here, conflict monitoring (Carter et al. 1998, Botvinick et al. 1999, Bush et al. 2000). By detecting conflict, the ACC is able to signal to the PFC that additional control needs to be allocated to a task. It has also been suggested that the upgrading of the ACC would have been critical for enabling animals to generate a 'virtual group' of spatially-dispersed individuals since it is linked to generating a sense of self in relation to others ("the troop in the head": Skoyles & Sagan 2002). In this respect, it is intriguing that spindle cells, a class of large projection neurons found principally in the ACC region, are found only in apes and humans and not in monkeys (Nimchinsky et al. 1999). Allman et al. (2001) have speculated that these cells are involved in coordinating widely distributed neural activity involving emotion and cognition, fitting well with our speculations on the need for greater cognitive control within a dispersed hominoid market place.

12.13 Implications for primate cognition and cooperation

Having introduced the notion of improved cognitive control as the key to coping with a dispersed market place, we can predict how cognitive abilities should differ between monkeys and apes. In essence, apes should possess an analogical reasoning ability that monkeys lack, show greater abilities to solve problems that require the completion of sub-tasks while keeping an overall goal in mind, better inhibition of pre-potent responses, increased planning abilities and finally, an ability to construct and sequence longer causal chains of events. Apes are known to show higher performance than monkeys in both causal (Limongelli et al. 1994, Visalberghi & Limongelli 1995) and analogical reasoning tasks (Thompson & Oden 2001), but there have been, as yet, few attempts to test for differences in the latter two abilities. Anecdotal evidence suggests that monkeys show extreme 'myopia for the future' (Roberts 2002), while recent work reveals that apes show extended memory for accumulated quantity (Beran & Beran 2004). Most importantly, we can also predict that, compared to apes, monkeys will show no evidence of generating 'contingency plans' for future events.

This has implications for the nature of monkey trading within a market, bringing us back to our initial arguments concerning the value of grooming to female primates. If our hypothesis is correct, and monkeys are unable to plan for the future, then grooming should only be exchanged for something immediately obtainable (like access to infants) or something that does not require any monitoring of checks and balances over time. This cognitive perspective therefore provides a further reason why coalitionary support is unlikely to be traded for grooming, at least among monkeys, because the need for support is unpredictable and highly variable across time. Coalitionary support may thus be needed immediately leaving no time for support to be 'bought' from others. However, the 'myopia' of monkeys means that they will be unable to plan ahead and groom potential partners before they are needed. In any case, this would be a wasted effort due to the myopia of the partners themselves who may fail to retain the relevant information regarding the price paid. Coalitions are thus most likely to occur when there are immediate and direct benefits for the females taking part, as seems to be the case at Amboseli (Silk et al. 2004a), rather than as a result of trading favors over time in a reciprocal manner (see also Stevens & Hauser 2004 who argue for similar cognitive limitations on reciprocal altruism).

Trading for something like feeding tolerance is different, both because it may be immediately obtainable and because some form of 'attitudinal reciprocity' can work as a mechanism (De Waal 2000c); regular grooming may change the general attitude of the groomee towards the groomer, putting them in a more relaxed state about the groomee, so that grooming could have a long-term effect with little loss of value over time and without requiring extensive 'book-keeping'. The same may be true for helping behavior (Lazaro-Perea et al. 2004). However, it seems unlikely that reducing tension in a partner would have the effect of increasing the willingness of such a partner to take aggressive risks on another's behalf and engage in coalitionary support (we thank Ronald Noë for pointing this out).

A focus on tolerance, how it is traded and the timeframe over which it operates, is the logical next step in our analyses, since it has important implications for our assumptions about what can and cannot be traded. If the cognitive timeframe over which baboons operate is fairly short, animals will be unable to groom too far in advance to achieve their goals. If so, then grooming "is not a hard currency but chocolate money that melts away" (R. Noë, pers. com.). Determining whether baboons are dealing in hard cash or perishable goods, and how this affects exchange rates over time, is an important goal for the future.

12.14 Summary and conclusions

Data from baboons, and an increasing number of other primate species, support the notion that primate groups represent 'biological markets', within which individuals 'trade commodities' with each other (e.g. grooming, tolerance, helping behavior) according to the laws of supply and demand. Grooming reciprocity among female chacma baboons is driven by market forces generated by the ecological and competitive circumstances under which they live, so that levels of cooperation vary across both space and time. Females also interchange grooming for tolerance around infants, with the 'price' of grooming set by the local supply of infants as economic theory predicts. Thus, the dynamic, individual-based approach of BM theory, with its emphasis on partner choice, is a much more appropriate framework within which to analyze primate cooperation than alternative models, like those based on the iterated Prisoner's Dilemma (Axelrod 1986).

Market-based theories can also shed light on other aspects of primate sociality, including the evolution of primate cognition. Monitoring a social market place that is in a constant state of flux requires high levels of cognitive and behavioral flexibility, but does not require that primates have to be especially 'Machiavellian' in their attitude to others. Differences between ape and monkey market places in terms of the spatial and temporal dispersion of individuals, and the timeframe of social decision-making provide us with a plausible and testable hypothesis concerning the evolution of primate social intelligence.

This, in turn, has implications for human evolutionary psychology and, specifically, the notion of 'massive modularity', the idea that selection has produced a mind comprised of computational algorithms designed to solve specific recurring problems. If, as we suggest, primate groups, including those of humans (La Cerra & Bingham 1998), constitute market places, then a massively modular psychology seems unlikely; the contingency inherent in a market means that what constitutes a fair trade today may actually be a dodgy deal tomorrow. A computational cheat-detection module, for example, triggered by certain conditions such as 'taking the benefit without paying the cost' (Cosmides & Tooby 1992) will be doomed to giving the wrong answer most of the time, because the truth of such a statement is entirely contingent on the state of the market. As La Cerra & Bingham (1998) point out, a more flexible form of decision-making is needed under such circumstances; one that can cope with these ever-changing contingencies and one for which the human PFC is well designed.

In line with this, it is clear from the work of Gächter & Herrmann (this volume) and Millinski (this volume) that human decisions regarding cooperation and cheating are contingent on the context in which individuals find themselves. While these may be emotionally-mediated actions, as opposed to perfectly rational ones, they are not automatic, involuntary or mandatory as a modular response would require. Nor do these decision 'mistakes' reflect the operation of ancient decision-making mechanisms selected for in small kin-based groups; if female baboons, who live in small kin-based groups, can differentiate among their kin according to the services they have to offer, as our work demonstrates, then it seems unreasonable to expect human decisions to be based on a much more crude rule of thumb. Rather, our decisions are the creative, flexible and contingent responses of a primate well versed in the workings of a biological market, with a flexible mind and brain to match.

Acknowledgments

We thank Peter Kappeler for inviting us to take part in the excellent Göttinger Freilandtage on Cooperation. We are also very grateful to Claudia Fichtel who, most importantly, made sure we actually made it to Göttingen, and was immensely patient regarding our ever-changing travel plans. Our work in the Drakensberg and De Hoop has been supported by the NRF (South Africa), The National Geographic Society, the University of Liverpool Research Development Fund and The Leverhulme Trust. We thank the Natal Parks Board and Cape Nature Conservation for permission to work in their reserves, plus our collaborator, Drew Rendall, and all the other members of the De Hoop Baboon project, past and present, for all their efforts with the baboons. We also thank Filippo Aureli, Sarah Brosnan, Peter Hammerstein, Colleen Schaffner and, in particular, Bernard Chapais for some very interesting discussions of the ideas presented in this paper. Finally, we are extremely grateful to Ronald Noë and Peter Kappeler for their clever, insightful and, in Ronald's case, forthright comments on an earlier draft of this chapter.

Chapter 13

Digging for the roots of trading

RONALD NOË

13.1 Introduction

Cooperative behavior is commonplace in human social interactions. It is easy to recognize equivalent forms of behavior of non-human animals, such as mutual support among kin and cooperative hunting. Other forms, such as trading and large-scale collective action, are perhaps not uniquely human, but are much more widespread among humans than among non-humans. Here, I use the term trading as shorthand for interactions in which individuals exchange goods and services; bartering, vendor-customer and employer-employee interactions and so forth. Trading may not be recognized by everybody as a typical cooperative interaction, but it has the hallmarks of cooperation: (i) two or more individuals exchange goods and services in such a way that the participants involved are usually better off after the interaction, than before it, and (ii) the participants have to invest something in the interaction without a full guarantee of net gain.

In this paper, I want to reflect on the evolutionary roots of human cooperative behavior in all its forms, with the exception of cooperation among close relatives. I start this discussion without knowing whether humans pursue the same strategies and use the same toolbox of mechanisms to implement those strategies when engaged in different forms of cooperation. Are, for example, the same mechanisms involved when two neighbors build a fence, when all inhabitants of a valley build a common irrigation system and when people trade goods at the weekly village market? The same question can also be asked when comparing different species. Do cooperatively hunting humans follow the same strategies towards their companions as cooperatively hunting lions? Does a customer use the same mechanisms to get what he wants from his barber as a reef fish does when he visits a cleaner wrasse? The most likely answer is that some basic mechanisms are common to all forms of cooperation while others are specific to a limited set of cooperative interactions only. The main question I want to ask in this chapter is thus: can the mechanisms we use in cooperation and trading be traced across species borders and down to the roots of our particular phylogeny? A related question is whether cooperation and trading are fundamentally different phenomena or merely represent different ends of a continuum. Tracing mechanisms back down the phylogenetic tree implies that I will only be considering evolved mechanisms, such that other descendants of the same remote ancestor living today might also use them, if confronted with comparable problems. In other words, I am looking for mechanisms that have evolved under natural selection, rather than cultural selection, which is a strong force in the evolution of human behavior (Richerson et al. 2003). A further task is therefore to try to distinguish naturally-selected from culturally-selected mechanisms.

13.2
A comparison between cooperation and trading

Not everybody will accept that trading among humans can and should be compared to, for example, pollination; an interaction in which plants pay insects with nectar for the transportation of their gametes. For me, however, the similarities are striking and, together with several colleagues, I therefore introduced the 'biological market' paradigm (see Box 1). Our main goal was to point out the analogies between cooperation among non-human organisms and trading among humans in order to pave the way for the introduction of theoretical insights derived from economics into the field of behavioral ecology. Before I try to identify mechanisms used in cooperation and trading that are homologous, i.e. can be traced back to a common origin, I will first make clear what I understand by the terms cooperation and trading. I then proceed by making an inventory of analogies between human trading and cooperation among non-human organisms.

13.2.1
What is cooperation?

Intuitively, most of us have an idea what is meant by the term 'cooperation' but when it comes to precise definitions, it is apparent that the term is used to cover a wide range of behaviors (Noë 2005, in press). I use the term 'cooperation' broadly for all activities that as a rule result in net benefit to both the actor and the recipient(s). In the following, I will concentrate on interactions in which one or both parties have to invest under uncertainty, without making any distinction between (intra-specific) cooperation, (inter-specific) mutualism and symbiosis. The only thing that counts theoretically is whether cooperating individuals are sufficiently closely related that their strategies can be explained by kin selection (Hamilton 1964). Nor do I distinguish mutualism (immediate benefits to both participants) from reciprocity or reciprocal altruism (delayed benefits received in an alternating manner). I am less interested in the delay between investments and eventual returns, because I consider this to be only one of several factors that determine the level of control that participants exert over their partners (see Noë 2005, in press for a more detailed discussion). I will examine both one-shot and repeated interactions, although the majority of my examples will be of the latter kind.

13.2.2 Models of cooperation

Models of cooperation can be divided into those focusing on partner control and those focusing on partner choice (Bshary & Noë 2003). Models of partner control take the formation of cooperating partnerships for granted and concentrate on the mechanisms that each participant uses to prevent being cheated by their partner. Bob Trivers (1971) was one of the first to propose the use of the two-player Iterated Prisoner's Dilemma (IPD) as a paradigm for what he called 'reciprocal altruism', although it required some adaptations to deal with asynchronous choices (for the 'alternating' PD see Frean 1994, Nowak & Sigmund 1994, Hauert & Schuster 1998, Neill 2001).

Partner choice models include extensions of the IPD-model (Dugatkin & Wilson 1991, Batali & Kitcher 1995, Ashlock et al. 1996, Roberts 1998) and the biological markets paradigm (Noë & Hammerstein 1994, 1995; see Box 1). In a two-player IPD model, a player sanctions an uncooperative partner by aborting the relationship, thereby losing the advantages of cooperation in the process. The price paid for imposing sanctions on a partner can be considerably reduced, however, if one switches to another partner, even if the latter is less profitable. In my papers on biological markets, I did not make a distinction between choices made on the basis of intrinsic attributes of the partner itself ('attributes-based partner choice') and choice on the basis of characteristics of the commodity offered by the partner ('commodity-based partner choice'). However, this distinction becomes important when one reflects on the mechanisms involved. Take, for example, a cleaner fish that chooses between two clients that present themselves simultaneously (see Bshary 2001 or Bshary & Noë 2003 for a description). The client can choose on the basis of the amount of resources carried by the client, for which he can take body size as a proxy, or he can choose on the basis of characteristics of the client that are independent of its parasite load (predatory or not; resident or floater; aggressive in previous interactions etc.)

▶ *Box 1.*
Biological markets

In a series of papers, my colleagues and I have pointed out the analogies between the cooperation between unrelated individuals, reproductive behavior and human trading (Noë et al. 1991, Noë 1992, 2001, Noë & Hammerstein 1994, 1995, Bshary & Noë 2003). Our main purpose has been to stimulate the development of new models for cooperation based on knowledge accumulated in two well-developed fields: (i) sexual selection theory and (ii) economics. Peter Hammerstein and I coined the phrase 'biological markets', because the common denominators of the three fields are reminiscent of human economic markets: exchange of services and goods, choice of partners, competition by outbidding etc.
The biological market paradigm stresses some aspects of cooperation that were ignored in earlier models (reviewed by Sachs et al. 2004), notably partner choice

and partner switching, competition in the form of outbidding among potential partners, the division of benefits and the exchange rates of commodities. Models based on the 2-player Iterated Prisoner's Dilemma (IPD) and related paradigms, which include reciprocal altruism (Trivers 1971), put the problem of partner control under the magnifying glass, assuming that possible cheating by the partner poses the greatest challenge to a cooperating individual and therefore to the evolution of cooperation itself (see Dugatkin 1997 and Sachs et al. 2004 for reviews). Thus, the biological market paradigm emphasizes the context in which cooperative interactions take place, while IPD-models emphasize the dynamics of repeated interactions between pairs of individuals.
Biological market theory shows its economic character in the prediction that changes in the supply-demand ratio should result in clearly specified directional shifts in the division of benefits communally gained by cooperation or in the exchange values of goods and services. In an analogy to sexual selection theory, biological market theory predicts that partner choice can lead to selection for specific traits. We assumed, therefore, that the same skills that are known to play a role in mate selection would also be important in the selection of cooperation partners: (i) judging the partner's quality, (ii) a memory for the partner's quality and location, (iii) searching strategies, (iv) judging the honesty of signals and so on. 'Market selection' can run counter to sexual selection, for example, when dominant males accept only satellite males that do not show the exuberant ornaments typical for the males of the species (Noë & Hammerstein 1994, Greene et al. 2000). In recent years, a number of empirical studies have shown that the biological markets approach leads to new insights in a variety of studies of intraspecific cooperation (grooming markets in primates: Barrett et al. 1999, Henzi & Barrett 1999, 2002, Barrett & Henzi 2001, this volume, Leinfelder et al. 2001, Payne et al. 2003, Lazaro-Perea 2004, Manson et al. 2004) and inter-specific mutualism (nutrient exchange mutualisms in mycorrhiza: Schwartz & Hoeksema 1998, Hoeksema & Schwartz 2001, Hoekesema & Kummel 2003; cleaner fish-client mutualism Bshary 2001, Bshary & Grutter 2002a, 2002b, Bshary & Noë 2003), and interactions between groups of different species of primates (Eckardt & Zuberbühler 2004). For further examples, see reviews by Bronstein 1998, Hoeksema & Bruna 2000, Noë 2001, Wilkinson & Sherratt 2001, Simms & Taylor 2002, Bshary & Noë 2003, Sachs et al. 2004, Bshary & Bronstein, in press).

13.2.3 What is trading?

The following strike me as typical attributes of trading interactions: (i) There is an exchange of goods and/or services. In advanced forms of trading, one party may use tokens of value (clams, money etc.) in exchange for the goods/services, or different tokens of value (dollars, euros) are exchanged for themselves. (ii) The goods and services traded have an exchange value that fluctuates with supply and demand. In advanced forms of trading, the value of different goods and services can have a 'market value' expressed in a common currency. (iii) Choice among trading partners and their goods or services is the main mechanism that causes exchange rates to follow changes in supply and demand. (iv) Trading can take place between total strangers and in one-off interactions.

13.2.4 Human trading compared to examples of non-human cooperation

Human economic interactions are very rich in form and I will not attempt to cover all relevant aspects here. Instead, I intend to classify different forms of trading in a manner that corresponds loosely to the categories of cooperative and mutualistic interactions observed in nature. The purpose of this list is to give the reader a feel for the similarities between the two. Both human and biological markets can be classified by the degree of lopsidedness in the freedom of choice possessed by different classes of trader. Many causes of asymmetry reflect the idiosyncrasies of specific markets, but two general factors can be identified: (i) differences in mobility and (ii) in size. Broadly speaking, more mobile traders possess a wider array of options, unless they are so much smaller than their trading partners that they have to pay a high price to move out of the partner's sphere of influence. Neither the human trading nor the non-human cooperation categories are mutually exclusive and the classification of some examples is therefore arbitrary.

1. **Bartering.** In its simplest form, different goods or services are directly exchanged against each other during short interactions. Traders play symmetrical roles; each of them can choose the other as a partner and initiate a transaction. Interactions take place in the larger context of similar interactions by the same and other traders. Shifts in supply and demand alter the exchange ratio between two commodities in a predictable direction in the long term, but there is little left to haggle about when two traders interact. Biological examples: non-specific pollination and seed dispersal interactions, which are both food-for-transport barters with large numbers of different individuals and species in both camps.
2. **Shopkeepers-customers.** Asymmetrical interactions in which a usually small class of traders exchanges goods or services with a usually large class of customers. The customers can exert choice more easily than the shopkeeper thanks to their mobility, but each of them contributes only a small portion to the latter's 'wealth'. Biological examples: (i) Cleaner fish with clients that roam over a wide area ('floaters'; Bshary & Noë 2003). (ii) Baboon mothers trading grooming for permission to touch infants (Henzi & Barrett 2002, Barrett & Henzi, this volume). (iii) The obligate and species-specific pollination mutualisms between yuccas and the yucca moths (James et al. 1994, Pellmyr & Huth 1994, Marr & Pellmyr 2003).
3. **Large employers-employees.** In contrast to the category above, these are asymmetrical interactions in which many individuals offer services in return for goods (money) provided by a few. The latter are usually sessile, but nevertheless able to exert choice by controlling a scarce commodity locally. The 'employees' have to pay a high cost to reach another employer. One could

think of a village with a single big factory or coal mine. Biological examples: (i) Ants in several ant-protection mutualisms. Each colony, acting as a single trader, exchanges protection against nutrients (e.g. nectar provided by lycaenid larvae, Axèn 2000; honeydew produced by homopterans, Fischer et al. 2001; food bodies growing on plants, Fischer et al. 2002; or housing facilities, e.g. domatia provided by plants, Izzo & Vasconcelos 2002). (ii) Plants that control the exchange of nutrients with much more numerous and smaller individuals by excluding those that provide small quantities; for example: mycchorrizal fungi (Schwartz & Hoeksema 1998, Hoeksema & Schwartz 2001, Hoekesema & Kummel 2003), soil bacteria such as rhizobia (West et al. 2002, Denison et al. 2003, Kiers et al. 2003; see also reviews by Agrawal 2001 and Simms & Taylor 2002). (iii) Symbiont choice by fungus-growing ants (Mueller et al. 2004).

4. **Business partnerships.** These are usually symmetrical relationships in which two or more parties produce a commodity that is exchanged with a third party; i.e., there are two interconnected cooperative interactions at different levels. Example: An architect and a contractor who construct a house together for a client. The contribution of each individual partner can be very different in quality and quantity, but the important characteristic is that all contributions are needed to produce the commodity to be traded. Biological examples: cooperative displays by two or more conspecific males to attract females; for example: manakins (McDonald 1989a, 1989b) and ruffs (van Rhijn 1973, 1983). Complex three-way mutualisms also fall into this category; for example between (i) leaf-cutter ants, (ii) the fungus grown by the ants on their gardens of leaf cuttings and (iii) the bacteria that keep the fungus gardens free from a virulent parasitic fungus (Currie et al. 1999).

Human economic transactions, like other cooperative interactions, can also be classified along another dimension: (i) isolated 'one-off' interactions, (ii) repeated interactions and (iii) (semi-)permanent relationships.

13.2.5
Is trading a form of cooperation?

My conclusion is that trading cannot be considered as merely a special category of cooperation, although the two phenomena strongly overlap. This is because some forms of modern human trading have become so emancipated from direct, dyadic interactions between individuals that the connection to cooperation is lost; cooperation is not the first term that comes to mind, for example, with respect to computer algorithms that buy and sell stock automatically when certain price levels are reached. I also think it makes sense to use the term trading for interactions among non-human organisms when these are clearly likely to be influenced by 'market effects' (Noë et al. 1991). This helps to distinguish them from other forms of cooperation in which supply and demand do not play any role, such as many forms of collective action (Nunn & Lewis 2001, Ostrom 2001), the cooperation between genes (Hoekstra 2003) and cells (Michod 2003,

Szathmary & Wolpert 2003), and the acceptance of a mutually respected border between territorial neighbors (Whitehead 1987, Hyman 2002).

13.2.6 Cultural versus natural selection

In their very inspiring paper on emotions as mechanisms used by boundedly rational agents, Muramatsu & Hanoch (2005) state "...we think that some emotional programs have been shaped by natural selection to help individuals resolve adaptive problems observed as far back as the Pleistocene era" (p. 213). Similar remarks can be found in other texts written by behavioral economists, evolutionary psychologists, biological anthropologists and their hybrids (see e.g. Tooby & Cosmides 1995, p. 1189 and Richerson et al. 2003, p. 367). However, the reference to the Pleistocene as the period during which crucial elements of human cooperative behavior evolved strikes me as odd. The Pleistocene, which runs from about 1.8 million years till about 12000 years before present, is the era of the more 'advanced' hominids, such as *Homo erectus* (*ergaster*), *H. heidelbergensis*, *H. neanderthalensis* and of course *H. sapiens*.

Given that human cooperative behavior is very likely to be a mosaic of behavioral traits produced by the actions of both natural and cultural selection, and that strategies produced by natural selection can easily be much older than the species in which they are expressed, I therefore have a hunch that some of the mechanisms that play a crucial role in cooperation are rooted much deeper in our phylogeny than the Pleistocene. Indeed, we may have to think in terms of primate phylogeny as a whole, i.e. a history stretching back more than 80 million years (Tavare et al. 2002), and perhaps much further back still. Behavioral strategies under cultural evolution, in contrast, are likely to be very recent. Most of the strategies relevant today probably evolved in the Holocene, i.e. in the last 12000 years (Richerson & Boyd 2001), although some forms of cooperation and trading, such as adhering to traffic rules or auctioning via Ebay, are very recent indeed and are clearly new strategies that have developed to cope with such situations. I am not suggesting that nothing interesting happened during the Pleistocene, but I do think that present day human cooperative behavior can be divided into elements that were selected by a process of individual natural selection starting long before the Pleistocene and elements that were selected under a process of cultural group selection that occurred mainly after the Pleistocene. The Pleistocene may, however, have been the period in which some behavioral elements typical for trading and bargaining evolved.

In this chapter, I want to concentrate on mechanisms that have evolved under natural selection, and which I suspect to have deep phylogenetic roots. I will concentrate on forms of cooperation and trading in which each actor typically has multiple **dyadic** encounters with multiple partners, because I assume that cooperation in which many unrelated individuals act simultaneously is largely limited to humans (but see Nunn & Lewis 2001) and that the mechanisms used are largely a product of cultural evolution (the 'cultural group selection hypothesis'; Boyd & Richerson 1985, 2002, Bowles & Gintis 2003, Richerson et al. 2003, Panchanathan & Boyd 2004).

First, after a brief discussion of homology and units of selection, I make an inventory of problems that are common to many forms of trading and cooperation. Thereafter, I discuss the likelihood that the mechanisms used to solve these problems are homologous. Finally, I consider the consequences of our evolutionary history for our present-day trading behavior. Is our behavior fully adapted to our role in economic life or do we see sub-optimal behavior that can be explained by phylogenetic inertia?

13.2.7 Analogies and homologies

Two traits or characteristics are called 'analogous' when natural selection hits upon the same solution independently in different phylogenetic lineages, a process called convergence. For example, wings of bumblebees, birds and bats are all used for flying, but have very different structures. Two structures are called 'homologous' when they share the same evolutionary history, i.e. the same 'Bauplan', which is the case, for example for the anterior extremities (front legs, wings, arms) of baleen whales, bears, birds and bats. I deliberately introduced confusion by mentioning birds and bats in both contexts in order to make a further point: bird and bat wings are homologous in the sense that each bone in the bird wing has a homologous counterpart in the bat wing, but the essential airfoil is formed in a very different way. This point is worth making clear because this kind of ambiguity can also lead to confusion when we consider mechanisms used in cooperation and trading (see Box 2).

Both analogies and homologies are useful in the pursuit of answers to evolutionary questions. For example, looking for analogous solutions to the same evolutionary problem can allow us to identify a common denominator that can be used as the basis of common models; this is the message of our previous papers on 'biological markets' (Box 1). A second incentive is to identify systems that can be used as models for more complex forms of cooperation. For example, the mycorrhiza markets, which have been explored by Schwartz & Hoeksema (1998), Hoeksema & Schwartz (2001), and Kummel & Salant (in prep.) may turn out to be good alternatives to computer simulations if one wants to model complex human markets. Plants and fungi exchange nutrients in mycorrhiza that are easily quantified in both field and laboratory conditions, and the interaction is sensitive to changes in supply and demand. The third reason, and the one which drives me to look closer at analogies, is that similar-looking strategies used by different species could turn out to be implemented using homologous mechanisms.

13.2.8 Units of selection and modularity of the brain

A discussion of the evolution of behavioral strategies specific to cooperation only makes sense when one accepts that there are specific units of behavior (and/or the psychological mechanisms that implement them) that are adapted to fulfill specific functions. In other words, one has to accept that the mind is 'modular' in the sense of Fodor's seminal book of 1983 (see Whiten & Byrne 1997, Todd & Gigerenzer 2000, Barrett et al. 2002a and references therein). Note that 'a module of the mind' is not necessarily the equivalent of 'a nucleus in the brain'. The modules that I have in mind are what Gigerenzer (1997) calls 'domain-specific'; i.e., modules that evolved as units with a single function connected to a single ultimate cause, such as repeatedly dealing with untrustworthy cooperation partners. Such modules are, however, much harder to identify than morphological traits, which is one reason I use the latter more frequently in examples. I follow Gigerenzer and colleagues in assuming domain-specific modules have evolved for those functions that require fast decisions using rules of thumb. Moreover, in cases in which it is plausible that our ancestors were confronted by similar problems, I assume that the corresponding module was shaped by natural selection deep in the past. I used the term 'similar' and not 'the same', because I also assume that modules can be emancipated from the narrow purpose they served when they first evolved (the 'proper' domain *sensu* Gigerenzer 1997) to a wider use (Gigerenzer's 'actual' domain).

In principle, it is possible to trace the use of a certain strategy across many species boundaries, all the way back to animals with very little brain or even no brain at all. Natural selection had time enough, in most cases, to fine-tune conditional strategies in such a way that they work well in most of the circumstances that the species encounters. There is little reason why natural selection would have replaced such mechanisms with more sophisticated cognitive ones, if they still work reasonably well for their more brainy descendants. Selection has to overcome considerable friction to drive evolution from a 'hardwired' strategy to a 'cognitive' strategy, because this implies traveling from one peak to another in Wright's (1932) 'adaptive landscape'. Even in animals with big brains, including humans, there will be selection against the use of complicated strategies that are costly in time and processing power, if a simple rule-of-thumb can do the trick (Gigerenzer 1997). It turns out that people use 'fast and frugal' simple heuristics in many circumstances where economic theory had predicted sophisticated and, above all, rational decisions. The outcome is not necessarily worse than the outcome of rational decision processes and may well be better under many circumstances, although at times they are bound to lead to sub-optimal choices (contributions in Gigerenzer et al. 1999, Todd & Gigerenzer 2000, 2003).

13.3 Choice and control: the basic problems of cooperation

13.3.1 Partner choice

The first problem a would-be cooperator needs to solve is picking a partner. He should then be ready to switch to another partner, if this is likely to lead to an increase in net benefit. On biological and mating markets, profitability translates to net fitness gain; on economic markets, to net utility gain. In proximate terms, this translates into items like a net amount of energy, reduced predation risk, more surviving offspring, an amount of money and so forth, depending on which commodities are traded. The other side of the coin is competition over potential partners. When a competitor cannot be excluded by brute force, the favors of an attractive partner must be won by placing a higher bid than all other competitors. A skilled competitor should outbid the competition without making concessions that are unnecessarily high. Thus, choosers and bidders have to be able to estimate the relative market value of themselves and their potential partners. The most important factors involved are the following:

- **Comparison of offers.** Partner choice requires an estimate of relative future benefit to be expected from interactions with different partners. This is relatively easy when the partners can be compared directly and when the commodities they offer are visible and tangible. An example is an ant colony interacting with different species of aphids that all offer honeydew in exchange for protection. Aphids occur in aggregations that can be considered as single traders when they are clones produced by parthenogenetic reproduction, as is often the case. The ants most frequently visit those aphids that have most to offer by laying pheromone trails, which correspond in strength to the amount of honeydew produced (Fischer et al. 2001). Moreover, the ant colony can compare partners belonging to different taxa that produce the same commodity using the same method; for example, aphids and plants that each offer sugar-rich rewards (Engel et al. 2001).

  The ants quantify the difference between commodities offered using a direct, analogue method; more food translates in to more pheromone on the trail. Not all quantitative comparisons will be this straightforward, however. Some animals potentially must be able to compare quantities that differ in the number of items, volume, energy content and so forth. In many cases, these quantities cannot be compared directly, because they are offered by different partners in different locations and at different times. We know that monkeys have difficulty discriminating between food resources that differ in the number of items, even if they are offered (almost) simultaneously (Hauser et al. 2000, Stevens & Hauser 2004). On the other hand, most animals cope fairly well with the problem of choosing between food patches of different quality (see Sugrue et al. 2004 for a recent neurobiological study). In fact, this is what the ants in the aphid-protection example are doing, as are many other animals involved in food-based cooperative exchange. The close connection between foraging and commodity-based choice in the many food-related cooperative relationships observed in nature leads me to hypothesize that commodity-based choice mechanisms are likely to be homologous to mechanisms used in the selection of food items, even in cases in which commodities are other tangible items besides food.
- **Honesty of advertisements of offers.** If signals are used to advertise commodities that cannot be assessed directly, selection is likely to put a premium on paying attention to honest advertisements only. The problem of honest signaling has been studied extensively in the context of sexual selection (Zahavi 1975, 1977a, Grafen 1990, Johnstone 1995), but hardly at all in the context of biological markets. In the social sciences, this idea is widely known as 'costly signalling' or 'signalling theory'. In economics, the idea goes all

the way back to the 'conspicuous consumption' of Veblen (1899; see also the reference to Spence 1973 in Bowles & Hammerstein 2003). An example of honest advertising within a biological market is the petal disk (corolla) of flowers. The potential of cheating by a plant is high when the anatomies of flower and pollinator allow for the transfer of pollen before the pollinator can assess the reward (Bell 1986, Thakar et al. 2003). I discuss this example at some length, because the potential of cheating is a factor in many cooperative interactions.

A mechanism that ensures the honesty of the petal disk signal has been described by Blarer et al. (2002); bumble bees learn the association between the size of the petal disk and quantity of the reward again and again for each new population of flowers visited. This mechanism can only work when flowers occur in large aggregations and are genetically identical, e.g. multiple flowers on single plants or trees, or stands of clones. However, neither this mechanism, nor the mechanism I proposed myself, which was based on honest signaling theory (Noë 2001), can explain why cheaters with large petal disks and low amounts of nectar could not invade populations of genetically heterogeneous individuals. The only answer I can think of is that nectar quantity directly influences stay times and thus the amount of pollen transferred to the pollinator's body, or that pollinators can assess nectar quantities directly once they land on the flower. Both of these options render 'false advertising' pointless.

In many cases, the benefits obtained in interactions with different partners in the past will provide the best proxy indicator for future benefits. The sexually selected parallel of this is repeated courtship feeding before mating, which can be used by females to assess the ability of a male to feed their future offspring (e.g. Wiggins & Morris 1986). The potential for deceit is obvious and the 'Concorde fallacy' (Dawkins & Carlisle 1976) looms, but a mate that is not even able to bring in food before mating will almost certainly be a lousy caretaker after mating. The direct benefits that females receive, often before actual mating takes place, make this a low-cost form of mate sampling. An individual choosing between two cooperation partners faces a similar problem: information about a past difference between the two partners may not be worth much, but is better than no information at all and at least it provides information about potential contributions. During periods in which two competitors try to outbid each other, they both may be forced to produce their commodity at the maximum possible level. Their output from such periods thus provides reliable information about their long-term potential. Attractive offers at the beginning of long-lasting trading relationships are commonplace. For example, one should always closely inspect the small print in price tariff listings in the advertisements of internet and mobile phone providers; the large print only gives the price for the first few months of service.

- **Cost of sampling and discounting the future.** The variation in benefit between partners must be large enough to make an investment in sampling worthwhile; if differences are very small, the cost of sampling does not outweigh the benefits of making a better choice. The costs of sampling depend in the main on the decision rule used. Search times using relaxed strategies like 'accept the first partner that provides a benefit over a certain threshold' can be considerably shorter than more ambitious strategies like 'best-of-n' with a large n. The cost of sampling is modest when potential partners are aggregated in time and space, and can be compared directly. Sampling problems have been modeled extensively in the sexual selection literature (reviewed by Harvey & Bradbury 1991 and Gibson & Langen 1996) and in the vast optimal foraging literature (Stephens & Krebs 1986). Similar considerations play a role in the choice of consumer goods; for example, clients of supermarkets tend to search longer in order to compare prices of different brands for more expensive products (Oliveira-Castro 2003). Shopping has in fact been likened to foraging directly (Rajala & Hantula 2000, DiClemente & Hantula 2003, Smith & Hantula 2003).

Discounting the future can have a strong effect on sampling rules and partner choices: most hungry animals are known to prefer receiving a small quantity of food immediately over waiting or searching for a larger quantity (Green et al. 2004, Stephens et al. 2004 and references therein). Similar results have been found for human subjects (Myerson et al. 2003). This means that partners that provide a small quantity immediately are likely to be preferred over partners that provide large quantities after a delay (Stephens 2000, Stephens et al. 2002). In terms of the sampling strategies mentioned above, this is equivalent to using a threshold strategy with a rather low threshold.

- **Estimating market value of self.** The most successful human traders can fine-tune their strategies to their own market value. This seems a tall order for non-human traders. Estimating exchange rates is much easier when the two commodities exchanged can be valued in a common currency. Sampling the market by tapping 'communal knowledge' can also help a lot. A hunter coming out of the forest can ask around and easily discover the current value of a dead tapir in terms of turnips. In an analysis of 'lonely hearts' advertisements, Pawłovski & Dunbar (1999, 2001) showed that both men and women have a keen perception of their own value on the mating market. This shows that humans are able to estimate their relative market value in a market in which values are not expressed in money. The advantage they have compared to most non-humans is that they can gather a lot of information about the other traders in the same market in a short time. In biological systems, selection can lead to conditional strategies that are tuned to shifts in the market as long as these are predictable. Plants, for example, could theoretically 'decide' to put out more reward for pollinators after harsh winters, if such weather conditions hit pollinators harder than plants (Noë 2001). More research on non-human traders is needed before anything more conclusive about this topic can be said, however.
- **Adjusting to market value.** There is a difference between increasing an offer to obtain more of the partner's commodity and increasing an offer out of fear of losing the partner. For example, lyceanid larvae increase the amount of nectar they produce to reward ants for protecting them in reaction to increased predation risk (Leimar & Axèn 1993, Agrawal & Fordyce 2000). They also increase the amount of reward when ant attendance is too low, but then

decrease it again when there are many ants around them (Leimar & Axèn 1993). The larvae thus balance ant attendance and predation pressure without receiving any direct information about the amount of nectar produced by conspecifics or about other sources of sugars available to the ants.

How do baboon mothers adjust to the value of their infant in the baby market? The lower the number of babies in a group, the longer baboon mothers can apparently demand to be groomed before they allow another female to touch their newborn (Henzi & Barrett 2002, Barrett & Henzi, this volume). The question is how they manage to turn their higher market value into longer grooming? Systematic observations are still underway, but I guess, based on my own experience with baboons, that would-be handlers try to touch the infant once in a while and that mothers block her advances until the moment is reached that the groomer stops grooming and starts showing a lack of interest in the infant. The mother would in that way be able to fathom the motivation of the handler. The motivation to handle a particular infant will depend on the number of other options the handler has and/or the number of infants she has handled in the recent past. An underlying mechanism was suggested to me by Louise Barrett (pers. com.), namely that grooming promotes the production of β-endorphins (Barrett & Henzi, this volume), which leads to a reduction in tension. The fewer mothers there are in a group, the more they are surrounded by would-be handlers. This means that they might start at higher levels of stress and thus need more grooming to be relaxed enough to accept the handling of their infants.

Playing off partners. The option of switching partners is a double-edged sword; a new partner can bring more profit, but the old one may also yield more when facing the threat of being deserted. A credible signal that a switch is imminent helps to increase the pressure considerably. Baboon males can signal their intention to switch allies by the way in which they interfere in their partner's conflicts. The most obvious signals we have seen are (i) males that turned away while their ally begged for help (signaled by 'head-flagging' and screaming) and (ii) switching allegiance during a conflict between two potential partners (Noë & Sluijter 1995, pers. obs.). The latter conflicts were markedly different in their duration, intensity and form from 'normal' conflicts with multiple males. The baboon mothers mentioned above can play-off handlers directly when two of the latter approach a mother-infant pair simultaneously. The mother can simply hold on to her baby until one of the two stops grooming. She can also invite further potential handlers, while being groomed by a single handler. Baboons have several facial expressions for such invitations at a distance.

13.3.2
Partner control

The second problem is to control for the continuation of net benefit from ongoing partnerships. Fear of being cheated by the partner should keep each trader on edge. And since one's partner is driven by the same fears, the other side of that coin is that one has to overcome mistrust by the partner. Building trust is thus another useful skill. This can be achieved by sending reliable signals that commodities will be (and continue to be) delivered. The obvious alternative is to deceive the partner, giving rise to selection for another pair of antagonistic skills; deception and the detection of deception.

Theoretical treatises of cooperation are dominated by the 'cheating' problem. Cheating can take many forms from subtly reducing the value of the commodity offered to not delivering the commodity at all. Expectations may be based on advertisements (e.g. petals, billboards), a pattern of taking turns in giving and receiving, or past interactions. Strategies that make participants less vulnerable to being cheated are thus seen as crucial for stable cooperation. Many such strategies have been proposed: ending the relationship; all kinds of sanctions (running from physical damage to damage to the cheater's reputation); reducing risks by offering the commodity exchanged in small parcels and so forth (see reviews by Dugatkin 1997, 2002b and Sachs et al. 2004)

'Cheating' between mutualistic partners belonging to different species is not necessarily controlled by behavioral counter-strategies. Co-evolution between two species can also have resulted in morphological structures that make it hard to exploit the partner. Take for example the interaction between figs and fig-wasps. The wasps are essential for the pollination of figs, but lay their eggs in the ovaries of the figs, which are then lost to the plant. Figs therefore have to defend themselves against overexploitation by the wasps. Several species of fig tree have ovaries that cannot be reached by the wasps (reviewed in Cook & Rasplus 2003). Other 'transport for food' and 'protection for food' mutualisms offer further examples of cooperation in which for one or both partners little scope is left for cheating after a long process of co-evolution. In many cases, the food reward cannot be reached without taking on the load to be transported, but subtle cheating is possible in some cases, as discussed above under 'honest advertising'. Selection for this sort of hardware defense can be accelerated under pressure of parasites that exploit the mutualism. There is often a thin line between mutualism and parasitism anyway (Bronstein 2001, Johnstone & Bshary 2002).

In human trading, we can also find such hardware solutions against cheating, such as burglar alarms and safes. However, thieves are similar to the parasites of mutualisms, rather than to the participants. What, for example, prevents cheating in simple human transactions like buying a loaf of bread? What prevents the customer from walking off without paying? Some mechanisms are typically human: the baker may call the police or may damage the reputation of the customer. Such mechanisms do not work well, however, when the customer visits the baker and his community only once. However, the customer may also decide to pay, even if he is a total stranger and the baker has no telephone. He may do so because he fears physical sanctions by the baker. This may even be true when the baker is much weaker than the customer; the risk of injury may nevertheless outweigh the benefit of getting a free loaf. The customer may also not know whether the baker has a weapon or can call for allies. The latter forms of cheating control backed up by physical sanctions can also be found throughout the animal kingdom. Using sanctions blends seamlessly with using harassment or punishment to control partners (see Noë 2005, in press and references therein for an explanation of these phenomena and their difference). Punishment also forms an important ingredient of 'strong reciprocity' (Fehr & Fischbacher 2003, 2004, Fehr & Gächter 2002, Gächter & Herrmann, this volume) but, in contrast to punishment used by animals, it is thought to have evolved under group selection during recent cultural evolution in humans. Yet, another very human explanation is that the customer adheres to internalized 'norms' (Gintis 2003, Young 2003), which may be adapted to a life in a close-knit community. Whether or not strong reciprocity and such norms are adapted or not to one-off interactions is currently a hotly debated issue (Fehr & Henrich 2003)

13.4 Potentially homologous mechanisms

13.4.1 Homology at two levels: strategies and mechanisms

Theoreticians, both biologists and economists, attempt to determine which strategies individuals should use to get most out of cooperation or trade. In order to identify the critical elements of the problem, the real-life situation is reduced to a bare-bones model, usually based on a theoretical game. The same game, for example the Iterated Prisoner's Dilemma, may turn out to be a paradigm that closely resembles cooperation among both baboons and bacteria.

Strategies are rather abstract algorithms that prescribe what to do in a manner that is conditional upon the situation at hand. Species as diverse as baboons and bacteria are assumed to play the same conditional strategies; for example, 'PAVLOV' or 'TIT-FOR-TAT', when confronted with the same kind of problem. However, the actual mechanisms that these species use to implement the strategy are likely to be very different. Baboons may acquire a strategy largely through learning, whereas bacteria use largely 'hard-wired' mechanisms produced over many generations by the action of natural selection. Thus, while the strategies used by two species can very well be analogous, the mechanisms used to implement them need be neither homologous nor analogous.

It is also possible that the same mechanisms are used to implement different strategies, which are then employed in different behavioral domains. For example, certain brain areas are highly sensitive to symmetry. This symmetry-detecting mechanism can be used both to select mates with symmetrical faces and select prey with asymmetrical bodies (Sasaki et al. 2005).

13.4.2 Homologies can be intra-specific or inter-specific

Two mechanisms used in cooperation by different species are homologous when they derive from a cooperative mechanism used by a common ancestor ('vertical homology'). 'Horizontal homology' is a term I will use to refer to all cases in which members of the same species apparently use the same mechanism in different contexts (see Box 2). 'Horizontally homologous' can be replaced by 'identical' if it can be shown that exactly the same neurons, neurotransmitters, learning processes etc. are involved. I prefer to use the former, more fuzzy, term for two reasons. First, I want to avoid a futile debate over which of the building blocks of a strategy should be identified as separate mechanisms. Second, mechanisms may be based on the same morphological and physiological substrate, but contain elements of learning that may result in differences in the details between behavioral domains. Hybrids between the two forms of homology are also possible; for example, two species may have inherited a mechanism that plays a role in mate choice from a common ancestor and this mechanism may have developed into a mechanism used in cooperation in each lineage independently.

Two mechanisms are likely to be vertically homologous when two species that descended from a recent common ancestor use the same mechanism under comparable circumstances. Macaques and humans, for example, both use the same brain nuclei to recognize symmetry (Sasaki et al. 2005).

What are plausible cases of horizontally homologous mechanisms? An almost trivial example would be a bird species in which females prefer to eat red fruit and prefer to mate with males with a red breast. In both cases, the same photo-pigments and neurons in the visual cortex are implicated in seeing red. It is likely that selection for finding red fruits resulted in a sensory bias in the visual system which males can then exploit, a phenomenon known as 'sensory exploitation' (Basolo 1990, Ryan et al. 1990). Other plausible sources of mechanisms (pre-adaptations or exaptations) used in cooperation, apart from foraging behavior, are mate choice mechanisms (for mechanisms used in the choice of cooperation partners) and the mutual control of mates in species with bi-parental care (for the mutual control among cooperation partners).

Interesting homologies, both vertical and horizontal, can be found in fMRI-studies in which human and non-human subjects show activity in the same brain nuclei, such as the nucleus accumbens, which, in humans, is activated in response to a very diverse range of stimuli, for example pretty faces of the opposite sex (Aharon et al. 2001), money (Knutson et al. 2001) and sports cars Erk et al. (2002). It remains to be seen, however, whether it is the same populations of neurons or different intermingled populations which are implicated in each case.

▶ *Box 2.*
Horizontal and vertical homologies

Fig. 13.1 illustrates the idea of horizontal and vertical homologies. Imagine three species A, B and C that belong to the same lineage (A being the oldest, or 'ancestral', species). All three species use the same choice strategy, say 'best-of-n' (see below). Of the many mechanisms needed to implement such a strategy, one is 'sensitivity to the relevant signal of quality'. I use three qualities here: symmetry, color and quantity. It is perhaps useful to think of a certain population of neurons that are specifically sensitive to one of these characteristics (e.g. Sasaki et al. 2005 for symmetry; Nieder et al. 2002 for quantities). Species A prefers symmetry when

selecting both mates and cooperation partners, and uses mechanisms that are likely to be horizontally homologous. Species A prefers red items when foraging. This preference is inherited by species B, which makes it a plausible case of vertical homology. Species B also shows preference for red when selecting cooperative partners, a horizontal homology, etc.
I also illustrate one case where the use of identical mechanisms in the same domain by two species is unlikely to be homologous. Species A and C both prefer symmetrical cooperation partners, but the intermediate species B does not. Species A is likely to use a mechanism evolved in the context of mating and species C is likely to use one that evolved in the context of foraging. For the interpretation of evolutionary pathways of mechanisms, it is thus important to consider in which domain the strongest selection pressure was exerted (see Discussion in the main text).

	Choice in non-cooperative domains		Choice of cooperation partner	
	Mating	Foraging	Attribute-based	Commodity-based
Species A	*Sensitive to* ***symmetry***	*Sensitive to* ***red***	*Sensitive to* ***symmetry***	*Sensitive to* ***quantity***
Species B	*Sensitive to* ***symmetry***	*Sensitive to* ***red***	*Sensitive to* ***red***	*Sensitive to* ***quantity***
Species C	*Sensitive to* ***red***	*Sensitive to* ***symmetry***	*Sensitive to* ***symmetry***	*Sensitive to* ***quantity***

Possible 'vertical' homology between species

Possible 'horizontal' homology within species

Possible analogy between species

Whether or not two strategies are homologous is another story. Suppose each of the three species uses either ' best-of-n' or 'threshold' as choice strategy. An individual that uses 'best-of-n' will sample all individuals, whether food sources or commodities, that are found within a certain time or area and then return to the best one encountered. Using the 'threshold' strategy means that the individual will pick the first individual/resource/commodity that is better than a certain fixed threshold value. Assume further that all three mechanisms can be used in the implementation of both strategies and that each strategy-mechanism combination can have evolved independently in each domain for each species. This means that any pair of identical strategies or identical mechanisms can in principle be either analogous or homologous. Strategies can be considered homologous when all mechanisms involved in the implementation of the strategy are homologous. This becomes increasingly less likely with increasing distance between species, with increasing complexity of the mechanisms involved and with an increased contribution of learning.

13.4.3 The phylogeny of a mechanism may explain how well it is adapted to its present cooperative function

Depending on its phylogenetic history, a mechanism may be more or less specifically adapted to its task in the context of cooperation or trading (cf. Gigerenzer's 1997 discussion of the adaptation of modules of social intelligence to 'proper' and 'actual' domains). I see the following possibilities:

- The mechanism evolved *de novo* in the context of a specific form of cooperation. Examples are morphological adaptations to mutualistic interactions; for example, the nectar organ of lycaenid larvae, domatia and food-bodies of plants, all of which are rewards for insects that serve as protectors. Several mechanisms involved in the cooperative mating display of male ruffs, which are not found in closely related species, also belong to this category. Males belonging to different color morphs cooperate to attract females. The differences in the color of the ruff, as well as differences in behavior, are genetically determined (van Rhijn 1973, 1983).
- The mechanism evolved in a different context. A hypothetical example would be adaptations to bi-parental care, which act as pre-adaptations for cooperative hunting. Both forms of cooperation resemble a 'synergistic mutualism' game (Maynard-Smith 1983) in which it is better to compensate the inadequate input of the partner, at least partially, than to punish the partner by ending the cooperation instantly (for compensation in bi-parental care, see Smiseth & Moore 2004 and references therein). In both domains, partner control is a problem.
- The mechanism stems from another non-cooperative context, but is emancipated from that context and thereafter only used in (a specific form of) cooperation. Examples are grooming and preening behaviors, which have their roots in parental care, but are now largely emancipated from their function of removal of ecto-parasites in several animal groups, such as the primates, and can be used as payment for 'tolerance' (Barrett & Henzi 2001, this volume, Barrett et al. 2002b, Henzi & Barrett 2002).
- The mechanism evolved in a non-cooperative context and is now used in both its old function and in cooperation. Examples are the dancing approach and the caressing of clients by cleaner wrasses, both of which are also used in the mating context (see below under 'Trust').

13.4.4 Reproductive relationships

Reproduction and cooperation resemble each other in so many aspects that it is questionable whether one should see them as independent domains. Reproductive relationships can be seen as a special form of cooperation in which the partners, by definition, belong to the same species and in which kin selection plays no role. The vast sexual selection literature (reviewed in Andersson 1994) describes an enormous variation in reproductive relationships. The operational sex ratio (OSR), i.e. the relative numbers of males and females available for mating at a particular time, is the basic supply-demand parameter of mating markets (Emlen & Oring 1977, see also Noë & Hammerstein 1995).

The OSR is largely determined by the amount of time each sex is bound by the production of a batch of offspring. In species that mate on leks (display arenas), the relationships do not last much longer than the copulation itself and males return to the market immediately, whereas females may return only in the next season. In such skewed markets, sexual selection results in competitiveness in one sex (usually males) and choosiness in the other (usually females). Species with obligate bi-parental care are at the other extreme. The OSR is much more balanced and both sexes have an interest in choosing their mates carefully. The division of labor and sexual fidelity can be major sources of conflict and partner control is an important issue. In general, the longer the relationship lasts, the more the emphasis shifts from partner choice to partner control.

The distinction between 'mating markets' and 'economic markets' becomes blurred where differences in market value between the sexes are compensated for by goods and services offered by the members of the competing sex to the members of the choosy sex; for example, in the form of nuptial gifts, safe nesting locations, territories containing resources etc. Human mating markets provide good examples of the linkage between the two (reviewed in Barrett et al. 2002a) and, as I will argue below, this may partially explain why we behave in one of these markets as if we were acting in the other. Thus, for each mechanism used in cooperation, one should consider the possibility that it initially evolved under sexual selection.

13.4.5 Emotions

Emotions, such as fear, trust, envy and guilt, can be seen as "discrete mechanisms crafted by evolutionary processes" (Fessler & Haley 2003, p. 9) that play an important role in the implementation of cooperative strategies. There is a clear continuum between emotions felt by humans and emotions observed in non-human animals (Muramatsu & Hanoch 2005, Panksepp 2005, in press). Therefore, emotions are prime candidates for cooperation/trade-related mechanisms that have deep roots in our phylogenetic tree. It is not easy to pin down exactly what emotions are and people in different disciplines certainly have different ideas about what constitutes an emotion. In connection with human cooperation, Fessler & Haley (2003) see emotions as psychological attributes, shaped by natural selection, that "enhance the individual's ability to engage in, and profit from, cooperative enterprises" (p. 8). Fessler & Haley go on to discuss 13 different emotions that play a role in human cooperative interactions, but do not consider their phylogenetic roots. I will do just that, but discuss only a small subset: trust, fear, envy-jealousy and a complex of related emotions: sympathy, liking and wanting. I will discuss the latter under the heading 'choice', which is the driving force behind biological, mating and economic markets.

- **Fear of being cheated.** Cosmides & Tooby (1992) have argued that our modular mind contains a 'cheat detection module', which evolved specifically in the context of cooperation. However, no module drops out of the blue, evolutionary speaking. Two brain structures play a major role in second-guessing the immediate behavior of others: the amygdala and the ventro-medial prefrontal cortex (Adolphs 1999, 2003). The amygdala is also known for its role in the emotion of fear. It is therefore likely that natural selection took a shortcut from fear in general, via fear of aggression from conspecifics, to fear of being deceived by conspecifics.
- **Trust-distrust.** The amygdala is only part of the hypothetical cheat-detector module; one needs to recognize the danger of being deceived before one can fear it. Several nuclei, including the amygdala, are involved in the judgment of the untrustworthiness of faces (Winston et al. 2002, Adolphs 2003). Detection of cheating can be specific to the context of cooperation, but it is also possible that it is part of a more general sensitivity to 'deception'. Deception is a major problem for animals involved in agonistic interactions with conspecifics. The combatants have an interest in not giving away their next moves, while at the same time trying to detect the true intentions of the adversary (Enquist 1985). Reading the other's intentions correctly, from facial expressions or otherwise, can make all the difference. Such interactions are likely to have had more impact on the early evolution of an eventual cheat-detection module than cooperation.

  Trust is the other side of the cheating coin. One needs to trust other individuals at some point to make cooperation possible. In an experiment with human subjects, Zak and colleagues found that oxytocin levels rose when people experienced trust from anonymous partners in an 'investment game' and responded in a 'trustworthy' manner themselves (Zak 2004, Zak et al. 2004, submitted). Higher levels of oxytocin apparently reduce the anxiety caused by 'mistrusting' others, possibly acting via receptors found in the amygdala. Oxytocin is a hormone well-known for its role in regulating lactation and the enhancement of the bond between parents and offspring, as well as in breeding pairs in mammals (Carter 1998). A second hormone, vasopressin, is also important for pair bonding in mammals, notably in males (Lim et al. 2004 and references therein). Both oxytocin and vasopressin evolved by one amino-acid substitution from vasotocin, the hormone that plays a role in egg-laying and nest-building in some reptiles (Konner 2004).

  Familiar people are also likely to be trusted more than unfamiliar people. DeBruine (2002) found that players in a two-person trust game are more likely to trust players shown on a computer screen with faces manipulated to resemble themselves. This result conforms to kin selection theory (Hamilton

1964) with phenotypic matching as the mechanism used to recognize kin. Using fMRI in people playing repeated trust games, King-Casas et al. (2005) showed that the caudate nucleus plays a role both in the judgment of the fairness of the partner's offer and in the intention of self to return trust. Activity in the caudate nucleus, a structure connected to the dopamine system, which in turn is important in the 'wanting module' (see below under 'Choice') increasingly anticipated the intention to trust, showing evidence of a learning effect in a series of interactions with a specific partner.

Overcoming fear and mistrust is not only a problem for individuals eager to cooperate, but also for prospective sexual partners. Accounts of pair formation in animals are full of descriptions of 'appeasement' behavior and postures that are the exaggerated opposites of aggressive and threatening postures, for example turning away the dark face and bill in several gulls (Tinbergen 1959). Appeasement behavior, evolved in the context of mating, can be transferred to the context of cooperation, even between members of different species. The cleaner fish mutualism provides two examples: firstly, the zig-zag dances with which the cleaner *Labroidus dimidiatus* and its parasitic mimicry *Aspidontus taeniatus* approach their 'clients' (Grutter 2004). This dance, which is similar to the famous zig-zag dance of the stickleback, is used in the mating context in these species and several of their congeners (Wickler 1963). Secondly, cleaner fish (*Labroidus*) often calm their clients down by stroking their backs with their pelvic and pectoral fins (Bshary & Würth 2001). This too is a behavior that has also been observed in the context of mating (R. Bshary, pers. com.).

- **Envy and jealousy.** I follow Fessler & Haley (2003, p. 14) in their distinction between 'envy' as an emotion caused by a disparity in possession of a valued item and 'jealousy as the emotion experienced by individuals desiring to take the role of another individual in a social relationship. This distinction between envy and jealousy maps onto my distinction between 'commodity-based partner choice' and 'attributes-based partner choice', respectively. Both emotions can play a role in cooperation and trading, but in a different way. Jealousy can be a factor in alliance formation. Envy can be a mechanism that changes the way communally-produced resources are shared, for example meat after a cooperative hunt. 'Envy' may also drive individuals to react strongly when they feel they have been short-changed after cooperative interactions. The expectation of such passionate reactions may in turn stabilize cooperative relationships (Fessler & Haley 2003). Such strong reactions are described by Brosnan & de Waal (2003) and Brosnan et al. (2005). In their experiments, capuchin monkeys and chimpanzees would throw tantrums if they observed a conspecific in an adjacent cage obtain a preferred food item from an experimenter, when they themselves had only obtained a non-preferred food item from the same experimenter.

In their coverage of this work, the popular press often used the term 'envy'. The authors themselves avoided this label and used 'inequity avoidance' instead, probably for good reason. Showing envy in animals would be another nice example of a human-non-human emotional continuum, but I suspect it will be very hard to show the existence of envy beyond reasonable doubt. In the good old Lorentzian tradition, I use anecdotal observations of my cat and dog to explain why. When I stroke my cat, the dog comes over to me straight away and tries to wriggle herself between me and the cat. The dog is apparently afraid that my relationship with the cat will come at a cost to her relationship with me. I think I can call that being jealous. This would make sense for a group-living wolf; social relationships can have a lot of value, because they may mean support, grooming, tolerance etc. The cat could not care less if I stroke the dog or not; he belongs to a more solitary species. Now what happens at feeding time? When I fill the dog's bowl, the cat comes running too, and vice versa. When they are not fed almost simultaneously, they both show signs of frustration. My guess is that the same happens when a capuchin or chimpanzee expects grapes, but receives cucumber. The feeding of another individual acts as a stimulus that predicts the imminent arrival of a certain kind of food for the subject animal, because the two have been linked together contingently in the past: a classical conditioned response. The same basic idea was also expressed by Tim Clutton-Brock (pers. com.) during the conference on which this book is based. The tantrum thrown by Brosnan & De Waal's capuchin monkeys, for example, can be interpreted as a sign of frustrated expectations (Wynne 2004) and this frustration may be so strong that the cucumber is not eaten. It is therefore unnecessary to invoke either 'envy', or the more cautious term 'inequity avoidance' to explain the behavior of the primates and the pets.

The use of the word envy would perhaps be warranted if I were to feed the cat and discover that the dog also runs to the cat's bowl, even though she has a full bowl of her own. My dog goes to her own food bowl, however, even if it is still empty, and so does the cat if I feed the dog first. I would still not feel the need to use the word envy if the dog tried to steal the food from the cat, because her behavior is typical for situations of food competition in many species. The use of the word envy does not add any explanatory value, because my dog runs to any form of food at any place and tries to get it. In the primate case, envy (or inequity avoidance for that matter) would only be apparent if monkey A stopped eating the cucumbers he had been happily munching before monkey B was given grapes, even though A had never received any grapes himself simultaneously with any other monkey in his life, and had no chance getting B's grapes during the experiment.

13.4.6
Choice

Choice is the mechanism that makes markets turn. A major chunk of sexual selection works through mate choice and one cannot imagine the outbidding competition that typically occurs within markets if there were no choosing agents. Many, if not most, 'choices' in life are made by natural selection on behalf of the organism. Giraffes do not hunt zebras, and lions do not eat leaves in treetops. Nevertheless the life of any animal remains a string of choices. Where to go? What to eat? Who to mate with? Choice seems such a general mechanism that an attempt to identify a common denominator and a common origin may be futile.

Nevertheless, work on humans seems to point to a central 'choice module', which can in turn be divided into a 'wanting' module, active in the anticipation of pleasure, and a 'liking' module, active during pleasant experiences (Berridge 2003). The wanting module, which roughly coincides with the dopamine system, was shown to be active in different mammal species, but notably humans, in reaction to a wide range of stimuli (food: Berridge 1996, Pagnoni et al. 2002, Arana et al. 2003; beautiful faces: Kampe et al. 2001, Aharon et al. 2001; pornographic material: Bocher et al. 2001, Karamara et al. 2002; sports cars: Erk et al. 2002; money: Knutson et al. 2001a, 2001b). Berridge (1996) showed that, as far as food preferences are concerned, the wanting module can be clearly distinguished from the liking module, which is linked to opioids rather than dopamine. According to Berridge (2003), this separation holds for other stimuli as well, such as monetary rewards in humans.

Anatomically, as well as functionally, these two hypothetical modules are closely connected and partially overlap. The 'wanting' part could be considered as the genuine choice module, but the 'liking' module is bound to be important for learning which choices are worth repeating. Tremblay & Schultz (1999) (see also comment by Watanabe 1999) and Arana et al. (2003) showed that the orbitofrontal cortex (the ventral part of the prefrontal cortex) plays a role in the relative choice of food items; i.e., a mechanism needed to perform a sampling strategy such as 'best-of-n'. This area of the brain is related to the liking module rather than the wanting module, which seems logical when the association with learned tastes determines the choices made. The fact that the choice module governs the selection of food items suggests that this module is rather archaic. I imagine that it evolved first in the context of foraging, then developed further in the context of mating and is now implicated in all kinds of choices, including economic ones like the purchase of consumer goods.

Mate choice and economic decisions may be related because they partially rely on the same mechanisms, but they are also directly connected, as I argued above. According to sexual selection theory, an asymmetry between the sexes in the direct investment in the offspring can be compensated by indirect investments. The sex with the larger direct investment is usually in a position to base mate choice on commodities offered such as nuptial gifts, high quality territories, safe nesting places etc. Economic decisions play a major role in mate choice in most human cultures (reviewed in Barrett et al. 2002a, chapter 8). The economic consequences of the pair bond play a role notably in the choice of long-term partners, for example the historical Krummhörn population described by Voland and colleagues (Voland & Engel 1990, Voland & Dunbar 1995, Voland 2000). This is quite different from the choice of partners for 'one-night stands', where the physical qualities of the partner play a much larger role. The latter is known as the 'good genes' explanation in sexual selection. However, one-night stands can also take the form of prostitution with an obvious economic component. Thus, within mating markets, mate choice can be based on the mate's qualities and/or on the basis of commodities offered by the mate. Above, I have proposed a similar distinction between attributes-based partner choice and commodity-based partner choice on biological markets. The question now is, do the different choice criteria evoke entirely different or partially overlapping mechanisms in the selection of sexual partners and does this apply to the selection of cooperation and trading partners too?

So far, I have assumed that there is a single 'choice module' with two parts, one for 'liking' and the other for 'wanting'. This runs parallel with the idea that there is a single decision-making centre in the brain that may or may not be homologous between humans and macaques (Rorie & Newsome 2005). Liking food is, however, a rather different emotion from liking a conspecific. Above, I proposed a split between two modes of partner choice, because I suspect that they are at least partly based on different mechanisms. 'Liking a conspecific' should be further separated into several sub-categories, depending on whether the conspecific is a sexual partner, a parent of one's offspring, a relative, or a frequent coalition partner. In human terms, we are talking about emotions like sexual desire, romantic love, sympathy and so forth. Trust plays a role in all of these, of course, so some remarks made above apply here too.

In the psychological literature, 'friendship' tends to be compared 'horizontally' with other human relationships, notably of a romantic or sexual nature (e.g. Sprecher & Regan 2002). The term friendship has also been used by several primatologists (Smuts 1985, Cords 1997, Hemelrijk et al. 1999, Palombit et al. 2001, Silk 2002). It seems worthwhile, therefore, to think in terms of phylogeny and to consider the possible homology between the feelings for an ally in non-human species and the feelings for friends and business partners in humans. De Waal (2000) has described what he calls 'attitudinal reciprocity' as a mechanism that stabilizes cooperation; instead of detailed, quantitative 'book-keeping' of past interactions, he envisages a more fuzzy, qualitative building up of a certain attitude, one could say sympathy, towards those group members with whom one has had positive interactions.

13.5 Is human trading behavior well-adapted?

Until recently, most economists considered economic decisions by humans to be the result of rational cognitive processes, and many perhaps still do. However, a number of recent studies show that our 'rationality' is rather limited (Chase et al. 1998, Colman 2003a, 2003b, Todd & Gigerenzer 2000, 2003, contributions in Gigerenzer et al. 1999). Gigerenzer et al.'s 'fast and frugal' algorithms, mentioned above, are assumed to be products of natural selection, which implies that they are optimally adapted to the circumstances in which their effect on fitness was, or is, the greatest. The process of natural selection inevitably incurs some inertia, making it likely that decision rules do not always follow fast cultural changes. One example is the recent surge in eating disorders that seem to be a mal-adaptation to a world of plenty. One should, however, not expect to see mal-adaptations all over the place. Even if genetic evolution was too slow, cultural evolution and learning processes might have papered over some of the cracks.

Sub-optimal decisions in modern circumstances may therefore be telltale signs that they are made using archaic mechanisms. The question is, sub-optimal for what? Utility maximization may have been the proximate mechanism for fitness maximization for most of human history, but that connection now seems mostly lost for those living in industrialized societies. People appear to seek compromises between the two, but compromises are not necessarily optimal solutions. Take my own guild, the academics. We postpone having kids till it is often too late; most of us could have done better in a material sense if we had learned an honest trade and we only have to look at the half-life of journal citations to see that we will not do very well in the cultural evolution theater either. Nevertheless, we spoil weekends with beautiful weather in order to finish book chapters. I will, however, limit myself here to behavior that seems irrational in the eyes of classical economists and propose some bold hypotheses about sub-optimality in trading behavior:

1. Suppose we take decisions about trading goods using mechanisms from foraging, does that explain sub-optimal consumer behavior?
2. Suppose we choose our cooperation partners using mechanisms from mating behavior, does this lead to sub-optimal choices by employers when recruiting new personnel?
3. A related question, I will not try to answer in keeping with my promise to limit myself to cooperation and trading among unrelated individuals is: Suppose our trust in trading partners is based on mechanisms that evolved under kin selection, such as the resemblance of the partner's face to faces of family members, does that hamper our trade with unfamiliar individuals?

13.5.1 Consumer behavior

Above we have seen that an eventual choice module would affect our choices in a wide range of domains. The hypothesis I put forward is that the selection pressure on this module would have been, and probably still is, strongest in activities with large implications for fitness, such as partner choice, parental behavior and foraging behavior. Consequently, the module would be less adapted to choices with less existential value, such as the purchase of luxury goods and gadgets. I also predict gender differences in choice behavior, because sexual selection would drive male and female behavior in different directions. There is a catch, however. Buying, and showing off with, luxury goods can influence mate acquisition. Some items are bought to impress the other sex, to adorn the body to attract the other sex or, explicitly to buy the favors of the other sex. I therefore propose to distinguish at least three categories of consumer behavior for which different predictions can be formulated.

A. Commodities for maintenance (e.g. food, housing, insurances etc.). In economical terms, this is a plain consumer choice problem (Frank 1994), which can be tackled with the theory of rational consumer choice. Products are chosen depending on their quality, price, cost of purchase (including cost of sampling and traveling) etc. A gender difference is not predicted. Biologists would be inclined to apply optimal foraging theory and assume that decision rules are at least partially a product of natural selection. Optimal foraging theory has indeed been applied to consumer behavior by some economists and it turns out that the behavior of customers can be described pretty well by this biological theory (Rajala & Hantula 2000, DiClemente & Hantula 2003, Smith & Hantula 2005).

B. Commodities that play a role in mate choice. A broad category that includes all relatively expensive possessions visible to the prospective mate during the process of mate choice. These can be items that are normally in the 'maintenance' category, but acquire a second function as signals to potential mates. I predict that many goods bought impulsively fall into this category and that consumer behavior under these circumstances is not what an economist would consider rational, unless he somehow succeeds in squeezing the acquisition of a mate into a utility function. In biological terms, sexual selection reigns. *H. sapiens* is a species with reciprocal mate choice, but the criteria of choice are rather different between the sexes. A strong gender difference is therefore predicted. Two sub-categories can further be distinguished (Buston & Emlen 2003, Borgerhoff-Mulder 2004):
   a. Choice of a short-term partner (one-night stands)
   b. Choice of a long-term partner (marriage)

C. Commodities that represent parental investment. The roles of parents differ, but the interests of parents coincide. A moderate gender difference is predicted.

There is a surprising lack of research in consumer preferences by gender (Moss & Colman 2001), but some consistent patterns emerge. Apart from factors related to mate choice, there are, of course, a number of other explanations for gender differences in consumer preferences, for example consistent differences in technical knowledge of the items bought, in purchase power, or in sensory acuteness. An example: women rarely suffer from color-blindness in contrast to men and have a stronger preference for more colorful consumer goods (Moss & Colman 2001). Note that it is an open question whether color-blindness, or better dichromatism, has adaptive value or is a sad consequence of having a Y-chromosome (see Dominy et al. 2001). In addition, there may be social causes for a gender difference, such as consistent differences in social relationships with vendors and shopkeepers. All such differences would not compromise the rational choice hypothesis. Other differences are clearly related to the role an item plays in the mating market, however. Impulsive buying by women has, in general, more to do with appearance; that of men more with status (Dittmar et al. 1995, 1996). Men are more interested in sports cars than in 'useful' cars, such as a family-friendly hatchback. We have seen above that the same brain nuclei are activated when men see sports cars, sexual stimuli, beautiful female faces and money. Buying a sports car rather than a car suited for plain transport, may enable men to give a handicap signal of wealth (Erk et al. 2002). The latter authors compare the sports car to the peacock's tail. This is a risky comparison as humans have bi-paternal care, while peacocks do not. Thus, the fancy car is perhaps a good signal for women seeking 'good genes', or even direct material reward, during a one-night stand, but a woman looking for a good caretaker may do better by avoiding the dandy with a tendency to spend money on sports cars, rather than the education of his children.

Conclusion: if we use our skills evolved in the domains of foraging and paternal care when purchasing goods and services, we remain pretty close to what economists would consider rational behavior. When it comes to buying commodities that have an impact on our mating behavior, however, we may mix up fitness maximization with utility maximization. Notably, in an era of birth control we may end up behaving irrationally in both domains.

13.5.2 Hiring by employers

The employment market has been likened to a 'marriage market' by economists ever since the classic papers on matching models by Gale & Shapley (1962) and Becker (1973). Animal mating markets, in turn, have been compared to human economic markets by biologists (Noë & Hammerstein 1995, Miller & Todd 1998). It seems obvious that several mechanisms used on both markets are identical, which is confirmed by the fact that the same people tend to be successful in both (Harper 2000). However, there is not only an analogy between the problems that traders on either market need to solve, but apparently the same mechanisms are also used in spite of the fact that this can lead to sub-optimal choices in the employment market. Good-looking people are more successful in their careers; they are preferentially hired as employees, get higher salaries etc. This phenomenon is known as 'the beauty premium'. A citation from Aharon et al. (2001): "The strong motivational influence of beauty has been shown in studies of labor markets suggesting that there is a 'beauty premium' and 'plainness penalty' (Hamermesh & Biddle 1994) such that attractive individuals are more likely to be hired, promoted, and to earn higher salaries than unattractive individuals (Marlowe et al. 1996, Frieze et al. 1990, 1991)".

The phenomenon can be explained partially by rational economic behavior of the employer; by preferring signs of health, he gets a better employee, everything else being equal. There are also examples of businesses that gain directly as a result of having good-looking employees (Pfann et al. 2000). The 'sexual selection' hypothesis would also predict an 'ugliness premium' for people hired by same-sex staff managers, but I have found no proof of that in the literature. Nevertheless, it seems to me that in many cases an employer behaves irrationally, when he ignores reliable signals such as CVs and follows his sexually-selected hunches.

13.6 Summary

The trading of commodities, like goods, services and information between human beings, is almost certainly the most widespread form of cooperation between unrelated members of the same species on earth. Trading would not take place if it did not normally result in a net benefit for all participants, but the potential for conflict over exchange rates looms large. It is therefore crucial for each trader to have a number of mechanisms at his disposal that ensure optimal profit. In this chapter, I have speculated about the evolutionary roots of such mechanisms. The main question I asked was whether we can trace mechanisms used in cooperation and trading by modern humans back to homologous mechanisms used by ancestral species. I speculated that, for some mechanisms, many species borders must be crossed in order to arrive at their evolutionary origin.

Two sets of mechanisms can be bracketed together as 'partner choice mechanisms' and 'partner control mechanisms'. Choice is what makes markets turn; it forms the link between supply-demand ratios and exchange rates of commodities. Partner choice can be based on certain attributes of the potential partners themselves ('attribute-based partner choice') or on the basis of the goods or services they offer ('commodity-based partner choice'). Partner control becomes important only after trading relationships have formed; it ensures continuing profit from ongoing relationships.

Mechanisms used in cooperation can be homologous in two different ways. The same mechanism can be used by multiple living species that share a common ancestor. It is then reasonable to assume that the mechanism evolved for the first time in that common ancestor or even earlier. I baptized this classical form of homology 'vertical homology' in order to distinguish it from another form of homology, which I called 'horizontal homology', the use of the same mechanism in different domains by members of the same species. Within domains, I referred to behavioral complexes such as 'foraging' and 'reproduction' that are under more or less independent selective forces.

In order to identify candidate homologies, I first considered analogies, mechanisms that resemble each other and are used by members of different species cooperating in comparable circumstances or mechanisms used by the same individuals in cooperation and in other domains. I then speculated as to whether some of these analogies could also be homologies. I identified some domains that are likely to have spawned pre-adaptations ('exaptations') for mechanisms used in cooperation. Mechanisms used in foraging are, for example, likely to be used in commodity-based partner choice as well and skills evolved in connection with the choice of sexual partners are likely to be used in attribute-based mate choice.

In the case of vertical homologies of mechanisms used by humans in cooperation and trade, one has to split mechanisms into those evolved under natural selection and those evolved under cultural selection. The mechanisms I am interested in, those that evolved under natural selection, are more likely to have evolved long before bipedal primates turned the planet into a dangerous place to live. I tried to identify candidate mechanisms by looking at emotions, such as fear, trust and jealousy, which are shared by humans and animals and by looking more closely at mechanisms used in partner choice. A number of studies, among them several using advanced techniques like fMRI, have identified sub-cortical brain areas that are active in both cooperative and non-cooperative contexts in human as well as non-human primates.

The homology between a mechanism used by modern humans in trading and a mechanism used by a distantly-related species in cooperation shows best when the former is not optimally adapted to its modern use. Sub-optimal adaptation of a genetically-determined mechanism is likely when it has to adapt to a fast-moving target or when it is used in different domains. In the latter case, the mechanism is likely to be adapted to the domain in which it has, or recently had, its biggest impact on fitness. I identified such apparent 'mal-adaptations' in two forms of human economic interaction: (i) consumer behavior and (ii) the hiring of personnel.

Acknowledgments

I would like to thank Peter Kappeler and his fellow organizers of the "4. Göttinger Freilandtage" for inviting me to a very inspiring meeting. I owe Peter, Carel van Schaik, Louise Barrett and an anonymous reviewer for their constructive comments on an earlier version that was even more chaotic than this one. Louise also greatly improved language and flow. I would finally like to thank Tim Clutton-Brock for pointing out the unsolved problem of price setting in the baby market.

Part VI
Cooperation in Humans

Chapter 14

Reputation, personal identity and cooperation in a social dilemma

Manfred Milinski

14.1 Introduction

Many problems of human society, such as overexploiting fish stock or the difficulty of sustaining the global climate, are problems of achieving cooperation. When individuals, groups or states are free to overuse a public good, they usually overuse it. Thus, public goods are at risk of collapsing, which happens to health insurance systems, fish stock and most probably the global climate (Hasselmann et al. 2003). This problem is known as the 'tragedy of the commons' (Hardin 1968, 1998). Social and political scientists, economists and recently evolutionary biologists have studied this issue intensively (e.g. Ledyard 1995, Ostrom 1999, Fehr & Fischbacher 2003, Nowak & Sigmund 2004). Several potential solutions to this social dilemma have been proposed and/or shown in experiments with human subjects: (i) punishment of uncooperative group members (Boyd & Richerson 1992, Gintis 2000, Sigmund et al. 2001, Fehr & Gächter 2000b, 2002), (ii) costly signaling with altruistic acts (Gintis et al. 2001), (iii) voluntary participation in the public goods game (Hauert et al. 2002, Semmann et al. 2003), (iv) a kind of interaction with both indirect reciprocity situations (see 14.3; Milinski et al. 2002a, Semmann et al. 2004, 2005) and a trust game (Barclay 2004). There are several examples from human societies where the social dilemma has been successfully avoided, at least for some time, by mechanisms such as control of access to public goods by the local community (Berkes et al. 1989). Nevertheless, tragedies of the commons are usually found to be tragedies.

The 'tragedy of the commons' problem can be formalized as a 'public goods game' (e.g. Ledyard 1995). The classic public goods game consists of four players, who are given the opportunity to contribute money into a public pool. The content of the pool is doubled, divided by the number of players and evenly paid to all players, irrespective of their contributions. The social dilemma lies in the conflict between the group and the individual's interest. The group does best when all players cooperate. However, a rational individual should never contribute anything, because each money unit paid into the pool yields, doubled and then divided by four, only a return of a half-unit to the contributor. Interestingly, public goods games played by human subjects usually start with unpredicted cooperation but cooperation collapses within a few rounds. Fig. 14.1 gives an example of groups of six human subjects each playing the public goods game for eight consecutive rounds. Had they all cooperated during all eight rounds,

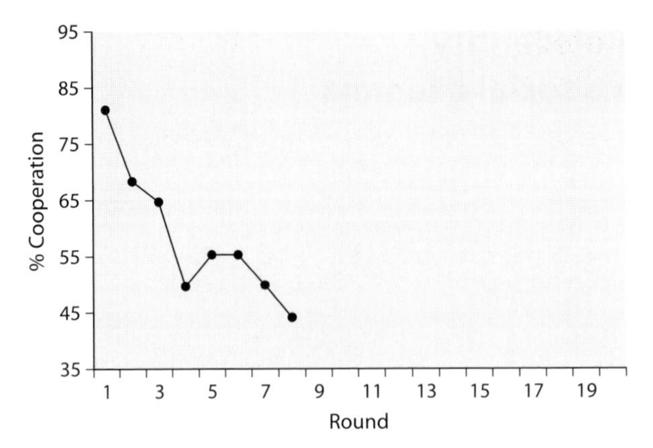

Fig. 14.1. Percentage of cooperation ('YES') per group of six subjects in each round of the public goods game. From Milinski et al. (2002a).

everybody would have gained a net income of 8 Euros, but they actually earned much less; hence, the social dilemma.

14.2 Indirect reciprocity: give and you shall receive

In direct reciprocity (Trivers 1971, Axelrod & Hamilton 1981), donor and recipient of an altruistic act meet repeatedly in alternating roles. If the cost to the donor is smaller than the gain to the recipient, both have a net benefit in the long run. Recently, theorists (Nowak & Sigmund 1998a, 1998b, Lotem et al. 1999, Fishman 2003, Mohtashemi & Mui 2003) have shown that cooperation can also evolve through indirect reciprocity (Alexander 1987, Zahavi 1991); "give and you shall receive", as the bible says. By helping others, who do not have the possibility of returning the help to the donor in the future, people build up good reputation or a positive image score, whereas refusing to help damages the reputation (Table 14.1). If one helps those who have helped others, one helps those who have a reputation for helping. Third parties reciprocate the altruistic act.

Nowak & Sigmund (1998a, 1998b) provided the first formal proof of the potential evolution of cooperation through indirect reciprocity. They assumed that each player has an image score, which is increased by each act of helping and

Table 14.1. The idea of indirect reciprocity:.

1) A observes that B helps C
2) A helps B

is decreased by each act of withhold helping. Now, a rule is needed that defines which minimum image score B has to have so that A would help him. Such rules for helping ranged from −5 to +6. A −5 strategy helps everybody who has an image score of at least −5, which means almost unconditional helping. A +6 strategy helps everybody who has an image score of at least +6, which means almost unconditional defecting. A zero-strategy is maximally discriminating; it helps everybody who has a zero or positive image score. Nowak & Sigmund (1998a) found that this most discriminating strategy dominated the population after 150 generations in their computer simulations.

We designed an experiment to test whether students would give money to others even if they know that there is no hope for direct reciprocity. If they do give money, would they help preferentially those who have a positive image score? Wedekind & Milinski (2000) tested eight groups with 10 students each from a first year biology course at the university of Bern.

All the subjects were anonymous throughout and after the game. Each subject was randomly assigned a number and a box with two button keys, one for 'YES', which would light a green lamp on the central desk, and one for 'NO' which would light a red lamp. A special procedure insured that nobody could hear who was responding. Each subject was provided with a starting account. The benefit of giving was always 4 Swiss Francs for the receiver, whereas the cost for the donor was 2 Swiss Francs. Each group played six rounds. Each subject played once per round both as a 'donor' and as a 'receiver'. The students knew that they would never meet the same player again with alternated roles; direct reciprocity was excluded. Each pair of players was randomly chosen, and the one who was chosen as a 'donor' was plugged to the lamps. The donor's decision (Would you give 2 Swiss Francs to player no. x? 'YES' or 'NO'?) was written down on a protocol sheet that was fixed to the blackboard behind the operator (left donor, right receiver). So everybody could see all of the previous choices in each interaction. Then the next pair was chosen and two other protocol sheets were displayed. After the game, everybody received her or his money in a way that did not disclose the players' identity.

The receivers' history of giving had a significant effect on the donors' decisions whether to give money; the image score of receivers who were given money was on average higher than the score of those who got nothing in all but one of the eight groups (Fig. 14.2; Wedekind & Milinski 2000). Furthermore, those who gave rarely, supported only players that had a very high image score. Other experimental studies found similar results (Bolton et al. 2001, Milinski et al. 2001, 2002b, Seinen & Schram 2001, Wedekind & Braithwaite 2002, Semmann et al. 2004, 2005). Hence, human subjects who have been helpful in the past are more likely to receive help from others through indirect reciprocity.

The success of the image scoring strategy (Nowak & Sigmund 1998a, 1998b) has been questioned by other theorists (Leimar & Hammerstein 2002, Panchanathan & Boyd 2003) who showed, by using a more complex population structure, that Sugden's (1986) 'standing' strategy is superior in establishing cooperation through indirect reciprocity. They argued that if according to the image scoring strategy a player would correctly refuse to help an individual with a low score, thereby reducing his own score, this player would suffer from not being helped

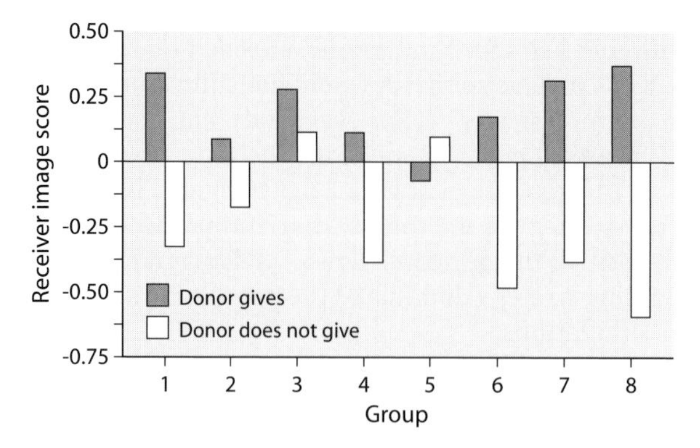

Fig. 14.2. Grey columns show the average image score of all the receivers who received something and white columns show the average image score of all the receivers who did not receive anything in each of the eight groups. Data are shown as deviations from the means per group and round to correct for group and round effects. From Wedekind & Milinski (2000).

thereafter. In Sugden's (1986) model, everybody is initially in good standing. An individual loses good standing by failing to help a recipient in good standing, whereas failing to help recipients who lack good standing does not damage the standing of a potential donor. This appears to make sense intuitively. Recently, both Brandt & Sigmund (2004) and Ohtsuki & Iwasa (2004) have done an in depth analysis of these two and several other updating rules for the indirect reciprocity game, and found standing strategies to be more successful than scoring strategies in most cases.

Now, it would be interesting to know, which strategy human subjects adopt when they cooperate in indirect reciprocity games. Since the results of previous experimental studies are compatible with either strategy, Milinski et al. (2001) addressed the question of standing versus scoring in a specific experiment. Each group included a secretly instructed NO-player who always refused to help. Strategies of either type would always refuse to give aid to such a player. The decisive question was, would their potential donors in turn penalize these players? Image scorers would penalize them, standing players would not. Fig. 14.3 shows that these donors of the NO-players were penalized almost as often as predicted for scoring but much more than expected from standing. Moreover, they appeared to compensate for 'NOs' to the NO-player by fewer 'NOs' to others, which would not make sense if they expect their co-players to follow a standing strategy. Providing the subjects with first plus second order information ('much information') compared with first order information only ('little information') appeared to have no obvious effect on the donors' strategy (Fig. 14.3), probably because all players had directly observed all previous interactions. Bolton et al. (2002) obtained similar results.

Why do humans not behave according to 'standing' as theory suggests? In order to decide whether a potential receiver is in good standing after he had refused help three times, one needs to know whether his last three receivers had been in good or bad standing, which implies that one must also know whether

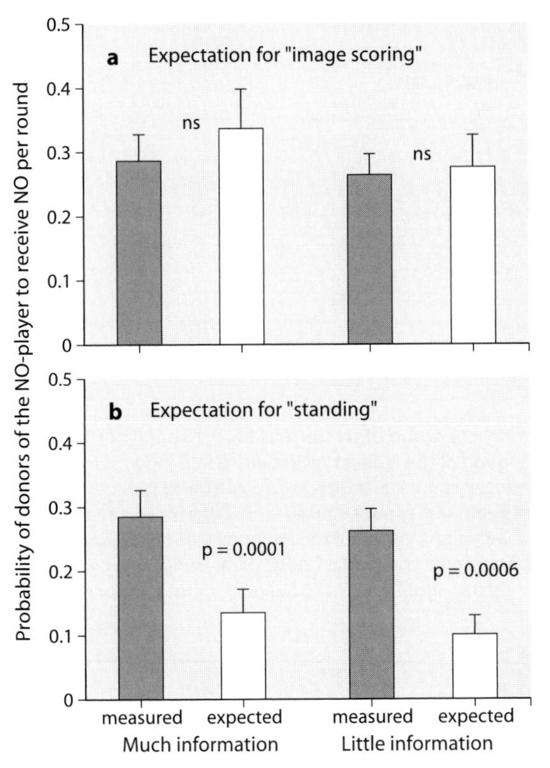

Fig. 14.3. Mean (+SE) probabilities of donors of NO-players receiving NO per round (during 16 rounds) of 12 groups with much information and 11 groups with little information of six subjects plus one NO-player each. (**a**) Expectation for image scoring; measured probabilities (grey bars) are compared with expected probabilities (white bars). (**b**) Expectation for standing. From Milinski et al. (2001).

their receivers had been in good or bad standing, which implies... etc. The evolution of a less demanding strategy such as image scoring may therefore reflect the constraints imposed by our limited working memory capacity (Milinski & Wedekind 1998). Because theorists mostly restricted their analysis to binary scores, taking only the score or standing of each player's last round into account (i.e. having a plus or minus score, good or bad standing), neither standing's memory problems nor scoring's advantage among cooperators could be fully expressed; in a mostly cooperative environment, image scorers would not be penalized for refusing help to rare defectors because their image score remains positive.

14.3 Reputation helps solve the 'tragedy of the commons'

Because individuals and countries often participate in several social games simultaneously, the interaction of these games may provide a sophisticated way by which to maintain the public resource. In order to gain profit in indirect reciprocity situations, it is important to build up and maintain a high reputation (image score or standing). If public goods and indirect reciprocity situations alternate, it would be worth not damaging one's reputation in a public goods situation by withholding help, because one meets the same co-players in both situ-

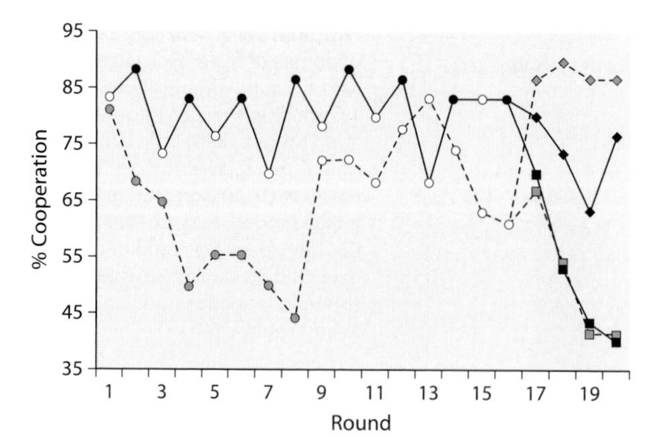

Fig. 14.4. Percentage of cooperation ('YES') per group of six subjects in each round of the public goods game (filled symbols) and in each round of the indirect reciprocity game (open symbols). In one treatment, the groups alternated between rounds of indirect reciprocity and rounds of public goods until round 16 (black); in the other treatment, groups started with eight consecutive rounds of the public goods game and continued with eight rounds of the indirect reciprocity game (grey); in rounds 17-20, groups of both treatments played the public goods game, which was either announced, "from now on only this type of game until the end" (squares), or not announced (diamonds). From Milinski et al. (2002a).

ations again. To test the potential interaction between public goods and indirect reciprocity games, Milinski et al. (2002a) performed an experiment with groups of six students, each of the university of Hamburg. Nine groups first played eight rounds of the public goods game followed by eight rounds of the indirect reciprocity game; 10 other groups played indirect reciprocity rounds alternated with public goods rounds. Both the rules of the game and the complete game itself were projected onto a big screen that everybody could see. Thus, everybody had complete information about everything except for the real identity of the players. All players had individual pseudonyms that lasted throughout the games, as in Milinski et al. (2001). Cooperation declined as usual in groups that began with eight rounds of public goods and built up during the subsequent indirect reciprocity rounds. However, when rounds of both games were alternated, the high starting level of cooperation was maintained during all eight rounds of public goods (Fig. 14.4). A bad reputation from not contributing to the public pool was recognized in the indirect reciprocity game where players refused to support such individuals. However, they supported individuals who had contributed to the public pool. Through this transfer of reputation between games, cooperation was maintained throughout the experiment. Cooperation in the public goods game paid off. Groups that alternated rounds of indirect reciprocity and public goods games, and thus were more cooperative in the public goods games, earned significantly more money during the eight rounds of the public goods game than did groups that played the two games in blocks of eight rounds each. The commons became productive and could be harvested.

Four additional rounds of public goods directly tested the hypothesis that interaction with the indirect reciprocity game keeps up cooperation in the public goods game. Groups in both treatments played public goods in rounds 17-20. Every second group was told before round 17 that from then on only public goods rounds would follow until the end of the game. In these groups, cooperation declined during the four public goods rounds, whereas cooperation was maintained when the risk of further rounds of indirect reciprocity was not excluded (Fig. 14.4). Obviously, refusing to give in the public goods game reduced the reputation of a player to a similar extent as if this person had refused to give in the indirect reciprocity game; his potential donor in the next round of indirect reciprocity just followed the rules for indirect reciprocity and refused to give to someone with a low image score.

Reputation can therefore maintain all-around contributions to public goods, and does so in the absence of a special punishing rule or motivation. The potential donor actually saves money by refusing to give, whereas punishing would be costly. Also, a recent theoretical analysis suggests that reputation is essential for fostering social behavior among selfish agents (Sigmund et al. 2001). In another empirical study, Wedekind & Braithwaite (2002) suggested that costly investment in reputation pays off in a subsequent direct reciprocity game (two-persons Prisoner's Dilemma), although cooperative persons being cooperative in both situations could have caused their result as well. Barclay (2004) showed that the need for becoming trustworthy in a future dyadic trust game maintained cooperation during the preceding five rounds of a public goods game. Here a kind of competitive altruism helped solve the tragedy of the commons.

14.4 Strategic investment in reputation

The previous results suggest that humans are aware of the possibility that their current behavior might affect their future gains depending on whether a situation where reputation pays off is likely to occur in the future or not. However, if this future situation will occur for sure but the subject knows that he/she will not be recognizable as the person that cooperated or did not cooperate in a previous game, will the person cooperate less than if he/she knows he/she will be recognized? In other words, do humans invest strategically in reputation when they know that their costly investment has a high probability of paying off in inevitable future interactions (Pollock & Dugatkin 1992, Ostrom 2003), and do they stop investing when this is not the case?

To test for strategic investment in reputation, Semmann et al. (2004) again alternated rounds of the public goods and the indirect reciprocity game but this time allowed for reputation transfer from the public goods to the indirect reciprocity game in some rounds but blocked this transfer in other rounds. They achieved this manipulation of the reputation transfer by providing the subjects each with two different new identities, i.e. two different pseudonyms. One name, the non-transferable name, was used only in some public good rounds, whereas the other name, the transferable name, was used in rounds of both games. Dur-

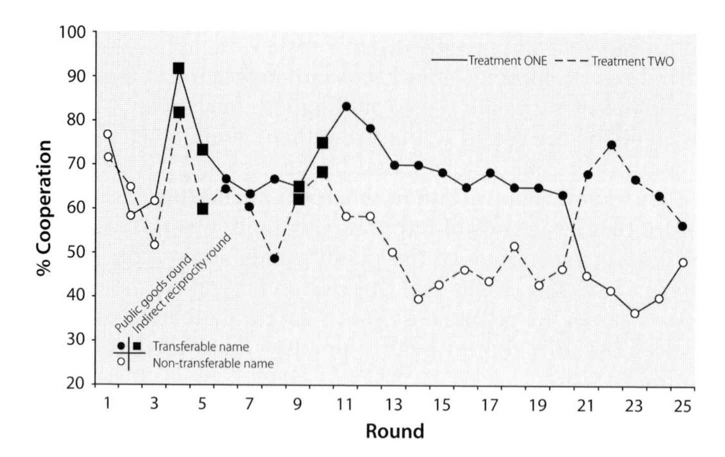

Fig. 14.5. For the public goods (PG) rounds (circles) and indirect reciprocity (IR) rounds (squares), the group mean 'YES' per round for both treatments are shown. In treatment 1 (black line), the groups played PG rounds, from round 11 to round 20 with their transferable name (filled symbols) and from round 21 to 25 with their non-transferable name (open symbols). In treatment 2 (dotted line), the groups played PG rounds, from round 11 to round 20 with their non-transferable name and from round 21 to 25 with their transferable name. The period from round 1 to 10 was identical in both treatments (three PG rounds played with the non-transferable name, two IR rounds with the transferable name, three PG rounds with the transferable name and two IR rounds with the transferable name. From Semmann et al. (2004).

ing each round of the public goods game, the subjects knew whether they played with their transferable or non-transferable name. Would they invest less in the public pool when playing with the non-transferable name?

After some training rounds during which all subjects experienced the alternation of public goods rounds with the non-transferable name and indirect reciprocity rounds, and the same with the transferable name, 10 groups with six subjects each played 10 consecutive public goods rounds with their transferable names followed by five rounds with their non-transferable names. To control for sequence effects, 10 other groups played their first 10 consecutive rounds of public goods with their non-transferable names followed by five rounds with their transferable names.

Fig. 14.5 shows that the level of cooperation was much higher during the rounds with the transferable name. Hence, the subjects used different strategies in the public goods game, conditional on whether the player knew that his decision would be either known or unknown in another social game. The knowledge of being recognized as the same individual in both games motivated players to invest in their reputation and thus sustain the public resource. The subjects earned significantly more money in public goods rounds with the transferable name than in rounds with the non-transferable name. If the circumstances render people recognizable, they will invest in the public good and because every one else does too, profit themselves. Humans strategically invest in various ways to preserve their good reputation within their own social group (Engelmann &

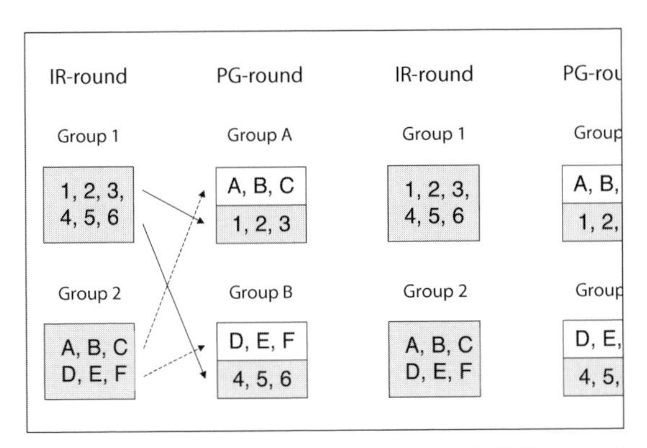

Fig. 14.6. Group composition in public goods (PG) and indirect reciprocity (IR) rounds. See text for details.

Fischbacher 2002). Also, in non-human animals, strategic reputation building seems possible. Bshary (2002) suggested that tactical deception in cleaner fish should occur if it pays to alter optimal behavior in a situation to induce responses in bystanders (clients), which will produce benefits during future interactions with these bystanders that exceed the momentary costs.

14.5 Reputation is valuable within and outside one's own social group

Players reward cooperative behavior of their group members in the public goods game and refuse to reward uncooperative behavior; this may be a kind of direct reciprocity (see also Engelmann & Fischbacher 2002). The interaction with indirect reciprocity should therefore sustain cooperation in public goods games only when the same players play in both games. The incentive for donating in public goods games would be direct reciprocity, not reputation building. We expect that cooperative players from another public goods group should not be rewarded in indirect reciprocity games.

To test this prediction, Semmann et al. (2005) had 12 students playing simultaneously, but in groups with six players each. The 12 subjects sat in one half circle watching a big screen. They were told that six of them would play together in one group for the pair (i.e. indirect reciprocity) game and the other six in another group for the same type of game. The same six people would play the pair game repeatedly and they could recognize their co-players by their pseudonyms. Every second round, they would play the group (i.e. public goods) game. For this game, three people of each group for the pair game would be combined with three people of the other pair game group (Fig. 14.6). Also, this group composition would remain stable whenever they played the group game. Again, they would recognize their co-players by their pseudonyms. Everybody could observe every interaction on the screen, and the sequence of 'YES' and 'NO' decisions of each potential receiver was displayed in the indirect reciprocity rounds. There were

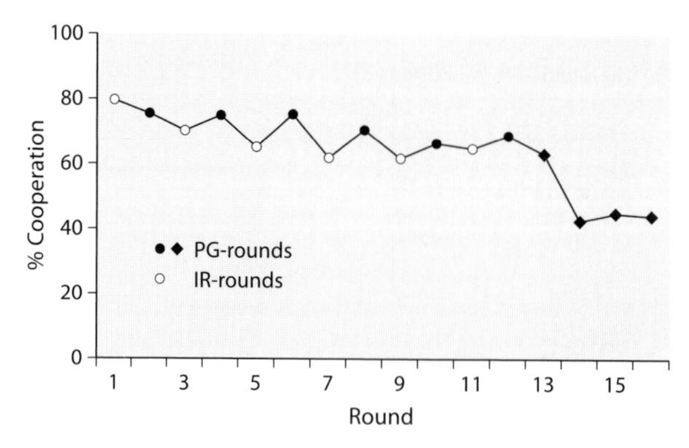

Fig. 14.7. Average level of cooperation ('YES') per round per group of six subjects each. Open circles represent indirect reciprocity (IR) rounds; black symbols represent public goods (PG) rounds. From Semmann et al. (2005).

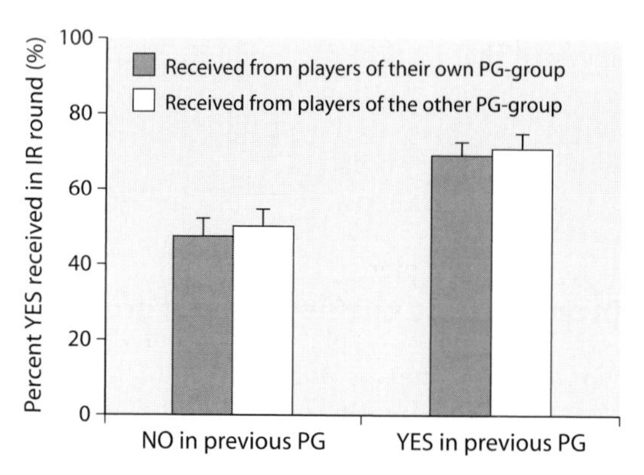

Fig. 14.8. Mean (+SE probability per group of receiving 'YES' in a round depending on whether the recipient had either given or not given in the previous public goods (PG) round, and on whether the recipient and the donor had played together or not together in the same PG group. From Semmann et al. (2005).

six rounds of each type of game. This procedure was explained several times in detail and the subjects were asked whether they had understood it completely.

Fig. 14.7 shows the overall results from all the 20 groups with 12 players each. The level of cooperation is again higher in public goods games than in indirect reciprocity games and it dropped after the announcement that there would be only group games until the end (black rhombs in Fig. 14.7). However, were group members and non-group members treated differently in indirect reciprocity rounds? Now one can test whether subjects coerce preferentially their co-players from their own public goods group.

Fig. 14.8 shows the percentage reward that subjects who played 'NO' or 'YES' in the public goods game received from members of their own public goods group (grey) and from members of the other public goods group (white) in the

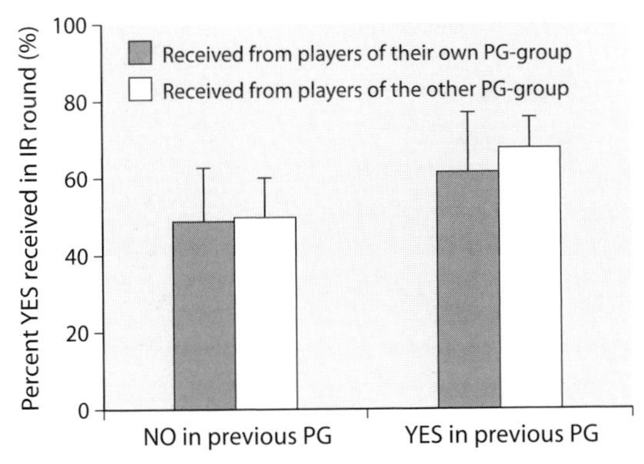

Fig. 14.9. Mean (+)SE probability per group to receive 'YES' in a round depending on whether the recipient had either given or not given in the previous public goods (PG) round, and on whether the recipient and the donor had played together or not together in the same PG group. Only decisions of participants who had answered in the questionnaire that they were conscious as to whether or not they had played together with the current receiver in the PG round and in addition had also used this information for their decision finding always or often are included. From Semmann et al. (2005).

indirect reciprocity game. YES-players received significantly more rewards than NO-players in either case. Surprisingly, donors did not take into account who came from their own group and who was a foreigner.

Had the players not understood the procedure of reshuffling players between groups? In anticipation of this potential problem, all players had been asked in a questionnaire after the experiment: did you realize that a person with whom you played in a 'pair-round' had/had not been in your 'group-round' (yes/no)? Did you use this information in the game (always/often/seldom/never)? The data were analyzed, again taking into account only those players who had responded 'YES' and 'always' or 'often'. These subjects should have both understood the procedure and used the information.

The new results are shown in Fig. 14.9; they look almost exactly the same as when all data were taken into account (Fig. 14.8). Obviously, the subjects did not treat own group members and foreign group members differently, even though they recognized them. This suggests that reputation is a label that is taken into account in social interactions irrespective of where this reputation was gained.

14.6 Gaining good reputation through donations to charity

If reputation is transferable between social groups, transporting the signal that a person is a valuable and trustworthy social partner, the same kind of reputation could be gained by giving to charity. It has to be insured, however, that people know about it. Milinski et al. (2002b) did an experiment, in which groups of six subjects each played several rounds of the indirect reciprocity game. After

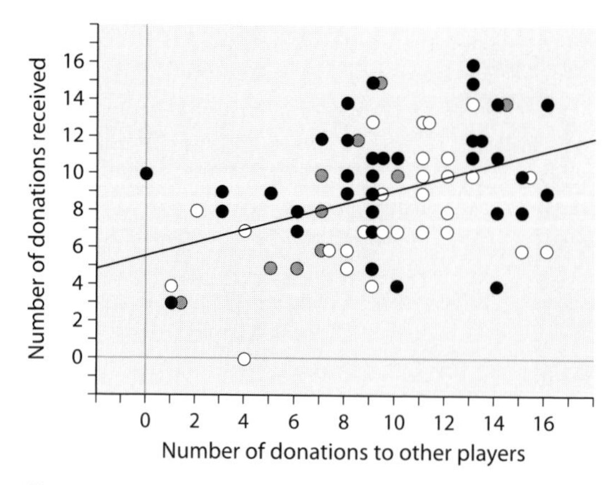

Fig. 14.10. Human subjects received money indirectly related to the amount they gave to others; i.e., the more they gave to others the more they received. The solid line depicts linear regression. Black circles are charitable donors (UNICEF) who gave more than the median, open circles are donors who gave less than the median and grey circles are median donors. From Milinski et al. (2002b).

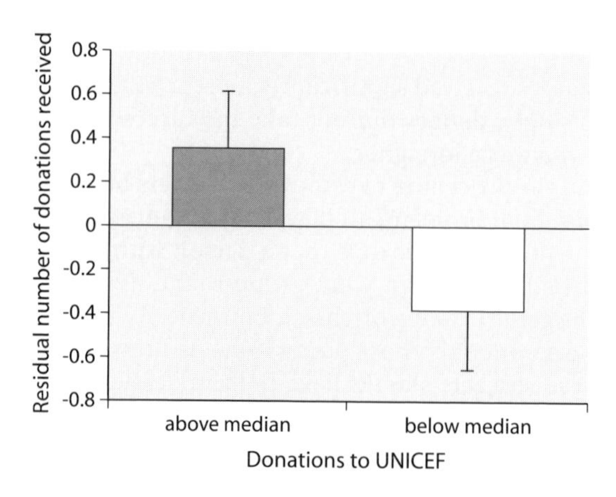

Fig. 14.11. Mean (+SE) residual number of donations per group of charitable donors (UNICEF) who gave more than the median (grey column) and who gave less than the median (white column) received from members of their group in indirect reciprocity rounds. From Milinski et al. (2002b).

each episode of being asked: "Do you want to give DM 2.50 to this person?" (if 'YES', the receiver obtained DM 4), the player was asked: "Do you want to give DM 2.50 to UNICEF?" (if 'YES', UNICEF obtained DM 4). It was made clear that the amount of money on UNICEF's account would be sent to UNICEF and every player would receive a copy of the receipt.

Fig. 14.10 shows that the more money subjects gave to players of their group, the more money they received from players of their group as expected for an indirect reciprocity game. The black dots depict players who donated more than the median amount to UNICEF and the open dots depict players who donated less than the median. There are more red dots above and more open dots below the regression line. This means that players of the group rewarded reputation gained by donating to UNICEF and withheld reward from persons who donated little or nothing to UNICEF. This can be shown more directly by comparing the residuals of black and open dots to the regression line (Fig. 14.11).

Even a reputation gained through giving money away from the group is rewarded in indirect reciprocity rounds. However, this only worked when donations to UNICEF were made 'in public', i.e. appeared on the screen. In a few control groups, in which the subjects knew that no information about their donations would be announced to the group, dramatically less money was donated privately to UNICEF. These results were expected based on the results of the previous experiment, illustrated in Figs. 14.8 and 14.9.

Relief organizations could probably collect more money if they would offer a way by which donations could be made in public. The Swiss radio station DRS has regularly asked people and companies to donate money ('Glückskette') whenever there was a catastrophe anywhere in the world, for example a flood in Bangladesh. The money had to be sent to a specific account that was observed by DRS. Every hour, just before the news, long lists of donations specifying the donor and the amount of money sent (e.g. a little boy has donated 2 Swiss Francs, a company 500) were read. Obviously, the radio station had found out that more money would be donated if donors would be named with their donations before the news. This is a very common procedure in the USA (Peter Kappeler, pers. com.).

14.7 Summary and conclusions

Many problems of human society, such as overexploiting fish stock or the difficulty of sustaining the global climate, are problems of achieving cooperation. When individuals, groups or states are free to overuse a public good, they usually overuse it. Thus, public goods are at risk of collapsing, which happens to health insurance systems, fish stock and probably the global climate. This problem that is known as the 'tragedy of the commons', has been studied intensively by social and political scientists and economists for decades, and recently by evolutionary biologists. Except for allowing for punishing defectors, no scenario that strongly facilitates a cooperative solution of the tragedy of the commons has been demonstrated yet. We could show that an unexpectedly efficient solution of the problem can be achieved when personal reputation, which is important for other social interactions such as gaining support through indirect reciprocity ("give and you shall receive"), is at stake in the public goods situation. When this interaction is allowed for, the public good is not only sustained but also provides all participants with a high payoff. Humans strategically invest in various ways to preserve their good reputation within their own social group; the subjects used different strategies in the public goods game conditional on whether the player knew that his/her decision would be either known or unknown in another social game. The knowledge of being recognized as the same individual in both games motivated players to invest in their reputation. Reputation gained in social interactions is transferable to other social groups, where it seems to be valued just as highly as within one's own social group. If reputation signals that a person is a valuable and trustworthy social partner, the same kind of reputation can be gained by giving to charity. Humans rewarded reputation gained by donating to UNICEF 'in public' and withheld reward from persons who donated little or nothing to UNICEF. Relief organizations could probably collect more money if they would offer a way by which donations could be made in public, which is already practiced in some countries. A good reputation is a valuable currency, which can be accumulated during observable actions. Direct observation, gossip and modern telecommunication can transmit the signal. Many social interactions are based on a person's trustworthiness, which has a name, reputation.

Acknowledgments

I thank Dirk Semmann and Derk Wachsmuth for help with preparing the figures and Peter Kappeler for his patience.

Chapter 15

Human cooperation from an economic perspective

Simon Gächter, Benedikt Herrmann

15.1 Introduction

Many important economic and social situations are characterized by a conflict of interest between individual and group benefits. The 'tragedy of the commons' (Hardin 1968) is probably one of the best known examples of this problem. Each individual farmer has an incentive to put as many cattle on the common meadow as possible. The tragic consequence may be overgrazing from which all farmers suffer. Collectively, all farmers would be better off if they were able to constrain the number of cattle that simultaneously graze on the commons. Yet, each individual farmer is better off by letting his cattle graze. A similar tension between individual and collective rationality is typical in such diverse areas like warfare, cooperative hunting and foraging, environmental protection, tax compliance, voting, the participation in collective actions like demonstrations and strikes, the voluntary provision of public goods, donations to charities, teamwork, collusion between firms, embargos and consumer boycotts, and so on.

While the logic of self-interest is straightforward, the facts seem to be at odds with theoretical predictions derived under the joint assumptions of rationality and selfishness. At the societal level, our societies have achieved a degree of cooperation and division of labor among genetically unrelated individuals that is unprecedented in the animal kingdom (see Seabright 2004 for a recent account). At a lower level, the fact that people even in anonymous situations vote, take part in collective actions, often manage not to overuse common resources, care for the environment, mostly do not evade taxes on a large scale, donate to public radio, as well as to charities, etc. suggests that the strict self-interest hypothesis is inconsistent with the degree of cooperation that we observe around us.

How can we explain this? This paper presents evidence from systematic experimental investigations on how people solve cooperation problems. Laboratory experiments are probably the best tool for studying cooperation. The reason is that in the field many factors are operative at the same time. The laboratory allows for a degree of control that is not feasible in the field. In all the laboratory experiments that we will discuss below participants, depending on their decisions, earned considerable amounts of money. Thus, the laboratory allows observing real economic behavior under controlled circumstances (see Friedman & Sunder 1994 for an introduction to methods in experimental economics and Kagel & Roth 1995 for an overview of important results).

In the next section, we will introduce two prototypical cooperation games that have been extensively investigated in experiments: (i) the 'Prisoner's Dilemma' (PD) and (ii) the 'public goods experiment'. These games are simple and contain the essence of the cooperation problems introduced above. Many of them are structured such that purely selfish individuals would not cooperate in these games. Yet, we will show that there is substantial cooperation even in completely anonymous one-shot situations. This finding has been termed 'altruistic cooperation' or 'altruistic rewarding' (e.g. Fehr & Fischbacher 2003) because apparently some people are prepared to benefit others by cooperating. Yet, most of this 'altruistic cooperation' takes the form of 'conditional cooperation'; people cooperate if others cooperate as well. 'Altruistic rewarding' has also been observed in other contexts (for surveys, see Fehr & Gächter 2000a and Camerer 2003, chapter 2).

One of the most important insights from the laboratory experiments is that in the absence of extrinsic incentives like reputation, social (dis-)approval and punishment, cooperation is fragile. Cooperation almost inevitably breaks down in repeated interactions. The reason is that conditional cooperators can only avoid being exploited by the free riders if they stop cooperating themselves. The lack of targeted punishment leaves the cooperators with the only option they have, stopping cooperation.

In Section 15.3, we will look at reputation, communication and social approval. These important mechanisms are frequently available in reality and may help to sustain cooperation. Reputation mechanisms have recently gained a lot of attention. It turns out that reputation can have a strong cooperation-enhancing effect. The same holds for communication. Similarly, there is also experimental evidence that social approval can lead to a substantial increase in cooperation.

Section 15.4 presents evidence that shows that many people are prepared to engage in altruistic cooperation but also in 'altruistic punishment'. They do this even in anonymous one-shot situations in which future benefits from reciprocal altruism (Trivers 1971), indirect reciprocity and reputation (Alexander 1987, Nowak & Sigmund 1998), signaling (Zahavi & Zahavi 1997, Gintis et al. 2001) and kinship (Hamilton 1964) are excluded by the experimental design. This punishment is altruistic, because it is costly to the individual and beneficial for someone else who interacts with the punished (and now well-behaved) individual in the future.

Section 15.5 discusses the role of emotions as a proximate mechanism that can explain altruistic punishment. Section 15.6 looks at evolutionary explanations for the observed behavior. Section 15.7 presents a summary and some concluding remarks.

15.2 Some stylized facts on cooperation

We start our discussion with a brief presentation about what is known about factors influencing cooperation and free riding. The most important vehicles for

Table 15.1. The Prisoner's Dilemma. The amounts in each cell refer to the players' payoff. In each cell, the left payoff refers to player 1's payoff, and the right payoff to player 2's payoff.

		Player 2	
		Cooperate	**Defect**
Player 1	Cooperate	€80, €80	€0, €100
	Defect	€100, €0	€35, €35

studying cooperation problems in controlled laboratory experiments are the PD and the 'public goods experiment'.

Table 15.1 illustrates the prototype cooperation game, the famous PD (see Poundstone 1992 for an illuminating discussion of this game). In the game of Table 15.1, two players are told that they can choose simultaneously between 'Cooperate' and 'Defect'. If both choose 'Cooperate', both earn 80 Euros each. If player 1, for instance, chooses 'Defect', while player 2 chooses 'Cooperate', player 1 earns 100 Euros, while player 2 gets nothing. If player 1 cooperates and player 2 defects, player 1 will earn nothing and player 2 will earn 100 Euros. If both players defect, they earn 35 Euros each.

If this game is played only once, selfishness predicts no cooperation. Yet, if 'the shadow of the future' is important, i.e, if players interact for an unknown length of time, and if people are not too impatient and therefore care for the future, then strategic cooperation becomes possible, because defection can be punished by withholding future cooperation and even more complicated punishment strategies (e.g. Fudenberg & Maskin 1986). The most famous idea is probably reciprocal altruism (Trivers 1971) and the related strategy of 'tit-for-tat', which turned out to be a very successful strategy in an 'evolutionary contest' where strategies played against each other in a computer simulation (Axelrod & Hamilton 1981). Its essence is the idea that favors are reciprocated ("I'll scratch your back if you'll scratch mine") and that unhelpful behavior is reciprocated by withholding future help.

Yet, the assumption that players are forced together for an unknown number of interactions may not hold (see Hammerstein 2003b for an extensive critique). In reality, people stop interacting with disliked partners and change social groups. Moreover, throughout (evolutionary) history, social groups were frequently disbanded by warfare, famine and other catastrophes (e.g. Knauft 1991, Gintis 2000, Fehr & Henrich 2003). These arguments suggest that studying short-term cooperation games is worthwhile. Moreover, though highly insightful, the studies by Axelrod & Hamilton (1981) are not about real behavior but are computer simulations. Therefore, we will turn next to some selected behavioral evidence on cooperation in finite PD games.

The PD game is probably one of the most extensively investigated games (see Rapoport & Chammah 1965, Colman 1999 and Ledyard 1995 for overviews on the

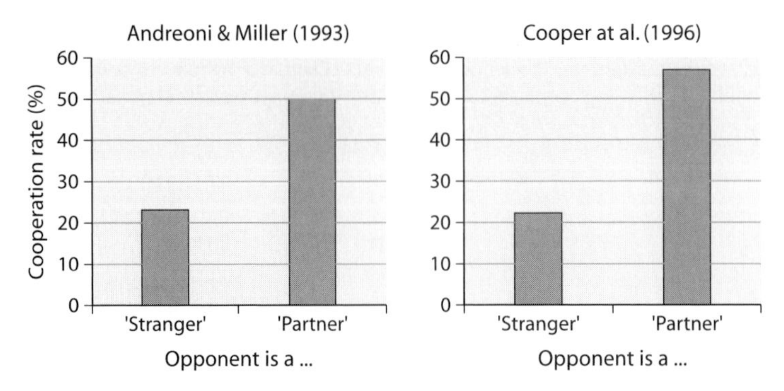

Fig. 15.1. Cooperation rates in the Prisoner's Dilemma. The figure shows the average cooperation rates from two studies, by Andreoni & Miller (1993) and Cooper et al. (1996), where players interacted for 10 periods, either with the same opponent ('Partner') or a randomly-matched opponent ('Stranger'). The prediction in both set-ups is a zero cooperation rate. Yet, in both set-ups, people cooperate, but substantially more in the 'Partner' than in the 'Stranger' set-up.

experimental evidence). Fig. 15.1 illustrates the results of two studies (by Cooper et al. 1996 and Andreoni & Miller 1993) in each of which the subjects played the game 10 times under two different conditions. In one condition, called the 'Stranger' condition, each player was matched with a new player in each of the 10 periods. In the second condition, the 'Partner' condition, the opponent stayed the same throughout all repetitions of the game. The subjects were informed about this. Thus, under the assumption of selfishness and rationality, all players in both conditions are predicted to defect. In the 'Stranger' condition, this prediction holds because each play of the game is against a new opponent and hence 'one-shot'. In the 'Partner' condition, the prediction holds with backward induction; in the last period, both players (who are assumed to be rational and selfish) will defect. Therefore, in the penultimate period, there is no incentive to cooperate, since players will surely defect in the last period. Hence, there is also no incentive to cooperate in the period prior to the penultimate one. Continuing this logic further implies that rational and selfish players will defect throughout. By contrast, if people are not completely sure that everyone is selfish, then it might pay to build up a reputation by cooperating if others cooperate until the final rounds, where a selfish player should defect for sure (see Kreps et al. 1982 for a game-theoretical explanation and Selten & Stoecker 1986 for a bounded rationality approach).

In both studies, the results in the 'Stranger' condition are at odds with this prediction. People cooperate on average in slightly more than 20% of the cases. To have a common future, if only for 10 rounds increases cooperation substantially. In the 'Partner' condition, the average cooperation rate is at least 50%. Thus: (i) people are prepared to cooperate even in one-shot games and (ii) the possibility of behaving strategically strongly increases cooperation.

Clark & Sefton (2001) studied an interesting variation of the game of Table 15.1. Instead of playing the game simultaneously, their subjects played the game sequentially, i.e. player 1 first made his or her choice, which was then observed by player 2 before deciding whether to cooperate or to defect. The subjects also played the game for 10 rounds in the 'Stranger' set-up. Clark & Sefton (2001) find that between 37% and 42% of the subjects cooperate conditionally on others' cooperation. Such conditional cooperation is also observed in two further treatments, 'double temptation', where the defection payoff was doubled, and 'double stakes', in which all payoffs were doubled. A statistical analysis shows that under 'double temptation', the fraction of conditional cooperation is reduced relative to the baseline, whereas 'double stakes' did not significantly affect the extent of conditional cooperation. Experiments on the sequential PD where the two players could also choose intermediate cooperation levels confirm the importance of conditional cooperation (e.g. Fehr et al. 1993, Fehr et al. 1997, Falk et al. 1999, Gächter & Falk 2002; see Fehr & Gächter 2000a for an overview).

These results are interesting, because the PD is such a simple and generic cooperation game. The fact that people cooperate (conditionally) even in one-shot games casts doubt on the selfishness assumption. The observation that there are strong effects of repeated interaction suggests that straightforward economic incentives are very helpful for successful cooperation. There can thus be no doubt that reciprocal altruism and the strategic gains from cooperation that come from repeated interactions are a powerful force in explaining real-world cooperation in small and stable groups. Yet, the success of reciprocal altruism in sustaining cooperation may be limited if groups become bigger. As has been shown theoretically (see Boyd & Richerson 1988), cooperation in the PD can only be sustained in groups larger than $n > 2$ if all other group members cooperated in the previous period. Thus, the basin of attraction for cooperation is very small because a few free riders can undermine cooperation. For this theoretical reason, it is worthwhile to move beyond dyadic relationships.

The most commonly used game for studying n-person cooperation problems is the public goods game. In contrast to a private good, a public good is a good which can be consumed even if one has not paid for it, or not contributed to its provision. Clean air, environmental quality and national security, but also collective reputations or team output are common examples of public goods.

An economic model of public goods provision is the public goods game. This game underlies many experiments that study cooperation for the provision of public goods. In a typical public goods experiment, four people form a group. All group members are endowed with 20 tokens. Each subject i has to decide independently how many tokens (between zero and 20) to contribute to a common project (the public good). The contributions of the whole group are summed up. The experimenter then multiplies the sum of contributions by 1.6 and distributes the resulting amount equally among the four group members. Thus each subject i's payoff is

$$\pi_i = 20 - g_i + \frac{1.6}{4} \sum_{j=1}^{4} g_j$$

The first term $(20 - g_i)$ indicates the payoff from the tokens not contributed to the public good (the 'private payoff'). The second term is the payoff from the public good. Each token contributed to the public good becomes worth 1.6 to-

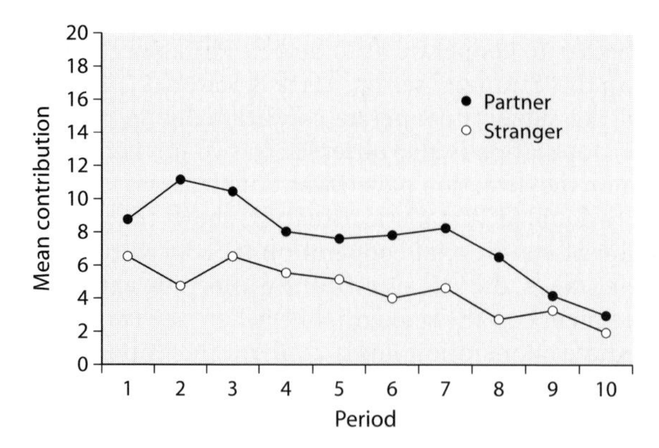

Fig. 15.2. Contributions to a public good in constant ('Partner') and randomly-changing groups ('Strangers') over 10 repetitions. Cooperation gains are maximized with full contributions (20 tokens). Selfishness predicts zero contributions. The figure shows that 'Partners' contribute more than 'Strangers' and that cooperation collapses in both treatments. From Fehr & Gächter (2000b).

kens. The resulting amount is distributed equally among the four group members, irrespective how much an individual has contributed. Thus, an individual benefits from the contributions of other group members, even if he or she has contributed nothing to the public good. Therefore, a rational and selfish individual has an incentive to keep all tokens for him- or herself, since the 'return' per token from the public good for him- or herself is only 0.4 (1.6/4), whereas it is one if he or she keeps the token. By contrast, the group as a whole is best off if everybody contributes all 20 tokens.

Since the public goods game is an n-person cooperation problem that is easy to implement and since it also reflects the tension between individual incentives and collective benefits, it has been frequently used in experimental studies (see Ledyard 1995 for an overview). Fig. 15.2 depicts a typical finding of a public goods experiment, where the exact same game is repeated 10 times and subjects know this. In each period, subjects receive 20 tokens and decide how many of them to keep or contribute to the public good. After each round, subjects are informed about what the other group members have contributed. Fig. 15.2 shows the resulting cooperation patterns in a 'Stranger' condition, where group members change randomly from round to round, and a 'Partner' condition, in which groups stay constant for all rounds.

Look at the 'Strangers' data first. Mean contributions start at about 6.5 tokens and decline to about two tokens in the 10 iterations of the public good game. In other words, by the end of the experiment, cooperation has almost entirely collapsed. As in the repeated PD, we find that cooperation in the 'Partner' condition is higher from the very beginning of the experiment. Yet, by the tenth round, cooperation has collapsed as well.

Fig. 15.2 illustrates two stylized facts from dozens of public goods experiments. First, as in the PD experiments reported above, 'Partners' contribute more than 'Strangers' (see Keser & van Winden 2000, and Andreoni & Croson 1998 for an overview). This result has also been found in other cooperation games (e.g.

Falk et al. 1999, Gächter & Falk 2002, Fehr & Fischbacher 2003). The significance of this and related findings is that people are immediately able to distinguish whether they are in a situation that requires strategic cooperation (the 'Partner' condition) or not (the 'Stranger' condition) and to adopt their behavior accordingly.

A second stylized fact is that cooperation is very fragile and tends to collapse with repeated interactions. Why is this so? One explanation is that people have to learn how to play this game. Since errors can only go in one direction, any erroneous decision looks like a contribution. Over time, people learn and commit fewer errors, which is why contributions decline (Palfrey & Prisbrey 1997). The problem with this explanation is that it is inconsistent with the fact that after a so-called 'restart' (after the tenth round, participants are told that they will play another 10 rounds), cooperation jumps up again and basically starts at the same level as in the first period. If learning would explain the decay in cooperation, then, after the restart, cooperation should have continued at the level at which cooperation was in the tenth round (see Andreoni 1988). A second explanation is that people are heterogeneous with respect to their cooperative inclinations. Some people are free riders who try to maximize their monetary income, irrespective of other group members' contribution. Other people are 'conditional cooperators', who cooperate if others cooperate.

To test this idea, Fischbacher et al. (2001) invented a design that allows measuring the 'type' of a player by observing each participant's contribution to the public good as a function of other group members' contributions. Specifically, subjects were asked to indicate for each possible average contribution of the other group members how much they would like to contribute to the public good. The payoff function is the same as in the other public goods experiments; i.e., incentives are such that, given others' average contribution, the monetary income is always highest if one contributes nothing. Thus, a free rider type will always contribute zero to the public good. A conditional cooperator type will increase his or her contribution in the average contribution of others.

Fig. 15.3 shows the results of experiments that applied the Fischbacher et al. (2001) design. Fig. 15.3 contains the pooled results of the experiments by Fischbacher et al. (2001) and Fischbacher & Gächter (2004) who conducted their experiments in Switzerland with n = 44 and n = 140 subjects, respectively. The authors also ran experiments with n = 148 subjects in various cities in Russia.

24% percent of the Swiss subjects turned out to be free riders who contribute nothing for all contributions of the other group members. In our Russian subject pools, this frequency is markedly lower. Only 7% turned out to be free riders. By contrast, the fraction of conditional cooperators who cooperate if others cooperate is strikingly similar in Russia and Switzerland. In Switzerland, 54% of the subjects show contributions that increase in others' contribution, whereas in Russia this is true for 57%. Fig. 15.3 shows the average contribution of all conditional cooperators. We find that not only the fraction of conditional cooperators, but also the average contribution schedules are very similar. The only difference is that our Russian subjects are prepared to contribute slightly more for a given contribution of the other group members than the Swiss subjects. A further remarkable result is that the average contribution schedule of conditional

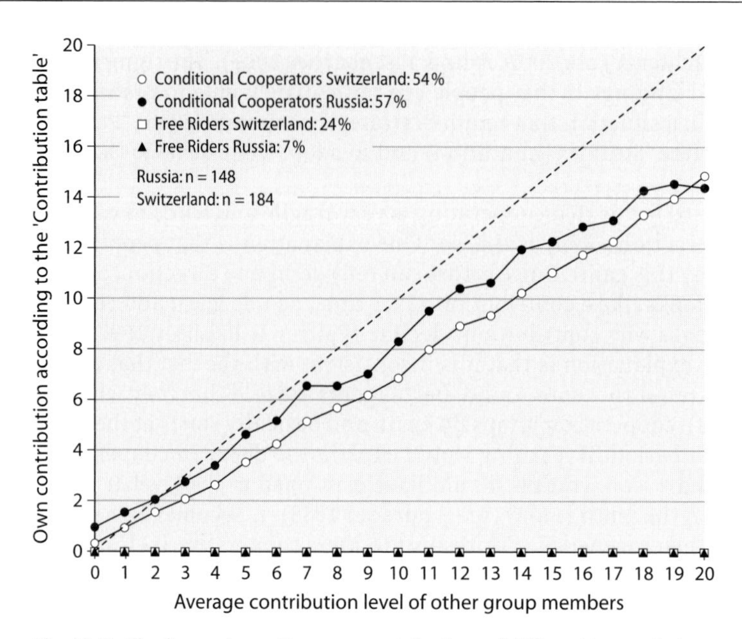

Fig. 15.3. The figure shows the mean contributions of different types of players to the public good as a function of other group members' average contributions. Free riders contribute nothing to the public good, irrespective of how much other group members contribute. In Switzerland [Russia], 24% [7%] of the subjects were free riders. Conditional cooperators increase their contributions the more others contribute. The graph 'Conditional Cooperators' is the average contribution of all subjects who report a contribution pattern that is increasing in other group members' contribution. In Switzerland [Russia], 54% [57%] of the subjects were conditional cooperators. From Fischbacher et al. (2001), Fischbacher & Gächter (2004) and new data from various places in Russia.

cooperators is 'self-servingly biased' because it is below the diagonal. Although conditional cooperators increase their contribution in the average contribution of the other group members, they do not fully match others' contribution.

How can the heterogeneity of types explain the fragility of cooperation that is so typical of repeatedly-played cooperation experiments (see Fig. 15.2)? The idea is simple. Conditional cooperators are prepared to cooperate if others cooperate. If they realize that others are taking a free ride, they reduce their contribution because they do not want to be 'suckered'. Moreover, even conditional cooperators have a 'self-serving bias'. Therefore, cooperation is bound to be fragile, even if most people are conditional cooperators (see Fischbacher & Gächter 2004 for a rigorous analysis).

Cooperation is even fragile if there is a leader who first decides on the contribution to the public good (e.g. Moxnes & van der Heijden 2003, Gächter & Renner 2004, Güth et al. 2004). This is remarkable, since one would expect that a leader should be able to utilize conditional cooperation by setting a good example. Yet, although conditional cooperation exists, free riding is there as well. Thus, the followers' cooperation is insufficient for inducing leaders to keep up their good example. Leaders get frustrated and stop setting a good example.

If it is indeed the mixture of types in a randomly-composed group that makes cooperation a fragile business, then an implication is that groups, where players know that others are of their type, should behave differently than randomly-composed groups. Specifically, conditional cooperators, who know that others are conditional cooperators as well, should find it easy to cooperate. To test this idea, Gächter & Thöni (in prep.) first had subjects play a one-shot public goods game. Then new groups were formed on the basis of the contribution to the public good in the one-shot game. The top cooperators were put in one group, the second to top in the next group and so on. After people had been sorted into the new groups, they were informed about this mechanism. Then they played the public goods experiment as 'Partners' for 10 rounds. It turned out that cooperators who knew that they were among other 'like-minded' cooperators, were able to maintain almost full cooperation until the final rounds. Surprisingly, even groups composed of free riders contributed to the public good. Yet, in stark contrast to the cooperator groups, cooperation among free riders entirely collapsed in the final period. Thus, they cooperated for purely strategic reasons and stopped doing so, when there was no future gain from cooperation anymore. The significance of this result is that the type composition and the knowledge of it (i.e. knowing that one is among like-minded players) matters strongly for the fragility of cooperation.

The experiments discussed so far looked at the most basic cooperation problem that exists in the absence of any extrinsic incentives, like reputation, social (dis-)approval and punishment. Any achieved cooperation must come from people's intrinsic readiness to cooperate, be it for strategic reasons and/or cooperative preferences. The results show that strategic incentives in a repeated interaction clearly help, but that cooperation is nevertheless fragile, with the exception of cooperators who know that they are among other like-minded cooperators. In the following, we look at evidence of how extrinsic incentives other than punishment mitigate the cooperation problem.

15.3 Reputation, communication and social approval

Humans often help each other or cooperate even if this act of altruism is not likely to be reciprocated. An important mechanism that may explain this kind of behavior in reality is reputation. One's behavior is often observed by third parties who may then decide to cooperate or not. Richard Alexander (1987) has coined the term 'indirect reciprocity' for such behavior, to distinguish it from direct reciprocity that occurs between two people. The idea is that helping someone, or refusing to help, changes ones social status, called 'image score'. People with a high image score are more likely to receive help from others: "Give and you shall receive". Game-theoretic analyses show that indirect reciprocity can be an evolutionary stable strategy (Nowak & Sigmund 1998). Seinen & Schram (in prep.), Engelmann & Fischbacher (2002) and Milinski and colleagues (see chapter 14) confirmed this experimentally; players with high image scores received more help than those with low image scores. Thus, indirect reciprocity is a mechanism that can sustain cooperation even in situations in which direct reciprocity is not feasible. Milinski et al. (2002b also showed that a good reputation helps a player also in other social activities that involve the same partners (see Milinski, chapter 14 and Panchanathan & Boyd 2004 for a theoretical model).

In addition to the straightforward economic incentives conferred by reputation, people in reality also sometimes have the chance to communicate about their cooperation problem. To the extent that such communication does not lead to binding agreements, but is merely 'cheap talk', it might not necessarily help to solve the cooperation problem. A free rider might well promise to cooperate but then go on and defect, if his or her cooperation cannot be enforced. Therefore, theoretically, it is not at all clear why communication should reduce free riding. However, casual evidence and intuition suggest that communication has an impact. Thus, there are competing hypotheses and the lab may be the judge. Dawes et al. (1977) and Isaac & Walker (1988) were among the first to study the role of communication in cooperation. In the public goods experiments of Isaac & Walker (1988), group members could talk between the 10 rounds of the game. In a control treatment, communication was not possible. In this latter treatment, again, cooperation collapsed during repeat play. When face-to-face communication was possible, cooperation was substantially higher relative to the control treatment. Almost full efficiency, even in the final rounds, was achieved. Bochet et al. (in prep.) found that even anonymous 'chat-room' communication can lead to very high cooperation rates. Thus, communication can be a very powerful device for sustaining cooperation (see also Brosig et al. 2003 and Sally 1995 for an overview).

Yet, it is not entirely clear why exactly communication works. If many people are conditional cooperators, communication may help coordinating on a certain cooperation level. However, communication, in particular if it is face to face, is also a highly loaded psychological process that creates social ties and disseminates social (dis-)approval. People might fear the disapproval of others or might want to win their approval.

Rege & Telle (2004) developed a very simple one-shot experiment to test for social approval effects. In the control experiment, subjects simply made an anonymous contribution decision. In the main treatment, a subject's decision was publicly but silently recorded on a blackboard. All other participants could see the decision. From a standard economic viewpoint, this treatment manipulation should be ineffective. However, contributions were substantially higher when they could be observed than when they were anonymous.

Gächter & Fehr (1999) also tested the influence of social approval on cooperation. In their experiments, groups of four played the game for 10 repetitions as 'partners'. There were four treatments. In the benchmark 'Anonymity treatment', contributions and group members were anonymous throughout. In the 'Social Exchange treatment', group members were informed that they would get to know each other at the end and that then they would also learn each other's individual contributions during the game. In the 'Group Identity treatment', group members were introduced to each other before they played the game. Thus, a group identity could be formed. At the end of the experiment, they left the building individually such that they could not meet each other. Subjects were aware of

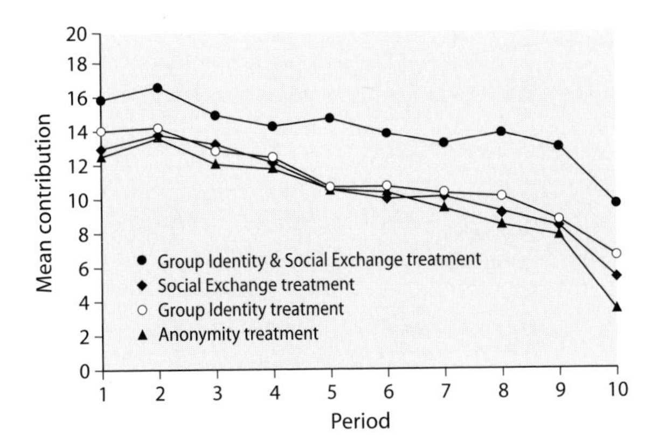

Fig. 15.4. The influence of group identity and social exchange on cooperation. The figure shows the mean contributions of 'Partners' to the public good. 'Group Identity' means that group members know each other's identity before they play; 'Social Exchange' means that subjects meet after the experiment to discuss what they did. In the 'Anonymity treatment', people neither meet before nor after the game. The figure shows that in this experiment only a combination of 'Group Identity' and 'Social Exchange' possibilities increases contributions. Cooperation is fragile in all treatments. From Gächter & Fehr (1999).

this. In the 'Group Identity & Social Exchange treatment', group members met before they played and where informed about the post-experimental meeting. In all four treatments, during the experiment subjects could not talk to each other. Between each round, they were only informed about the group average contribution. How can these treatments influence cooperation? Social exchange theory (e.g. Blau 1964) argues that people might exchange cooperation for social approval. Therefore, if people anticipate social approval effects, cooperation might be higher than under anonymity. Likewise, as suggested by psychological theory and previous evidence, group identity might increase cooperation (see for example Dawes et al. 1988). Fig. 15.4 shows the results.

The results show that, contrary to the hypotheses, neither group identity, nor social exchange alone, were able to increase cooperation. Only if both group identity and social exchange were possible did cooperation increase substantially relative to the anonymity benchmark. This result is consistent with the findings of Rege & Telle (2004). Yet, Rege & Telle (2004) only played their game once. The results from Fig. 15.4 show that social exchange, even if it increases cooperation, is not able to break the downward trend in cooperation. Cooperation is still very fragile, albeit at a higher level.

In summary, under appropriate circumstances, there is no doubt that reciprocal altruism, indirect reciprocity and reputation, and communication as well as social approval can enhance cooperation. A reason for observing higher cooperation when social approval is possible might be that the threat of disapproval of known group members induces higher cooperation rates. Thus, disapproval works like punishment. In fact, the group discussions at the end of the social exchange treatments often revealed quite some anger and frustration

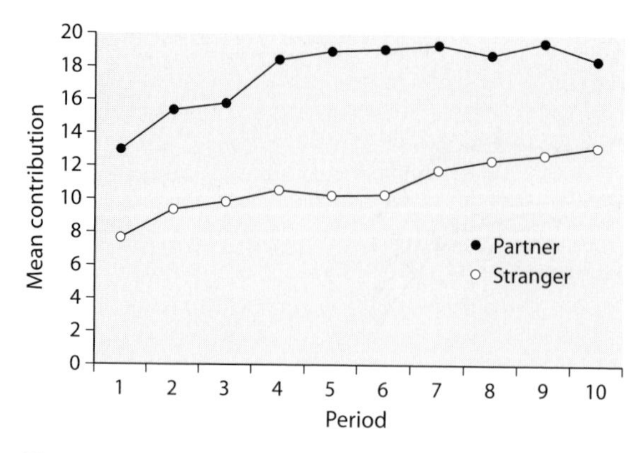

Fig. 15.5. Mean contributions to the public good in the presence of a punishment opportunity. The figure shows that contributions are substantially higher among 'Partners' than among 'Strangers'. A comparison with Fig. 15.2 shows that contributions to the public good are much higher and more stable when punishment is possible. From Fehr & Gächter (2000b).

towards the free riders. Since during the experiment 'social disapproval' could not be targeted at a free rider, it might not have been enough of a deterrent. In the next section, we therefore look at targeted punishment as a means to enhance cooperation.

15.4 Altruistic punishment and cooperation

Casual evidence as well as the observation reported above suggests that many people are in principle prepared to cooperate but want to avoid being the 'sucker' in social dilemma situations. Recall from Fig. 15.3 that roughly half of our subject pools are conditional cooperators who cooperate if others cooperate. If these people encounter a free rider in a typical anonymous standard public goods experiment, the only way to avoid being the 'sucker' is to withhold one's own cooperation. Since people typically strongly dislike being the 'sucker', they may be prepared to punish free riders if they could target them individually and even if it were costly for the punisher.

Yamagishi (1986) and Ostrom et al. (1992) were among the first to allow for punishment in interesting games. Yamagishi (1986) looked at people's willingness to provide a sanctioning system that itself is a public good. Ostrom et al. (1992) studied punishment in a common pool extraction system. Yet, these studies were not primarily interested in how people punish free riders. This was the focus of Fehr & Gächter (2000b) who developed an experimental design that allowed studying punishment in a public goods game. Specifically, after subjects had made their contributions to the public good, they entered a second stage, where they were informed about each individual group member's contribution. They could then assign up to 10 punishment points to each individual group member. Punishment was costly for the punishing subject and each punishment point received reduced the punished subject's income from the first stage by 10%. Fehr & Gächter (2000b) played this experiment under two treatment conditions, a 'Partner'-treatment, where group members knew that they would play the game with the same four group members for 10 periods, and the 'Stranger'-treatment, where group composition was changed from period to period. Fehr & Gächter (2000b) also ran control experiments in which punishment was not possible (see Fig. 15.2). Fig. 15.5 shows the results in the treatments with punishment.

As the comparison with Fig. 15.2 shows, contributions to the public good are strongly increased in the presence of a punishment opportunity. This is true for both the 'Partner'- and the 'Stranger'-treatment. In the case of the 'Partner'-treatment, contributions approach almost 100% of the endowment; in the 'Stranger'-treatment contributions amount to 60% of the endowment. Thus, again we see that 'Partners' contribute more than 'Strangers'. From the very first period onward, contributions are significantly higher in the 'Partner'-treatment than in the 'Stranger'-treatment.

A theoretically very important question concerns the relevance of future interactions. In the 'Partner'-treatment, the likelihood of future interaction is one; in the 'Stranger'-treatment, where groups are randomly re-matched, it is much smaller (depending on the size of the pool from which groups are re-matched), but still positive. An interesting benchmark case is the situation where the likelihood of future interaction is zero, i.e. groups play a one-shot game. This situation is interesting, because neither reciprocal altruism and indirect reciprocity, nor any other form of reputation building is possible, since they require some future interactions. Therefore, Fehr & Gächter (2002) set up a so-called 'Perfect Stranger' design where in each of the six repetitions all groups were composed of completely new members, and participants knew this. Subjects played both games with no punishment and games with punishment. Half of the subjects started with the no-punishment condition and then were introduced to the punishment condition. For the other half, this order was reversed. Fig. 15.6 contains the results on the cooperation rates achieved.

The results are very clear-cut. When punishment is not available, cooperation collapses, as in all previous experiments. The picture changes dramatically, when punishment is possible. For instance, in the experiments that started with the punishment option (labeled '1. punishment, 2. no punishment'), contributions in the very first period were significantly higher than in the experiment that started with the no punishment option. In the experiments where punishment was introduced in the second sequence, cooperation jumped up immediately. This is remarkable, because in this sequence subjects experienced a strong decline in the games with no punishment. Still, after punishment had been introduced, cooperation jumped up to a level that even exceeded cooperation in the very first period. In both sequences, cooperation in the presence of a punishment opportunity strongly increased over time. Thus, contrary to theoretical predictions, in the presence of punishment, cooperation can flourish even in purely one-shot interactions.

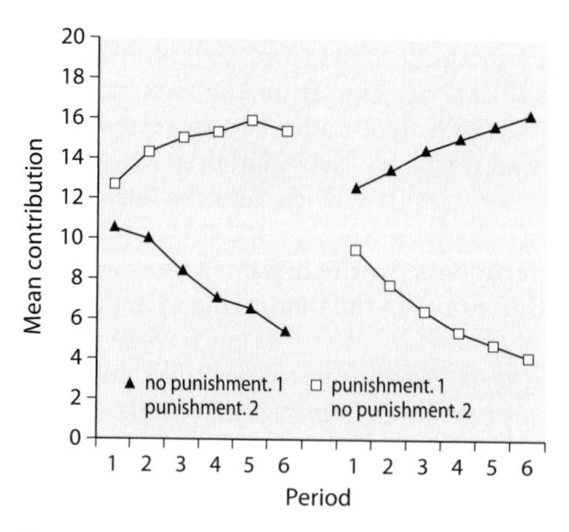

Fig. 15.6. Mean contributions to the public good among 'Perfect Strangers' in the absence and presence of a punishment option. In the sequence labeled "'1. no punishment, 2. punishment', subjects first played six rounds without the punishment option and were then introduced to an environment where they had a punishment option available in each of the following six rounds. In the sequence '1. punishment, 2. no punishment', subjects started in the game with punishment and were after the sixth round informed that there would be no punishment option in the next six rounds. The results show that contributions increase in the presence of punishment and decrease in its absence. From Fehr & Gächter (2002).

The reason why cooperation strongly increased in the presence of punishment is that cooperators were prepared to punish the free riders. Fig. 15.7 shows (separately for the 'Partner'-, the 'Stranger'- and the 'Perfect Stranger'-experiments in Zurich) the punishment expenditures for a given deviation from the other group members' average contributions. Fig. 15.7 also shows the punishment in a 'Partner'-experiment conducted in Samara (Russia). We will discuss this experiment below.

A couple of observations can be made from Fig. 15.7. First, the more a subject's contribution falls short of the average contribution of the other group members, the stronger is the punishment for the deviating group member. This is true in all treatments. Second, with the exception of very strong negative deviations (which comprise only a few cases, however) punishment is very similar between treatments. This is quite remarkable because cooperation levels differ strongly between the 'Partner', 'Stranger' and 'Perfect Stranger' treatments (compare Figs. 15.2, 15.5 and 15.6). In our view, this suggests that punishment is to a very large degree non-strategic. This view is also corroborated by the fact that the punishment pattern of Fig. 15.7 is temporally stable; i.e., some people are prepared to harm a free rider even in the final periods.

Why is punishment so successful in increasing cooperation? The most important reason is probably that it gives the selfish subjects, who care most about their individual payoff, a material incentive to cooperate. Since altruistic punishment is frequent, it apparently is a credible threat and induces selfish individuals to cooperate. It is exactly this feature that makes punishment altruistic;

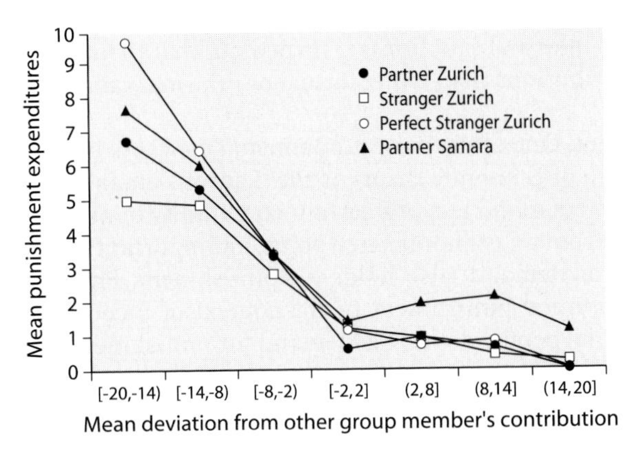

Fig. 15.7. Mean expenditures on punishment as a function of the deviation of the punished group member's cooperation from the average cooperation of the other members. The data are from experiments with 'Partners' and ('Perfect') 'Strangers' in Zurich and Samara. Each money unit spent on punishment reduced the income of the punished member by three money units. For example, group members spent 10 money units on punishing individuals whose contribution to the public good deviated between –20 and –14 units from the group average contribution. The data show that the more people free ride, the more altruistic punishment prevails. There is also some punishment of above-average contributors, in particular in the Samara subject pool. From Fehr & Gächter (2000b, 2002), Gächter et al. (2003).

a punished free rider might in his next encounter abstain from defecting, which benefits his or her future interaction partners.

By now, these results have been replicated by many researchers (see for example, Bowles et al. 2001, Sefton et al. 2002, Gächter et al. 2003, Masclet et al. 2003, Carpenter 2004, Carpenter et al. 2004, Falk et al. 2004, Gürerk et al. 2004, Anderson & Putterman, in prep., Bochet et al., in prep., Page et al. in prep., Carpenter in prep., Noussair & Tucker, in prep.). For lack of space, they cannot all be discussed here. We focus on three issues: (i) the perception of punishment, (ii) the demand for punishment, and (iii) cross-societal differences in norms of cooperation and punishment.

- **The perception of punishment.** A punishment may contain two messages. On the one hand, punishment directly inflicts a payoff reduction. On the other hand, punishment may also signal disapproval; i.e., it sends a message about socially inappropriate behavior. Both may be perceived as punishment and may therefore increase cooperation. Masclet et al. (2003) tested this intuition and studied 'formal and informal' sanctions. The structure of both formal and informal sanctions was the same as in Fehr & Gächter (2000b). Yet, while the formal sanctions were costly both for the punisher and the punished subjects, the informal sanctions were free; they neither caused costs for the punisher, nor the punished individual. Thus, they are tantamount to a symbolic disapproval. Consistent with the evidence on social approval effects reported above, it turned out that even informal sanctions were able to increase contributions. Yet, cooperation was more stable with formal than with informal punishment. In the experiments of Noussair & Tucker (in prep.), subjects could use both formal and informal sanctions. It turned out that their combination led to higher contributions than either formal or informal sanctions alone.

- **The demand for punishment.** One of the most fundamental concepts in economics that underlies much of economic theory is the 'Law of Demand', according to which people will demand less of a certain commodity or activity the higher its price. Thus, from an economic viewpoint, an important question is whether this 'Law of Demand' also holds for punishment. Fig. 15.7 and all papers that have studied punishment in the context of a cooperation game confirm that many people do have a 'demand for punishment', in the sense that they are willing to pay a certain amount of money to inflict punishment on others (i.e. they 'buy' punishment). The more a subject free rides, the higher is the demand for punishment. Yet, studying the 'Law of Demand' requires a systematic variation of the cost of punishment. This is what Anderson & Putterman (in prep.) and Carpenter (2004) did. Their subjects played the cooperation and punishment game in the 'Stranger' set-up to minimize strategic effects. In each of the games, subjects faced different costs for inflicting a punishment unit on the punished subject. The results confirm that people demand less punishment, for a given amount of free riding, the higher the costs of punishing are. Thus, the 'Law of Demand' holds for punishment.
- **Cross-societal differences.** Cross-societal differences in norms of fair sharing have recently attracted a lot of attention (e.g. Henrich et al. 2001, Oosterbeek et al. 2004). It is therefore an interesting question to what extent there are differences in cooperation and punishment norms. To examine this question, Gächter et al. (2003) ran experiments in Russia, where they exactly replicated the Zurich 'Partner'-experiments. Fig. 15.7 also contains the punishment pattern for the Samara subjects. We find that the punishment of free riders is very similar to that in Zurich. Yet, above-average contributors in Samara experienced substantially more punishment than their counterparts in Zurich. Fig. 15.8 looks at the consequences of such punishment for cooperation behavior.

A comparison with the 'Partner'-experiments in Fig. 15.5 yields a striking difference, in particular when a punishment option is available. In the exact same experiment, the Zurich subjects were able to achieve almost full cooperation. By contrast, the presence of a punishment option is only able to prevent the collapse of cooperation. The average cooperation the Samara subjects achieve is only 68% of the level the Zurich subjects manage to maintain. Another stark difference is that in the Zurich experiments the presence of a punishment option strongly increased cooperation relative to cooperation in the absence of punishment (compare Figs. 15.2 and 15.5). This is not at all the case in Samara. Here, cooperation is not statistically significantly higher when subjects have a punishment option available. A potential explanation lies in the punishment behavior. As was shown in Fig. 15.7, the Samara subjects often substantially punished the above-average cooperators. This probably scared them off and thereby prevented the average cooperation level from increasing.

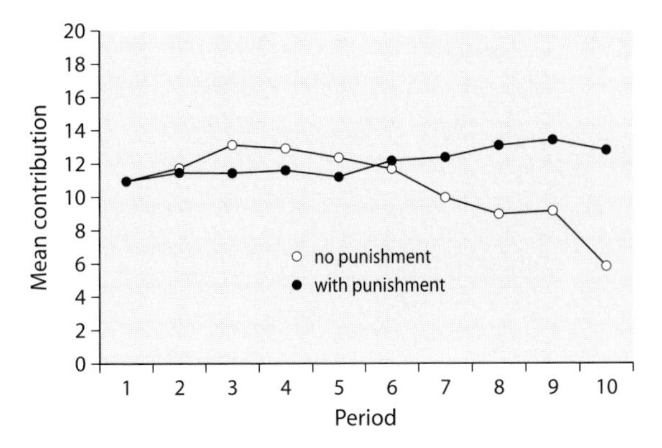

Fig. 15.8. The figure shows the mean contributions to the public goods in the absence and presence of punishment in a 10 times repeated 'Partner' experiment in Samara (Russia). In stark contrast to the results from Figs. 15.5 and 15.6, contributions are not significantly higher when punishment is possible. From Gächter et al. (2003).

In our view, the significance of this result is that different social groups may have widely differing social norms of cooperation, and in particular of punishment. This preliminary result suggests that it is worthwhile to understand logic and scope of cross-societal differences in norms of cooperation and punishment.

15.5 Emotions as a proximate mechanism

Given punishment, subjects' cooperation behavior looks quite rational. To avoid punishment, subjects cooperate in accordance with the group norm. Yet, why do people punish free riders in a one-shot context although this is costly? Emotions may play a decisive role here (Fessler & Haley 2003) and negative emotions, in particular, may provide a proximate explanation. Free riding may cause strong negative emotions among the cooperators and these emotions, in turn, may trigger the willingness to punish the free riders. If this conjecture is correct, we should observe particular emotional patterns in response to free riding. To elicit these patterns, the participants of the Fehr & Gächter (2002) experiments and the subjects in the Samara experiments were confronted with the following two hypothetical scenarios after the final period of the second treatment. The numbers in square brackets relate to the second scenario.

"You decide to invest 16 [5] francs to the project. The second group member invests 14 [3] and the third 18 [7] francs. Suppose the fourth member invests 2 [20] francs to the project. You now accidentally meet this member. Please indicate your feeling towards this person."

After they had read a scenario, subjects had to indicate the intensity of their anger and annoyance towards the fourth person (the free rider) on a seven-point

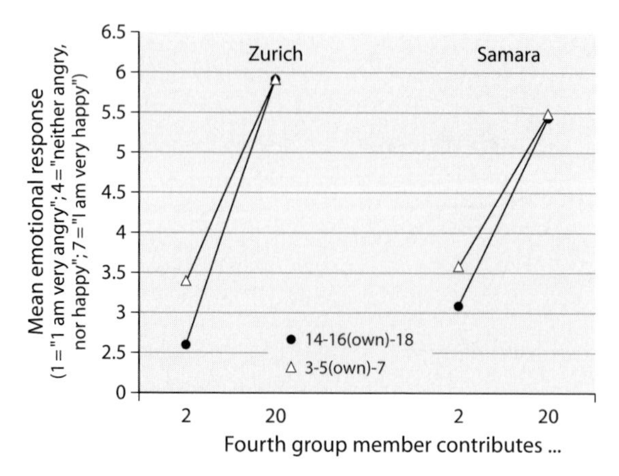

Fig. 15.9. Emotions as a proximate mechanism. The data are elicited in scenarios that describe own and others' contributions and then elicit one's own emotion toward a contribution of a 'fourth group member'. For instance, in Zurich, subjects who in the scenario contributed 16 tokens (whereas two others contributed 14 and 18 tokens, respectively) expressed an emotion score of 2.6 toward the fourth group member who only contributed 2 tokens. The emotion score is 5.9 if the fourth member contributes 20 tokens. The results show that people experience negative emotions toward a free rider more strongly the higher their own contribution level. The Samara subjects expressed less intensive emotions both toward the free rider and the high contributor. From Fehr & Gächter (2002; n = 240) and new results (n = 220).

scale (1 = 'very angry', 4 = 'neither angry nor happy', 7 = 'very happy'). The difference between Scenario 1 and 2 is that the other three persons in the group contribute relatively much in Scenario 1 and relatively little in Scenario 2. Fig. 15.9 documents the results for our experiments in Zurich and Samara.

Subjects report that they are angry if the fourth group member contributes less than they did. This effect is certainly more pronounced in the scenario where they contributed 16 than in the scenario where they contributed 5. The difference is highly significant, both in the Zurich and the Samara sample ($p < 0.001$, Mann-Whitney tests). When the fourth group member contributes more than the pivotal subject, then people report to be quite happy. Surprisingly, subjects are equally happy about the contribution of 20 of the fourth member both when they have contributed 5 or 16 tokens. In other words, the gain in happiness seems not to depend on the own contribution, whereas the intensity of the negative emotions strongly depends on the own contribution.

When we compare the Zurich subjects with the Samara subjects, we find qualitatively very similar results. Yet, a striking difference is that the Samara subjects reported significantly less intensive negative emotions towards the free rider (for an own contribution of 16) than the Zurich subjects. Likewise, for the Samarians, the reported positive emotions were also highly significantly less intense than for the Zurich subjects ($p < 0.0001$, Mann-Whitney tests). Thus, there seem to be strong cross-societal differences in the reported emotions.

Overall, the results suggest that free riding causes negative emotions. Moreover, the emotional pattern is consistent with the hypothesis that emotions trigger punishment. First, the majority of punishments are executed by above-average contributors and imposed on below-average contributors. Second, recall that punishment increases with the deviation of the free rider from the other members' average contribution. This is consistent with the observation that negative emotions are the more intense the more the free rider deviates from the others' average contribution. Third, evidence from neuroscientific experiments supports the interpretation that emotions trigger punishment. For instance, Sanfey et al. (2003) had their subjects play the ultimatum game, while the subjects' brains were scanned (using fMRI). The ultimatum game (invented by Güth et al. 1982) is a two-player game in which player 1 is asked to split an amount of money, say 10 Euros, between him- or herself and a player 2. Player 2 can only accept or reject the proposal. If he accepts, the offer is implemented; if he rejects, both get nothing. A rejection of a positive offer in the ultimatum game is also an instance of altruistic punishment. The brain scans showed that in the recipients who received an unfairly low offer by a human player 1, areas in the brain lit up that are related to negative emotions. When the unfair offer came from a computerized player 1, recipients were much less negatively aroused. Bosman & van Winden (2002) investigated the 'power-to-take game', which is related to the ultimatum game. They elicited self-reported emotions and found that unfair behavior triggers negative emotions that are correlated with punishment. de Quervain et al. (2004) studied neural activations of punishing subjects. They found that punishment activates the 'reward centre' of the brain; i.e., to punish is rewarding. Hence, the proverb "revenge is sweet". They were also able to show that subjects, for whom punishment was more rewarding, actually punished more. Taken together, these regularities are consistent with the view that emotions are an important factor in the process triggering altruistic punishment. Yet, more research is certainly needed here. The emerging field of neuroeconomics (see Camerer et al., in prep.) will certainly play an important role in this endeavor.

15.6
The evolution of strong reciprocity

The evidence presented above shows that many people, but not all, behave reciprocally. They reward nice behavior and punish misdeeds. Since this takes place even in one-shot games, this kind of reciprocity has been termed 'strong reciprocity' (e.g. Gintis 2000), to distinguish it from reciprocal altruism that occurs in repeated games. Reciprocal altruism is strategic reciprocity that can also be exhibited by a completely selfish individual, who would never cooperate or punish in a one-shot context. In economics, the kind of evidence presented in this chapter helped to pave the way for replacing the once ubiquitous selfishness assumption with more realistic assumptions about human's social preferences. A recent and very fruitful development in economic theory has been to take up the experimental evidence and model it. For instance, Fehr & Schmidt (1999) and Bolton & Ockenfels (2000) assume that people have a dislike for inequality. A free rider puts himself into a payoff advantage and inequality-averse people punish to reduce this inequality. Rabin (1993), Falk & Fischbacher (in prep.) and

Dufwenberg & Kirchsteiger (2004) assume that many people punish unkind intentions (to free ride reveals a greedy intention) and that they reward kind behavior (i.e. they cooperate to reward others' cooperation). Falk et al. (2004) show that intentions indeed play an important role in punishment since people also punish when they cannot diminish payoff inequities through punishment.

These new models, whose power extends beyond cooperation games, can be seen as proximate theories, but what explains the existence of strong reciprocity? Specifically, if sufficiently many people punish free riders sufficiently strongly, then free riders have no incentive to free ride anymore. Yet, why should anyone punish and not free ride on other's punishment, since altruistic punishment is just a second-order public good? The answer will probably be found in the evolutionary conditions of the human species that caused a propensity for strongly reciprocal behavior among a significant fraction of the population. The evidence presented suggests that strong reciprocity cannot easily be explained by kin selection (Hamilton 1964), reciprocal altruism (Trivers 1971, Axelrod & Hamilton 1981), indirect reciprocity (Alexander 1987, Nowak & Sigmund 1998) and by costly signaling theory (Zahavi & Zahavi 1997, Gintis et al. 2001).

In our view, one promising approach is 'gene-culture co-evolution' (Gintis 2000, Henrich & Boyd 2001, Bowles et al. 2003, Boyd et al. 2003, Gintis et al. 2003, Boyd & Richerson 2004). One line of reasoning (e.g. Boyd et al. 2003) goes as follows. Assume that in a population there are two behavioral types, cooperators and defectors. The cooperators incur a cost c to produce a benefit b that accrues to all group members. Defection is costless and produces no benefit. If the fraction of cooperators is x, then the expected payoff for cooperators is $bx - c$, whereas defectors get bx. Thus, the payoff difference is c, independent of the number of cooperators. Cooperators would always be at an evolutionary disadvantage under such circumstances. Now assume that there is a fraction y of 'punishers' who cooperate and punish defectors. Punishment reduces the payoff of the punished defector (by p) but also of the punishing subject (by k). The payoff of cooperators who cooperate but do not punish ('second-order free riders') is $b(x + y) - c$; the punished defectors get $b(x + y) - py$, and the punishers earn $b(x + y) - c - k(1 - x - y)$. If the cost of punishments exceed the costs of cooperation (i.e. if $py > c$), then cooperators have a higher fitness than defectors and the fitness disadvantage of punishers relative to the second-order free riders is $k(1 - x - y)$. Thus, punishment is altruistic and the cooperation and punishment game can have multiple equilibria.

This line of reasoning reveals two things. First, there is an important asymmetry between altruistic cooperation and punishment. In an environment without punishment, cooperators are always worse off than defectors, irrespective of how numerous they are. Second, by contrast to the first observation, the cost disadvantage of altruistic punishment declines as defection becomes infrequent because punishment is not needed anymore. The selection pressure against altruistic punishers is weak in this situation.

This latter observation suggests that within-group forces, like copying successful and frequent behavior (see Henrich & Boyd 2001) can stabilize cooperation. Boyd et al. (2003) formally investigate another mechanism, cultural group selection. Recall that in the presence of strong reciprocators the cooperation game may have multiple equilibria, equilibria which imply cooperation, and defection equilibria. Different groups may settle at different equilibria. Here, cultural group selection may come into play. The main idea is that groups with more cooperators are more likely to win inter-group conflicts and are less likely to become extinct, because they may better survive during famine, manage their common resources better etc. (see also Soltis et al. 1995). Therefore, this kind of group selection will tend to increase cooperation because groups who arrived at a cooperative equilibrium are more likely to survive. Moreover, cooperative groups will tend to have more punishers. Since the within-group selection effect is weak if there is a lot of cooperation, cultural group selection can support the evolution of altruistic punishment and maintain it, once it is common. To test this intuition rigorously, Boyd et al. (2003) developed a simple model and simulated it for important parameters, like group size, migration rates between groups and the cost of being punished. The parameters were chosen to mimic likely evolutionary conditions. The simulation results are very interesting because they show that cultural group selection can support altruistic punishment under a wide range of parameters. First, in the absence of punishment, group selection can only sustain cooperation in very small groups, whereas in the presence of punishment, high and stable cooperation rates can be achieved even in large groups. Second, higher migration rates between groups decrease cooperation rates. If the cost of being punished is small, then cooperation breaks down. This result is also consistent with the experimental evidence (see Anderson & Putterman, in prep. and Carpenter 2004). The significance of this and related models is to show that individual selection and cultural factors, like conformism and group selection may coexist (and not be incompatible as in purely gene-based models) and can explain why strong reciprocity may survive. Of course, further models that highlight the links between individual and cultural group selection should and will arise.

We conclude this section with a short discussion of frequent critiques that are leveled at evolutionary explanations of strong reciprocity (see Johnson et al. 2003, Fehr & Gächter 2003 and Fehr & Henrich 2003). One critique concerns group selection. According to the critics, strong reciprocity is merely a byproduct of reciprocal altruism, indirect reciprocity, or signaling. The skepticism against group selection arguments is probably founded in the view that genetic group selection is an implausible mechanism (see also Sober & Wilson 1998). Yet, as the above account of the Boyd et al. (2003) model should make clear, cultural group selection models work completely differently from genetic group selection models.

The second line of critique is that strong reciprocity is a 'mal-adaptation' (see e.g. Johnson et al. 2003). According to this argument, humans evolved in small and mostly stable groups and thereby acquired the psychology needed for sustaining cooperation. Thus, the human brain applies ancient cooperative heuristics even in modern environments, where they are mal-adaptive. Humans did not evolve to play one-shot games and therefore, when they are in a novel environment like a one-shot game in the experimental lab, they behave as if they were in a repeated game. In our view, this argument is problematic for two reasons. First, it is obvious that people did not evolve to play one-shot lab experiments and the strong reciprocity observed there does not represent adaptive behavior. Yet, lab experiments allow us to test to what extent people distinguish between one-shot and repeated games and to what extent they think strategically. As demonstrated repeatedly above (compare Figs. 15.1, 15.2 and 15.5, and the references therein), people cooperate substantially more with 'Partners' than with 'Strangers'. People also report stronger negative emotions when they are cheated by a 'Partner' than by a 'Stranger' (Fehr & Henrich 2003). Moreover, there is also systematic evidence that people respond strongly to increased costs of punishment; they punish less and therefore cooperate less (Fehr et al. 1997, Anderson & Putterman, in prep., Carpenter 2004). Second, research by anthropologists shows that group dispersal, migration and thereby the possibility of meeting strangers was quite common (see Fehr & Henrich 2003, in particular p. 69-76). Thus, vigilant individuals who are able to distinguish whether they deal with a 'Partner' or a 'Stranger' should have a fitness advantage.

Irrespective of one's take in this debate, one should notice that the phenomenon of strongly reciprocal behavior sheds new light on important economic issues (see Fehr & Gächter 2000a, Fehr et al. 2002 and Fehr & Fischbacher 2002). Even if strong reciprocity is a mal-adaptation, it is an important element in explaining patterns of human behavior.

15.7 Summary and conclusions

Humans have achieved a level of cooperation in large groups of genetically unrelated individuals that is outstanding in the animal kingdom. Understanding why this is so is a challenge for all social and behavioral sciences. A theoretically important question in all behavioral sciences is to establish to what extent the observed behavior can be explained by selfishness alone. People might cooperate for various (selfish) reasons. They might cooperate strategically to secure long-term benefits, to gain a favorable reputation in other social activities, to avoid social disapproval and punishment and to gain a high social status and approval. In reality, these motives are in most cases inextricably intertwined. In this paper, we have demonstrated that the experimental laboratory allows the researcher to separate motivations. The most important findings from experimental research are as follows:

- People cooperate even in one-shot PDs and public goods experiments.
- Relative to one-shot encounters, cooperation is strongly increased in stable groups.
- In the absence of communication and/or punishment, cooperation in randomly-composed groups is very fragile. Even stable groups cannot maintain cooperation.
- There seem to be two main types of players: (i) selfish free riders, who in one-shot experiments do not contribute to the public good but may cooperate strategically in repeated games and (ii) conditional cooperators who cooperate if others cooperate. In randomly-composed groups, the interaction of these two types of players explains why cooperation is fragile. The exception

to this rule is groups that are composed of like-minded cooperators, who know that the other group members share their cooperative attitude.

- Communication, possibilities for exchanging social (dis-)approval and reputation building substantially enhance cooperation. Yet, cooperation may still be fragile.
- Many people are prepared to punish free riders if they have the possibility to do so. Such punishment is often 'altruistic' because it can be observed even in one-shot games where the punishing subject does not benefit from induced cooperation. Altruistic punishment can substantially increase and stabilize cooperation.
- Negative emotions toward free riders may be a proximate mechanism that can explain altruistic punishment.

From a theoretical point of view, the most important observation is the existence of 'strong reciprocity', the fact that people are prepared to cooperate and to punish free riders even in anonymous one-shot encounters where there are no future interactions. While the existence of strong reciprocity can be considered an undisputed fact, evolutionary explanations are still open to debate.

Acknowledgments

We gratefully acknowledge financial support by the Grundlagenforschungsfonds of the University of St. Gallen through the research project "Soziale Interaktionen, Unternehmenskultur und Anreizgestaltung". We also thank Peter Kappeler, Carel van Schaik, and the participants of the Freilandtage in Göttingen 2003 for their very helpful comments.

References

Achenbach GG, Snowdon CT (1998) Response to sibling birth in juvenile cotton-top tamarins (*Saguinus oedipus*). Behaviour 135:845–862

Achenbach GG, Snowdon CT (2002) Costs of caregiving: weight loss in captive adult male cotton-top tamarins (*Saguinus oedipus*) following the birth of infants. Int J Primatol 23:179–190

Adolphs R (1999) Social cognition and the human brain. Trends Cogn Sci 3:469–479

Adolphs R (2003) Cognitive neuroscience of human social behaviour. Nat Rev Neurosci 4:165–178

Agrawal AA (2001) Nectar, nodules and cheaters. Trends Ecol Evol 16:123–124

Agrawal AA, Fordyce JA (2000) Induced indirect defence in a lycaenid-ant association: the regulation of a resource in a mutualism. Proc R Soc Lond B 267:1857–1861

Aharon I, Etcoff N, Ariely D, Chabris CF, O'Connor E, Breiter HC (2001) Beautiful faces have variable reward value: fMRI and behavioral evidence. Neuron 32:537–551

Alberts SC, Watts H, Altmann J (2003) Queuing and queue jumping: long- term patterns of reproductive skew in male savannah baboons, *Papio cynocephalus*. Anim Behav 65:821–840

Albon SD, Mitchell B, Staines BW (1983) Fertility and body weight in female red deer: a density dependent relationship. J Anim Ecol 52:969–980

Alexander RD (1974) The evolution of social behavior. Ann Rev Ecol Syst 5:325–383

Alexander RD (1979) Darwinism and Human Affairs. Univ of Washington Press, Seattle

Alexander RD (1987) The Biology of Moral Systems. De Gruyter, New York

Allee WC (1938) The Social Life of Animals. Henry Schuman, New York

Allee WC (1951) Cooperation among Animals. Henry Schuman, New York

Altmann SA (1962) A field study of the sociobiology of the rhesus monkey, *Macaca mulatta*. Ann NY Acad Sci 102:338–435

Allman JM, Hakeem A, Erwin JM, Nimchinsky E, Hof P (2001) The anterior cingulate cortex: the evolution of an interface between cognition and emotion. Ann NY Acad Sci 935:107–117

Altmann JM, Alberts SC (2003) Variability in reproductive success viewed from a life-history perspective in baboons. Am J Hum Biol 15:401–409

Anderson J (1984) Ethology and ecology of sleep in monkeys and apes. Adv Stud Behav14:165–229

Anderson CM, Putterman L (2005) Do non-strategic sanctions obey the law of demand? Games Econ Behav, in press

Andreoni J (1988) Why free ride? Strategies and learning in public goods experiments. J Publ Econ 37:291–304

Andreoni J, Croson R (1998) Partners versus strangers: random rematching in public goods experiments. In: Plott C, Smith V (eds) Handbook of Experimental Economic Results. Elsevier Press, Amsterdam

Andreoni J, Miller J (1993) Rational cooperation in the finitely repeated prisoner's dilemma: experimental evidence. Econ J 103:570–585

Anzenberger G, Mendoza SP, Mason WA (1986) Comparative studies of social behavior of *Callicebus* and *Saimiri*: behavioral and physiological responses of established pairs to unfamiliar pairs. Am J Primatol 11:37–51

Arana FS, Parkinson JA, Hinton E, Holland AJ, Owen AM, Roberts AC (2003) Dissociable contributions of the human amygdala and orbitofrontal cortex to incentive motivation and goal selection. J Neurosci 23:9632–9638

Armitage KB (1981) Sociality as a life history tactic of ground squirrels. Oecol 48:36–49

Armitage KB (1999) Evolution of sociality in ground squirrels. J Mammal 80:1–10

Arnold K, Aureli F (2005) Postconflict reconciliation. In: Campbell CJ, MacKinnon KC, Panger M, Fuentes A, Bearder S (eds) Primates in Perspective. Univ of Chicago Press, Chicago

Arnold K, Whiten A (2001) Post-conflict behaviour of wild chimpanzees (*Pan troglodytes schweinfurthii*) in the Budongo forest, Uganda. Behav 138:649–690

Ashlock D, Smucker MD, Stanley EA, Tesfatsion L (1996) Preferential partner selection in an evolutionary study of prisoner's dilemma. Biosystems 37:99–125

Aureli F (1992) Post-conflict behaviour among wild long-tailed macaques (*Macaca fascicularis*). Behav Ecol Sociobiol 31:329–337

Aureli F (1997) Post-conflict anxiety in nonhuman primates: the mediating role of emotion in conflict resolution. Aggr Behav 23:315–328

Aureli F, de Waal FBM (2000) Natural Conflict Resolution. Univ of California Press, Berkeley

Aureli F, Schaffner CM (2002) Relationship assessment through emotional mediation. Behaviour 139:309–420

Aureli F, Smucny DA (2000) The role of emotion in conflict and conflict resolution. In: Aureli F, de Waal FBM (eds) Natural Conflict Resolution. Univ of California Press, Berkeley, pp 199–224

Aureli F, van Schaik CP (1991) Post-conflict behaviour in long-tailed macaques (*Macaca fascicularis*). II. Coping with the uncertainty. Ethology 89:101–114

Aureli F, Whiten A (2003) Emotions - The mediation of behavioral flexibility. In: Maestripieri D (ed) Primate Psychology - The Mind and Behavior of Human and Nonhuman Primates. Harvard Univ Press, Cambridge/MA, pp 289–323

Aureli F, van Schaik CP, van Hooff JARAM (1989) Functional aspects of reconciliation among captive long-tailed macaques (*Macaca fascicularis*). Am J Primatol 19:39–51

Aureli F, Cozzolino R, Cordischi C, Scucchi S (1992) Kin-oriented redirection among Japanese macaques: an expression of a revenge system? Anim Behav 44:283–291

Aureli F, Das M, Veenema HC (1997) Differential kinship effect on reconciliation in three species of macaques (*Macaca fascicularis, M. fuscata* and *M. sylvanus*). J Comp Psychol 111:91–99

Aureli F, Preston SD, de Waal FBM (1999) Heart rate responses to social interactionsin free-moving rhesus monkeys: a pilot study. J Comp Psychol 113:59–65

Aureli F, Cords M, van Schaik CP (2002) Conflict resolution following aggression in gregarious animals: a predictive framework. Anim Behav 64:325–343

Axelrod R (1984) The Evolution of Cooperation. Basic Books, New York

Axelrod R, Hamilton W (1981) The evolution of cooperation. Science 211:1390–1396

Axén AH (2000) Variation in behavior of lycaenid larvae when attended by different ant species. Evol Ecol 14:611–625

Azzam AM (1999) Asymmetry, rigidity in farm retail price transmission. Am J Agric Econ 81:525–534

Bahuchet S (1985) Les Pygmés Aka et la Forêt Centrafricaine. Selaf, Paris

Baker AJ (1991) Evolution of the social system of the golden lion tamarin. PhD thesis, Univ of Maryland, College Park/MD

Baker AJ, Bales K, Dietz JM (2002) Mating system and group dynamics in golden lion tamarins (*Leontopithecus rosalia*). In: Kleiman DG, Rylands AB (eds) The Lion Tamarins of Brazil. Smithsonian Institution Press, Washington/DC, pp 188–212

Balshine-Earn S, Neat FC, Reid H, Taborsky M (1998) Paying to stay or paying to breed? Field evidence for direct benefits of helping behavior in a cooperatively breeding fish. Behav Ecol 9:432–438

Barclay P (2004) Trustworthiness and competitive altruism can also solve the "tragedy of the commons". Evol Hum Behav 25:209–220

Barnett SA (1965) Adaptation of mice to cold. Biol Rev 40:5–51

Barrett L, Henzi SP (2001) The utility of grooming in baboon troops. In: Noë R, van Hooff JARAM, Hammerstein P (eds) Economics in Nature – Social Dilemmas, Mate Choice and Biological Markets. Cambridge Univ Press, Cambridge, pp 119–145

Barrett L, Henzi SP (2002) Constraints on relationship formation among female primates. Behaviour 139:263–289

Barrett L, Henzi SP, Weingrill T, Lycett JE, Hill RA (1999) Market forces predict grooming reciprocity in female baboons. Proc R Soc Lond B 266:665–670

Barrett L, Henzi SP, Weingrill T, Lycett JE, Hill RA (2000) Female baboons do not raise the stakes but they give as good as they get. Anim Behav 59:763–770
Barrett L, Dunbar RIM, Lycett J (2002a) Human Evolutionary Psychology. Palgrave, Houndsmills
Barrett L, Gaynor D, Henzi SP (2002b) A dynamic interaction between aggression and grooming reciprocity among female chacma baboons. Anim Behav 63:1047–1053
Barrett L, Henzi SP, Dunbar RIM (2003) Primate cognition: from "what now?" to "what if?" Trends Cog Sci 11:494–197
Barton RA (1985) Grooming site preferences in primates and their functional implications. Int J Primatol 6:519–532
Barton RA, Byrne RW, Whiten A (1996) Ecology, feeding competition and social structure in baboons. Behav Ecol Sociobiol 38:321–329
Basolo AL (1990) Female preference predates the evolution of the sword in swordtail fish. Science 250:808–810
Batali J, Kitcher P (1995) Evolution of altruism in optional and compulsory games. J theor Biol 175:161–171
Becker GS (1973) A theory of marriage. Part I. J Polit Econ 81:813–846
Bednekoff PA (1997) Mutualism among safe, selfish sentinels: a dynamic game. Am Nat 150:373–392
Bélisle P, Chapais B (2001) Tolerated co-feeding in relation to degree of kinship in Japanese macaques. Behaviour 138:487–509
Bell G (1986) The evolution of empty flowers. J theor Biol 118:253–258
Bennett NC, Jarvis JUM, Faulkes CG, Millar RP (1993) LH responses to single doses of exogenous GnRH by freshly captured Damaraland mole-rats, *Crypotmys damerensis*. J Reprod Fertil 99:81–86
Bennett NC, Jarvis JUM, Millar R, Sasano H, Ntshinga KV (1994) Reproductive suppression in eusocial *Cryptomys damarensis* colonies – socially-induced infertility in females. J Zool Lond 233:617–630
Beran MJ, Beran MM (2004) Chimpanzees remember the results of one-by-one addition of food items to a set over extended time periods. Psychol Sci 15:94–99
Bercovitch FB (1988) Coalitions, cooperation and reproductive tactics among adult male baboons. Anim Behav 36:1198–1209
Bergmüller R, Taborsky M (2004) Experimental manipulation of helping in a cooperative breeder: helpers "pay to stay" by pre-emptive appeasement. Anim Behav 69:19–28
Berkes F, Feeny D, McCay BJ, Acheson JM (1989) The benefits of the commons. Nature 340:91–93
Berman CM (1983a) Matriline differences and infant development. In: Hinde RA (ed) Primate Social Relationships: An Integrated Approach. Sinauer Associates, Sunderland/MA, pp 132–134
Berman CM (1983b) Early differences in relationships between infants and other group members based on the mother's status: their possible relationship to peer-peer rank acquisition. In: Hinde RA (ed) Primate Social Relationships: An Integrated Approach. Sinauer Associates, Sunderland/MA, pp 154–156
Berman CM (1983c) Influence of close female relations on peer-peer rank acquisition. In: Hinde RA (ed) Primate Social Relationships: An Integrated Approach. Sinauer Associates, Sunderland/MA, pp 157–159
Berman CM (2004) Developmental aspects of kin bias in behavior. In: Chapais B, Berman C (eds) Kinship and Behavior in Primates. Oxford Univ Press, New York, pp 317–346
Bernstein IS (1991) The correlation between kinship and behaviour in non-human primates. In: Hepper PG (ed) Kin Recognition. Cambridge Univ Press, Cambridge, pp 6–29
Berridge KC (1996) Food reward: brain substrates of wanting and liking. Neurosci Biobehav Rev 20:1–25
Berridge KC (2003) Comparing the emotional brains of humans and other animals. In: Davidson RJ, Scherer KR, Goldsmith HH (eds) Handbook of Affective Sciences. Oxford Univ Press, pp 25–51
Berry RJ (1971) Life and death in an island population of the house mouse. Exp Geront 6:187–197
Berry RJ (1981a) Biology of the House Mouse. Academic Press, London
Berry RJ (1981b) Town mouse, country mouse: adaptation and adaptability in *Mus domesticus* (*M. musculus domesticus*). Mammal Rev 11:91–136

Bertram BCR (1978) Living in groups: predators and prey. In: Krebs JR, Davies NB (eds) Behavioural Ecology: An Evolutionary Approach. Blackwell, Oxford, pp 64–96
Bertram BCR (1992) The Ostrich Communal Nesting System. Princeton Univ Press, Princeton/NJ
Bettinger T, Walis J, Goodall J (1993) Male-male chimpanzee interactions from the alpha male's perpective. Am J Primatol 30:298–299
Bird RR, Smith EA, Bird DW (2001) The hunting handicap: costly signaling in human foraging strategies. Behav Ecol Sociobiol 50:9–19
Blarer A, Keasar T, Shmida A (2002) Possible mechanisms for the formation of flower size preferences by foraging bumblebees. Ethology 108:341–351
Blau P (1964) Exchange and Power in Social Life. Transaction Publishers, New Brunswick
Blouin M (2003) DNA-based methods for pedigree reconstruction and kinship analysis in natural populations. Trends Ecol Evol 18:503–511
Blount S (1995) When social outcomes aren't fair: the effect of causal attributions on preferences. Org Behav Hum Dec Proc 63:131–144
Blumstein DT, Armitage KB (1999) Cooperative breeding in marmots. Oikos 84:369–382
Blurton-Jones NG (1984) A selfish origin of human food sharing: tolerated theft. Ethol Sociobiol 5:1–3
Blurton-Jones NG (1987) Tolerated theft, suggestions about the ecology and evolution of sharing, hoarding, and scrounging. Soc Sci Info 26:31–54
Boccia ML, Reite M, Laudenslager M (1989) On the physiology of grooming in a pigtail macaque. Physiol Behav 45:667–670
Bocher M, Chisin R, Parag Y, Freedman N, Weil YM, Lester H, Mishani E, Bonne O (2001) Cerebral activation associated with sexual arousal in response to a pornographic clip: a 15O-H2O PET study in heterosexual men. NeuroImage 14:105–117
Bochet O, Page T, Putterman L (2005) Communication and punishment in voluntary contribution experiments. J Econ Behav Org, in press
Boesch C (1994) Cooperative hunting in wild chimpanzees. Anim Behav 48:653–667
Boesch C, Boesch H (1984) Mental map in wild chimpanzees: an analysis of hammer transports for nut cracking. Primates 25:160–170
Boesch C, Boesch H (1989) Hunting behavior of wild chimpanzees in the Tai National Park. Am J Phy Anthropol 78:547–573
Boesch C, Boesch-Achermann H (2000) The Chimpanzees of the Taï Forest: Behavioural Ecology and Evolution. Oxford Univ Press, Oxford
Boinski S (1987) Mating patterns in squirrel monkeys (*Saimiri oerstedi*): implications for seasonal sexual dimorphism. Behav Ecol Sociobiol 21:13–21
Boinski S (1994) Affiliation patterns among male Costa Rican squirrel monkeys. Behaviour 130:191–209
Bolton G, Ockenfels A (2000) A theory of equity, reciprocity and competition. Am Econ Rev 100:166–193
Bolton GE, Katok E, Ockenfels A (2001) What´s in a reputation? Indirect reciprocity in an image scoring game. Working paper, Smeal College of Business Administration, Penn State University
Boyd R, Richerson PJ (1992) Punishment allows the evolution of cooperation (or anything else) in sizable groups. Ethol Sociobiol 13:171–195
Boness DJ (1990) Fostering behavior in Hawaiian monk seals: is there a reproductive cost? Behav Ecol Sociobiol 27:113–140
Bonnie KE, de Waal FBM (2004) Primate social reciprocity and the origin of gratitude. In: Emmons RA, McCullough ME (eds) The Psychology of Gratitude. Oxford Univ Press, Oxford, pp 213–229
Borgerhoff Mulder M (2004) Are men and women really so different? Trends Ecol Evol 19:3–6
Boroditsky L (2000) Metaphoric structuring: understanding time through spatial metaphors. Cognition 75:1–28
Borries C (1993) Ecology of female social relationships: Hanuman langurs (*Presbytis entellus*) and the van Schaik model. Fol Primatol 61:21–30
Borries C, Sommer V, Srivastava A (1991) Dominance, age, and reproductive success in free-ranging female Hanuman langurs. Int J Primatol 12:231–257

Borries C, Sommer V, Srivastava A (1994) Weaving a tight social net: allogrooming in free-ranging female langurs (*Presbytis entellus*). Int J Primatol 15:421–443
Borries C, Launhardt K, Epplen C, Epplen JT, Winkler P (1999) Males as infant protectors in Hanuman langurs (*Presbytis entellus*) living in multimale groups – defence pattern, paternity and sexual behaviour. Behav Ecol Sociobiol 46:350–356
Bos R van den (1998) Post-conflict stress response in confined group-living cats (*Felis silvestris catus*). Appl Anim Behav Sci 59:323–330
Bosman R, van Winden F (2002) Emotional hazard in a power-to-take experiment. Econ J 112:147–169
Botvinick M, Nystrom LE, Fissell K, Carter CS, Cohen JD (1999) Conflict monitoring versus selection-for-action in ACC. Nature 402:179–181
Bowles S, Gintis H (2003) Origins of human cooperation. In: Hammerstein P (ed) Genetic and Cultural Evolution of Cooperation. MIT Press, Cambridge/MA, pp 429–443
Bowles S, Hammerstein P (2003) Does market theory apply to biology? In: Hammerstein P (ed) Genetic and Cultural Evolution of Cooperation. MIT Press, Cambridge/MA, pp 153–165
Bowles S, Carpenter J, Gintis H (2001) Mutual Monitoring in Teams: Theory and Evidence on the Importance of Residual Claimancy and Reciprocity. Mimeo, Univ of Massachusetts, Amherst
Bowles S, Fehr E, Gintis H (2003) Strong reciprocity may evolve with and without group selection. Theoret Primatol Proj Newslett 1, December issue
Boyce CCK, Boyce JLI (1988) Population biology of *Microtus arvalis*. I. Lifetime reproductive success of solitary and grouped breeding females. J Anim Ecol 57:711–722
Boyd R (1982) Density dependent mortality and the evolution of social behavior. Anim Behav 30:972–982
Boyd R (1992) The evolution of reciprocity when conditions vary. In: Harcourt AH, de Waal FBM (eds) Coalitions and Alliances in Humans and other Animals. Oxford Univ Press, Oxford, pp 493–510
Boyd R, Richerson PJ (1985) Culture and the Evolutionary Process. Univ of Chicago Press, Chicago
Boyd R, Richerson PJ (1988) The evolution of reciprocity in sizable groups. J theor Biol 132:337–356
Boyd R, Richerson PJ (1992) Punishment allows the evolution of cooperation (or anything else) in sizable groups. Ethol Sociobiol 13:171–195
Boyd R, Richerson PJ (2002) Group beneficial norms can spread rapidly in a structured population. J theor Biol 215:287–296
Boyd R, Richerson PJ (2004) Not by Genes Alone: How Culture Transformed Human Evolution. Univ of Chicago Press, Chicago
Boyd R, Gintis H, Bowles S, Richerson PJ (2003) Evolution of altruistic punishment. Proc Natl Acad Sci USA 100:3531–3535
Bradley BJ (1999) Levels of selection, altruism, and primate behavior. Q Rev Biol 74:171–194
Bradley BJ, Doran-Sheehy DM, Lukas D, Boesch C, Vigilant L (2004) Dispersed male networks in Western gorillas. Curr Biol 14:510–513
Brandt H, Sigmund K (2004) The logic of reprobation: assessment and action rules for indirect reciprocity. J theor Biol 231:475–486
Bronson FH (1979) The reproductive ecology of the house mouse. Q Rev Biol 54:265–299
Bronson FH (1989) Mammalian Reproductive Biology. Univ of Chicago Press, Chicago
Bronstein JL (1998) The contribution of ant plant-protection studies to our understanding of mutualism. Biotropica 30:150–161
Bronstein JL (2001) The exploitation of mutualisms. Ecol Lett 4:277–287
Brosig J, Ockenfels A, Weimann J (2003) The effect of communication media on cooperation. Germ Econ Rev 4:217–241
Brosnan SF (in prep) A sense of fairness in animals
Brosnan SF, de Waal FBM (2002) A proximate perspective on reciprocal altruism. Hum Nat 13:129–152
Brosnan SF, de Waal FBM (2003) Monkeys reject unequal pay. Nature 425:297–299
Brosnan SF, de Waal FBM (2004a) Social learning about value in capuchin monkeys, *Cebus apella*. J Comp Psychol, 118:133–139
Brosnan SF, de Waal FBM (2004b) A concept of value during experimental exchange in brown capuchin monkeys, *Cebus apella*. Fol Primatol, in press

Brosnan SF, de Waal FBM (2004c) Fair refusal by capuchin monkeys. Nature 428:140
Brosnan SF, Schiff HC, de Waal FBM (2004) Tolerance for inequity may increase with social closeness in chimpanzees. Proc R Soc Lond B 272:253–258
Brotherton PNM, Clutton-Brock TH, O'Riain MJ, Gaynor D, Sharpe L, Kansky R, McIlrath GM (2001) Offspring food allocation by parents and helpers in a cooperative mammal. Behav Ecol 12:590–599
Brown JL (1983) Cooperation: a biologist's dilemma. Adv Stud Behav 13:1–37
Brown JL (1987) Helping and Communal Breeding in Birds. Princeton Univ Press, Princeton/NJ
Brown JL, Brown ER, Brown SD, Dow DD (1982) Helpers: effects of experimental removal on reproductive success. Science 215:421–422
Bshary R (2001) The cleaner fish market. In: Noë R, van Hooff JARAM, Hammerstein P (eds) Economics in Nature. Social Dilemmas, Mate Choice and Biological Markets. Cambridge Univ Press, Cambridge, pp 146–172
Bshary R (2002) Biting cleaner fish use altruism to deceive image-scoring client fish. Proc R Soc Lond B 269:2087–2093
Bshary R, Bronstein JL (2005) Game structures in mutualistic interactions: what can the evidence tell us about the kind of models we need? Adv Stud Behav, in press
Bshary R, Grutter AS (2002a) Asymmetric cheating opportunities and partner control in a cleaner fish mutualism. Anim Behav 63:547–555
Bshary R, Grutter AS (2002b) Experimental evidence that partner choice is a driving force in the payoff distribution among cooperators or mutualists: the cleaner fish case. Ecol Lett 51:130–136
Bshary R, Noë R (2003) Biological markets: the ubiquitous influence of partner choice on the dynamics of cleaner fish-client reef fish interactions. In: Hammerstein P (ed) Genetic and Cultural Evolution of Cooperation. MIT Press, Cambridge/MA, pp 167–184
Bshary R, Würth M (2001) Cleaner fish *Labroides dimidiatus* manipulate client reef fish providing tactile stimulation. Proc R Soc Lond B 268:1495–1501
Buchan JC, Alberts SC, Silk JB, Altmann J (2003) True paternal care in a multi-male primate society. Nature 425:179–181
Bulger JB (1993) Dominance rank and access to estrous females in male savanna baboons. Behaviour 127:67–103
Bunte F, Peerlings J (2003) Asymmetric price transmission due to market power in the case of supply shocks. Agribusiness 19:19–28
Burnham TC (2002) Ultimatum games. In: Nadel L (ed) The Encyclopedia of Cognitive Science. Nature Publishing Group, London, pp 238–245
Burnham TC, Johnson D (2005) The biological and evolutionary logic of human cooperation. Anal Kritik 27, in press
Burt A, Trivers R (2005) Genes in Conflict: The Biology of Selfish Genetic Elements. Harvard University Press, Cambridge/MA
Bush G, Luu P, Posner MI (2000) Cognitive and emotional influences in anterior cingulate cortex. Trends Cogn Sci 4:215–222
Busse C (1977) Chimpanzee predation as a possible factor in the evolution of red colobus monkey social organization. Evolution 31:907–911
Busse C (1978) Do chimpanzees hunt cooperatively? Am Nat 112:767–770
Buston PM, Emlen ST (2003) Cognitive processes underlying human mate choice: the relationship between self-perception and mate preference in Western society. Proc Natl Acad Sci USA 100:8805–8810
Butovskaya M, Verbeek P, Ljungberg T, Lunardini A (2000) A multicultural view of peacemaking among young children. In: Aureli F, de Waal FBM (eds) Natural Conflict Resolution. Univ of California Press, Berkeley, pp 243–258
Butynski T (1990) Comparative ecology of blue monkeys (*Cercopithecus mitis*) in high- and low-density subpopulations. Ecol Monogr 60:1–26
Bygott D (1979) Agonistic behavior and dominance among wild chimpanzees. In: Hamburg D, McCown L (eds) The Great Apes. Benjamin-Cummings, Menlo Park, pp 405–427
Byrne RW, Whiten A (1988) Machiavellian Intelligence: Social Expertise and the Evolution of Intellect in Monkeys, Apes, and Humans. Clarendon Press, Oxford

Byrne RW, Whiten A, Henzi SP (1989) Social relationships of mountain baboons: leadership and affiliation in a nonfemale-bonded monkey. Am J Primatol 18:191–207
Caine NG (1993) Flexibility and co-operation as unifying themes in *Saguinus* social organization and behavior: the role of predation pressures. In: Rylands AB (ed) Marmosets and Tamarins: Systematic, Behavior, and Ecology. Oxford Univ Press, Oxford, pp 200–219
Caine NG, Potter MP, Mayer KE (1992) Sleeping site selection by captive tamarins (*Saguinus labiatus*). Ethology 90:63–71
Call J (2001) Chimpanzee social cognition. Trends Cogn Sci 5:388–393
Calow P (1979) The cost of reproduction – a physiological approach. Biol Rev 54:23–40
Camerer CF (2003) Behavioral Game Theory: Experiments in Strategic Interaction. Princeton Univ Press, Princeton/NJ
Camerer CF, Loewenstein G, Prelec D (2005) Neuroeconomics: how neuroscience can inform economics. J Econ Lit, in press
Caraco T, Brown JL (1986) A game between communal breeders: when is food-sharing stable? J theor Biol 118:379–393
Carlson AA, Young AJ, Russell AF, Bennett NC, McNeilly A, Clutton-Brock TH (2004) Hormonal correlates of dominance in meerkats (*Suricata suricatta*). Horm Behav 46:141–150
Carpenter J (2004) The Demand for Punishment. Mimeo, Middlebury College
Carpenter J (2005) Punishing free-riders: how group size affects mutual monitoring and the provision of public goods. Games Econ Behav, in press
Carpenter J, Matthews HP, Ong'ong'a O (2004) Why punish? Social reciprocity and the enforcement of prosocial norms. J Evol Econ 14:407–430
Carter CS (1998) Neuroendocrine perspectives on social attachment and love. Psychoneuroendocrinology 23:779–818
Carter CS, Braver TS, Barch DM, Botvinick MM, Noll D, Cohen JD (1998) ACC, error detection and online monitoring of performance. Science 280:747–749
Caro TM (1994) Cheetahs of the Serengeti plains: group living in an asocial species. Univ of Chicago Press, Chicago
Castles DL, Whiten A (1998) Post-conflict behaviour of wild olive baboons. II. Stress and self-directed behaviour. Ethology 104:148–160
Castles DL, Aureli F, de Waal FBM (1996) Variation in conciliatory tendency and relationship quality across groups of pigtail macaques. Anim Behav 52:389–403
Chalmeau R, Visalberghi E, Gallo A (1997) Capuchin monkeys (*Cebus apella*) fail to understand a cooperative task. Anim Behav 54:1215–1225
Chapais B (1983) Dominance, relatedness, and the structure of female relationships in rhesus monkeys. In: Hinde RA (ed) Primate Social Relationships: An Integrated Approach. Sinauer Associates, Sunderland/MA, pp 209–219
Chapais B (1986) Why do adult male and female rhesus monkeys affiliate during the birth season? In: Rawlins RG, Kessler MJ (eds) The Cayo Santiago Macaques: History, Behavior and Biology. State Univ of New York Press, Albany/NY, pp 173–200
Chapais B (1992) The role of alliances in social inheritance of rank among female primates. In: Harcourt AH, de Waal FBM (eds) Coalitions and Alliances in Humans and other Animals. Oxford Science Publications, Oxford, pp 29–59
Chapais B (1995) Alliances as a means of competition in primates: evolutionary, developmental, and cognitive aspects. Ybk Phys Anthropol 38:115–136
Chapais B (2001) Primate nepotism: what is the explanatory value of kin selection? Int J Primatol 22:203–229
Chapais B (2005) How kinship generates dominance structures: a comparative perspective. In: Thierry B, Singh M, Kaumanns W (eds) How Societies Arise: The Macaque Model. Cambridge Univ Press, Cambridge, in press
Chapais B, Bélisle P (2004) Constraints on kin selection in primate groups. In: Chapais B, Berman C (eds) Kinship and Behavior in Primates. Oxford Univ Press, New York, pp 365–386
Chapais B, Berman C (2003) Kinship and Behavior in Primates. Oxford Univ Press, New York
Chapais B, Berman C (2004) The kinship black box. In: Chapais B, Berman C (eds) Kinship and Behavior in Primates. Oxford Univ Press, New York, pp 3–11

Chapais B, Mignault C (1991) Homosexual incest avoidance among females in captive Japanese macaques. Am J Primatol 23:171–183
Chapais B, Schulman S (1980) An evolutionary model of female dominance relationships in primates. J theoret Biol 82:47–89
Chapais B, Girard M, Primi G (1991) Non-kin alliances and the stability of matrilineal dominance relations in Japanese macaques. Anim Behav 41:481–491
Chapais B, Prud'homme J, Teijeiro S (1994) Dominance competition among siblings in Japanese macaques: constraints on nepotism. Anim Behav 48:1335–1347
Chapais B, Gauthier C, Prud'homme J, Vasey P (1997) Relatedness threshold for nepotism in Japanese macaques. Anim Behav 53:1089–1101
Chapais B, Savard L, Gauthier C (2001) Kin selection and the distribution of altruism in relation to degree of kinship in Japanese macaques. Behav Ecol Sociobiol 49:493–502
Chase VM, Hertwig R, Gigerenzer G (1998) Visions of rationality. Trends Cogn Sci 2:206–214
Cheeseman LL, Mallinson PJ, Ryan J, Wilesmith JW (1993) Recolonisation by badgers in Gloucestershire. In: Hayden TJ (ed) The Badger. Royal Irish Academy, Dublin, pp 78–93
Cheney DL (1977) The acquisition of rank and the development of reciprocal alliances among free-ranging immature baboons. Behav Ecol Sociobiol 2:303–318
Cheney DL (1983a) Extrafamilial alliances among vervet monkeys. In: Hinde RA (ed) Primate Social Relationships: An Integrated Approach. Sinauer Associates, Sunderland/MA, pp 278–286
Cheney DL (1983b) Proximate and ultimate factors related to the distribution of male migration. In: Hinde RA (ed) Primate Social Relationships: An Integrated Approach. Sinauer Associates, Sunderland/MA, pp 241–249
Cheney DL, Seyfarth RM (1983) Non-random dispersal in free-ranging vervet monkeys: social and genetic consequences. Am Nat 122:392–412
Cheney DL, Seyfarth RM (1997) Reconciliatory grunts by dominant female baboons influence victims' behaviour. Anim Behav 54:409–418
Cheney DL, Seyfarth RM, Smuts BB (1986) Social relationships and social cognition in non-human primates. Science 234:1361–1366
Cheney DL, Seyfarth RM, Andelman SJ, Lee PC (1988) Reproductive success in vervet monkeys. In: Clutton-Brock TH (ed) Reproductive Success. Univ of Chicago Press, Chicago, pp 384–402
Cheney DL, Seyfarth RM, Silk JB (1995) The role of grunts in reconciling opponents and facilitating interactions among adult female baboons. Anim Behav 50:249–257
Cheney DL, Seyfarth RM, Fischer J, Beehner J, Berman T, Johnson SE, Kitchen DM, Palombit RA, Silk JB (2004) Factors affecting reproduction and mortality among baboons in the Okavango Delta, Botswana. Int J Primatol 25:401–428
Chism J (2000) Allocare patterns among cercopithecines. Fol Primatol 71:55–66
Clark AB (1978) Sex ratio and local resource competition in a prosimian primate. Science 201:163–165
Clark MS, Grote NK (2003) Close relationships. In: Millon T, Lerner MJ (eds) Handbook of Psychology: Personality and Social Psychology. John Wiley & Sons, New York, pp 447–461
Clark MS, Mills J (1979) Interpersonal attraction in exchange and communal relationships. J Pers Soc Psychol 37:12–24
Clark K, Sefton M (2001) The sequential prisoner's dilemma: evidence on reciprocation. Econ J 111:51–68
Clements KC, Stephens DW (1995) Testing models of non-kin cooperation: mutualism and the prisoner's dilemma. Anim Behav 50:527–535
Cleveland J, Snowdon CT (1984) Social development during the first twenty weeks in the cotton-top tamarin (*Saguinus o. oedipus*). Anim Behav 32:432–444
Clutton-Brock TH (1989a) Female transfer and inbreeding avoidance in social mammals. Nature 337:70–72
Clutton-Brock TH (1989b) Mammalian mating systems. Proc R Soc Lond B 236:339–372
Clutton-Brock TH (1991) The Evolution of Parental Care. Monographs in Behavior and Ecology. Princeton Univ Press, Princeton/NJ
Clutton-Brock TH (1998a) Reproductive skew, concessions and limited control. Trends Ecol Evol 13:288–292

Clutton-Brock TH (1998b) Reproductive skew: disentangling concessions from control. A reply to Emlen and Reeve. Trends Ecol Evol 13:459

Clutton-Brock TH (2002) Breeding together: kin selection and mutualism in cooperative vertebrates. Science 296:69–72

Clutton-Brock TH, Parker GA (1995) Punishment in animal societies. Nature 373:209–216

Clutton-Brock TH, Brotherton PNM, Smith R, McIlrath G, Kansky R, Gaynor D, O'Riain MJ, Skinner JD (1998a) Infanticide and expulsion of females in a cooperative mammal. Proc R Soc Lond B 265:2291–2295

Clutton-Brock TH, Gaynor D, Kansky R, MacColl ADC, McIlrath G, Chadwick P, Brotherton PNM, O'Riain JM, Manser M, Skinner JD (1998b) Costs of cooperative behaviour in suricates, *Suricata suricatta*. Proc R Soc Lond B 265:185–190

Clutton-Brock TH, MacColl ADC, Chadwick P, Gaynor D, Kansky R, Skinner JD (1999a) Reproduction and survival of suricates (*Suricata suricatta*) in the southern Kalahari. Afr J Ecol 37:69–80

Clutton-Brock TH, O'Riain MJ, Brotherton PNM, Gaynor D, Kansky R, Griffin AS, Manser M (1999b) Selfish sentinels in cooperative mammals. Science 284:1640–1644

Clutton-Brock TH, Gaynor D, McIlrath GM, MacColl ADC, Kansky R, Chadwick P, Manser M, Skinner JD, Brotherton PNM (1999c) Predation, group size and mortality in a cooperative mongoose, *Suricata suricatta*. J Anim Ecol 68:672–683

Clutton-Brock TH, Brotherton PNM, O'Riain MJ, Griffin AS, Gaynor D, Sharpe L, Kansky R, Manser M, McIlrath GM (2000) Individual contributions to babysitting in a cooperative mongoose, *Suricata suricatta*. Proc R Soc Lond B 267:301–305

Clutton-Brock TH, Russell AF, Sharpe L, Brotherton PNM, McIlrath GM, White S, Cameron EZ (2001a) Effects of helpers on juvenile development and survival in meerkats. Science 293:2446–2449

Clutton-Brock TH, Brotherton PNM, O'Riain MJ, Griffin AS, Gaynor D, Kansky R, Sharpe L, McIlrath GM (2001b) Contributions to cooperative rearing in meerkats. Anim Behav 61:705–710

Clutton-Brock TH, Brotherton PNM, Russell AF, O'Riain MJ, Gaynor D, Kansky R, Griffin A, Manser M, Sharpe L, McIlrath GM, Small T, Moss A, Monfort S (2001c) Cooperation, control and concession in meerkat groups. Science 291:478–481

Clutton-Brock TH, Russell AF, Sharpe LL, Young AJ, Balmforth Z, McIlrath GM (2002) Evolution and development of sex differences in cooperative behavior in meerkats. Science 297:253–256

Clutton-Brock TH, Russell AF, Sharpe LL (2003) Meerkat helpers do not specialize in particular activities. Anim Behav 66:531–540

Clutton-Brock TH, Russell AF, Sharpe LL (2004) Behavioural tactics of breeders in cooperative meerkats. Anim Behav 68:1029–1140

Cockburn A (1998) Evolution of helping behaviour in cooperatively breeding birds. Ann Rev Ecol System 29:141–177

Cole E (2004) Bone to Pick: Of Forgiveness, Reconciliation, Reparation and Revenge. Atria Books, New York

Colman AM (1999) Game Theory and its Applications in the Social and Biological Sciences. Routledge, London, New York

Colman AM (2003a) Cooperation, psychological game theory, and limitations of rationality in social interaction. Behav Brain Sci 26:139–198

Colman AM (2003b) Depth of strategic reasoning in games. Trends Cogn Sci 7:2–4

Colvin J (1983) Description of sibling and peer relationships among immature male rhesus monkeys. In: Hinde RA (ed) Primate Social Relationships: An Integrated Approach. Sinauer Associates, Sunderland/MA, pp 20–27

Combes S, Altmann J (2001) Status change during adulthood: life-history by-product or kin selection based on reproductive value? Proc R Soc Lond B 268:1367–1373

Connor RC (1986) Pseudo-reciprocity: investing in mutualism. Anim Behav 34:1562–1584

Connor RC (1995) Impala allogrooming and the parcelling model of reciprocity. Anim Behav 49:528–530

Constable JL, Ashley MV, Goodall, Pusey A (2001) Noninvasive paternity assignment in Gombe chimpanzees. Mol Ecol 10:1279–1300

Cook JM, Rasplus J-Y (2003) Mutualists with attitude: coevolving fig wasps and figs. Trends Ecol Evol 18:241–248

Cooney R, Bennett N (2000) Inbreeding avoidance and reproductive skew in a cooperative mammal. Proc R Soc Lond B 267:801–806
Cooper MA, Bernstein IS, Hemelrijk CK (2005) Reconciliation and relationship quality in Assamese macaques (*Macaca assameusis*). Am J Primatol 65:269–282
Cooper SM (1991) Optimal hunting group size: the need for lions to defend their kills against loss to spotted hyaenas. Afr J Ecol 29:130–136
Cooper R, DeJong D, Forsythe R, Ross T (1996) Cooperation without reputation: experimental evidence from prisoner's dilemma games. Games Econ Behav 12:187–218
Cords M (1992) Post-conflict reunions and reconciliation in long-tailed macaques. Anim Behav 44:57–61
Cords M (1997) Friendship, alliances, reciprocity and repair. In: Whiten A, Byrne RW (eds) Machiavellian Intelligence II. Cambridge Univ Press, Cambridge, pp 24–49
Cords M, Aureli F (1993) Patterns of reconciliation among juvenile long-tailed macaques. In: Pereira ME, Fairbanks LA (eds) Juvenile Primates: Life History, Development and Behavior. Oxford Univ Press, Oxford, pp 271–284
Cords M, Aureli F (1996) Reasons for reconciling. Evol Anthropol 5:42–45
Cords M, Aureli F (2000) Reconciliation and relationship qualities. In: Aureli F, de Waal FBM (eds) Natural Conflict Resolution. Univ of California Press, Berkeley, pp 177–198
Cords M, Killen M (1998) Conflict resolution in human and nonhuman primates. In: Langer J, Killen M (eds) Piaget, Evolution and Development. Lawrence Earlbaum Associates, Mahwah/NJ, pp 193–219
Cords M, Thurnheer S (1993) Reconciliation with valuable partners by long-tailed macaques. Ethology 93:315–325
Cosmides L, Tooby JH (1992) Cognitive adaptations for social exchange. In: Barkow JH, Cosmides L, Tooby JH (eds) The Adapted Mind. Oxford Univ Press, Oxford, pp 163–228
Courchamp F, Clutton-Brock T, Grenfell BT (2000a) Multipack dynamics and the Allee effect in the African wild dog *Lycaon pictus*. Anim Conserv 3:277–285
Courchamp F, Grenfell BT, Clutton-Brock TH (2000b) Impact of natural enemies on obligately cooperative breeders. Oikos 91:311–322
Crawford M (1937) The cooperative solving of problems by young chimpanzees. Comp Psychol Monogr 14:1–88
Creel SR (1990) How to measure inclusive fitness. Proc R Soc Lond B 241:229–231
Creel SR (2001) Social dominance and stress hormones. Trends Ecol Evol 16:491–497
Creel SR, Creel NM (1995) Communal hunting and pack size in African wild dogs, *Lycaon pictus*. Anim Behav 50:1325–1339
Creel SR, Creel NM (2002) The African Wild Dog. Princeton Univ Press, Princeton/NJ
Creel SR, Macdonald D (1995) Sociality, group size and reproductive suppression among carnivores. Adv Stud Behav24:203–257
Creel SR, Waser PM (1991) Failures of reproductive suppression in dwarf mongooses (*Helogale parvula*): accident or adaptation? Behav Ecol 2:7–15
Creel SR, Waser PM (1997) Variation in reproductive suppression among dwarf mongooses: interplay between mechanisms and evolution. In: Solomon NG, French JA (eds) Cooperative Breeding in Mammals. Cambridge Univ Press, Cambridge, pp 150–170
Crespi BJ, Yanega D (1995) The definition of eusociality. Behav Ecol 6:109–115
Crockett CM (1984) Emigration by female red howler monkeys and the case for female competition. In: Small MF (ed) Female Primates: Studied by Women Primatologists. Alan R Liss, New York, pp 159–173
Crockett CM, Pope TR (1993) Consequences for sex difference in dispersal for juvenile red howler monkeys. In: Pereira ME, Fairbanks LA (eds) Juvenile Primates: Life History, Development and Behavior. Oxford Univ Press, Oxford, pp 104–118
Crowcroft P, Rowe FP (1963) Social organization and territorial behaviour in the wild house mouse. Proc R Soc Lond B 140:517–531
Crozier R, Pamilo P (1996) Evolution of Social Insect Colonies: Sex Allocation and Kin Selection. Oxford Univ Press, New York

Currie CR, Mueller UG, Malloch D (1999) The agricultural pathology of ant fungus gardens. Proc Natl Acad Sci USA 96:7998–8002
Darwin C (1859) On the Origin of Species. Murray, London
Das M (1998) Conflict management and social stress in long-tailed macaques. PhD thesis, Univ of Utrecht/NL
Das M, Penke Z, van Hooff JARAM (1998) Post-conflict affiliation and stress-related behavior of long-tailed macaque aggressors. Int J Primatol 19:53–71
Datta SB (1983a) Relative power and the acquisition of rank. In: Hinde RA (ed) Primate Social Relationships: An Integrated Approach. Sinauer Associates, Sunderland/MA, pp 93–103
Datta SB (1983b) Relative power and the maintenance of dominance. In: Hinde RA (ed) Primate Social Relationships: An Integrated Approach. Sinauer Associates, Sunderland/MA, pp 103–112
Datta SB (1983c) Patterns of agonistic interference. In: Hinde RA (ed) Primate Social Relationships: An Integrated Approach. Sinauer Associates, Sunderland/MA, pp 289–297
Datta SB (1988) The acquisition of dominance among free-ranging rhesus monkey siblings. Anim Behav 36:754–772
Davidson RJ, Kabat-Zinn J, Schumacher J, Rosenkranz M, Muller D, Santorelli SF, Urbanowski F, Harrington A, Bonus K, Sheridan JF (2003) Alterations in brain and immune function produced by mindfulness meditation. Psychosom Med 65:564–570
Dawes RM, McTavish J, Shaklee H (1977) Behavior communication and assumptions about other people's behavior in a commons dilemma situation. J Pers Soc Psychol 35:1–11
Dawes RM, van de Kragt AJC, Orbell JM (1988) Not me or thee, but we: the importance of group identity in eliciting cooperation in dilemma situations – experimental manipulations. Acta Psychol 68:83–97
Dawkins R (1979) Twelve misunderstandings of kin selection. Z Tierpsychol 51:184–200
Dawkins R, Carlisle TR (1976) Parental investment, mate desertion and a fallacy. Nature 262:131–133
DeBruine LM (2002) Facial resemblance enhances trust. Proc R Soc Lond B 269:1307–1312
DeLong KT (1967) Population ecology of feral house mice. Ecology 48:611–634
Demaria C, Thierry B (2001) A comparative study of reconciliation in rhesus and Tonkean macaques. Behaviour 138:397–410
Denison RF, Bledshoe C, Kahn M, O'Gara F, Simms EL, Thomasow LS (2003) Cooperation in the rhizosphere and the "free rider" problem. Ecology 84:838–845
Di Bitteti MS (1997) Evidence for an important social role of grooming in a platyrrhine primate. Anim Behav 54:199–211
DiClemente DF, Hantula DA (2003) Optimal foraging online: increasing sensitivity to delay. Psychol Mark 20:785–809
Diedrichsen U (1993) Ethophysiologie der Energieallokation der Jungenaufzucht von Hausmäusen (*Mus domesticus*) in unterschiedlichen Sozialsystemen. Diploma thesis, Julius-Maximilians-University, Würzburg
Dietz JM (2004) Kinship structure and reproductive skew in cooperatively breeding primates. In: Chapais B, Berman C (eds) Kinship and Behavior in Primates. Oxford Univ Press, Oxford, pp 223–241
Dietz JM, Baker AJ (1993) Polygyny and female reproductive success in golden lion tamarins, *Leontopithecus rosalia*. Anim Behav 46:1067–1078
Di Fiore A, Rendall D (1994) Evolution of social organization: a reappraisal for primates by using phylogenetic methods. Proc Natl Acad Sci USA 91:9941–9945
Digby L (1995) Infant care, infanticide and female reproductive strategies in polygynous groups of common marmosets (*Callithrix jacchus*). Behav Ecol Sociobiol 37:51–61
Digby L (2001) Infanticide by female mammals: implications for the evolution of social systems. In: van Schaik CP, Janson CH (eds) Infanticide by Males and its Implications. Cambridge Univ Press, Cambridge, pp 423–446
Dittmar H, Beattie J, Friese S (1995) Gender identity and material symbols: objects and decision considerations in impulse purchases. J Econ Psychol 16:491–511
Dittmar H, Beattie J, Friese S (1996) Objects, decision considerations and self-image in men's and women's impulse purchases. Acta Psychol 93:187–206

Dobson FS, Jacquot C, Baudoin C (2000) An experimental test of kin association in the house mouse. Can J Zool 78:1806–1812

Doebeli M, Hauert C, Killingback T (2004) The evolutionary origin of cooperators and defectors. Science 306:859–862

Dominy NJ, Lucas PW, Osorio D, Yamashita N (2001) The sensory ecology of primate food perception. Evol Anthropol 10:171–186

Doncaster CP, Woodroffe R (1993) Den site can determine shape and size of badger territories: implications for group living. Oikos 66:88–93

Doran DM, Jungers WL, Sugiyama Y, Fleagle J, Heesy CP (2002) Multivariate and phylogenetic approaches to understanding chimpanzee and bonobo behavioral diversity. In: Boesch C, Hohmann G, Marchant LF (eds) Behavioral Diversity in Chimpanzees and Bonobos. Cambridge Univ Press, Cambridge, pp14–34

Dufwenberg M, Kirchsteiger G (2004) A theory of sequential reciprocity, games and economic. Behavior 47:268–298

Dugatkin LA (1997) Cooperation among Animals. An Evolutionary Perspective. Oxford Series in Ecology and Evolution, Oxford Univ Press, Oxford

Dugatkin LA (2002a) Animal cooperation among unrelated individuals. Naturwissenschaften 89:533–541

Dugatkin LA (2002b) Cooperation in animals: an evolutionary overview. Biol Philos 17:459–476

Dugatkin LA, Wilson DS (1991) ROVER: a strategy for exploiting cooperators in a patchy environment. Am Nat 138:687–701

Dunbar RIM (1984) Reproductive decisions an economic analysis of gelada baboon social strategies. Princeton Univ Press, Princeton/NJ

Dunbar RIM (1988) Primate Social Systems. Cornell Univ Press, Ithaca, New York

Dunbar RIM (1991) Functional significance of social grooming in primates. Fol Primatol 57:121–131

Dunbar RIM (1992) Social behaviour and evolutionary theory. In: Jones S, Martin R, Pilbeam R and D (eds) The Cambridge Encyclopedia of Human Evolution. Cambridge Univ Press, Cambridge, pp 145–149

Dunbar RIM (1998) The social brain hypothesis. Evol Anthropol 6:178–190

Dunbar RIM (2001) Brains on two legs: group size and the evolution of intelligence. In: de Waal FBM (ed) Tree of Origin. Harvard Univ Press, Cambridge/MA, pp 175–91

Dunbar RIM, Sharman M (1984) Is social grooming altruistic? Z Tierpsychol 64:163–173

Dunn PO, Cockburn A, Mulder RA (1995) Fairy wren helpers often care for young to which they are unrelated. Proc R Soc Lond B 259:339–343

Eberle M, Kappeler PM (2002) Mouse lemurs in space and time: a test of the socioecological model. Behav Ecol Sociobiol 51:131–139

Echeverria-Lozano G (2004) Aggression and conflict management in adult and juvenile Chacma baboons. PhD thesis, Univ of Liverpool/UK

Eckardt W, Zuberbühler K (2004) Cooperation and competition in two forest monkeys. Behav Ecol 15:400–411

Ekman J, Dickinson JL, Hatchwell B, Grieszer M (2004) Delayed dispersal. In: Koenig W, Dickinson J (eds) Ecology and Evolution of Cooperative Breeding in Birds. Cambridge Univ Press, Cambridge, pp 35–47

Emlen ST (1984) Cooperative breeding in birds and mammals. In: Krebs JR, Davies NB (eds) Behavioural Ecology: An Evolutionary Approach. Blackwell, Oxford, pp 305–339

Emlen ST (1991) Evolution of cooperative breeding in birds and mammals. In: Krebs JR, Davies NB (eds) Behavioural Ecology: An Evolutionary Approach. Blackwell, Oxford, pp 301–337

Emlen ST (1997) Predicting social dynamics in social vertebrates. In: Krebs JR, Davies NB (eds) Behavioural Ecology: An Evolutionary Approach. Blackwell, Oxford, pp 228–253

Emlen ST, Oring LW (1977) Ecology, sexual selection, and the evolution of mating systems. Science 197:215–223

Emlen ST, Wrege PH (1988) The role of kinship in helping decisions among white-fronted bee-eaters. Behav Ecol Sociobiol 23:305–315

Emlen ST, Wrege PH (1992) Parent-offspring conflict and the recruitment of helpers among bee-eaters. Nature 356:331–333

Emlen ST, Reeve HK, Keller L (1998) Reproductive skew: disentangling concessions from control. Trends Ecol Evol 13:458–459
Engel V, Fischer MK, Wäckers FL, Völkl W (2001) Interactions between extrafloral nectaries, aphids and ants: are there competition effects between plant and homopteran sugar sources? Oecologia 129:577–584
Engelmann D, Fischbacher U (2002) Indirect reciprocity and strategic reputation building in an experimental helping game. IEW Working Paper No.132, Univ of Zürich, Zürich
Enquist M (1985) Communication during aggressive interactions with particular reference to variation in choice of behaviour. Anim Behav 33:1152–1161
Erhart EM, Coelho AM, Bramblett CA (1997) Kin recognition by paternal half-siblings in captive *Papio cynocephalus*. Am J Primatol 43:147–157
Erk S, Spitzer M, Wunderlich AP, Galley L, Walter H (2002) Cultural objects modulate reward circuitry. NeuroReport 13:2499–2503
Fagen R (1981) Animal Play Behavior. Oxford Univ Press, Oxford
Fagen R (1993) Primate juveniles and primate play. In: Pereira ME, Fairbanks LA (eds) Juvenile Primates: Life History, Development and Behaviour. Oxford Univ Press, NewYork, pp 182–196
Fairbanks LA (1990) Reciprocal benefits of allomothering for female vervet monkeys. Anim Behav 40:553–562
Fairbanks LA (1993) Juvenile vervet monkeys: establishing relationships and practicing skills for the future. In: Pereira ME, Fairbanks LA (eds) Juvenile Primates. Oxford Univ Press, New York, pp 211–227
Fairbanks LA (2000) Maternal investment throughout the life span in Old World monkeys. In: Whitehead PF, Jolly CJ (eds) Old World Monkeys. Cambridge Univ Press, Cambridge, pp 341–367
Falk A, Fischbacher U (2005) A theory of reciprocity. Games Econ Behav, in press
Falk A, Fehr E, Fischbacher U (2004) Driving forces behind informal sanctions. IEW Working paper No. 59, Univ of Zürich, Zürich
Falk A, Gächter S, Kovács J (1999) Intrinsic motivation and extrinsic incentives in a repeated game with incomplete contracts. J Econ Psychol 20:251–284
Fanshawe JH, Fitzgibbon CD (1993) Factors influencing the hunting success of an African wild dog pack. Anim Behav 45:479–490
Faulkes CG, Abbott DH (1997) The physiology of a reproductive dictatorship: regulation of male and female reproduction by a single breeding female in colonies of naked mole-rats. In: Solomon NG, French JA (eds) Cooperative Breeding in Mammals. Cambridge Univ Press, Cambridge, pp 302–334
Faulkes CG, Abbott DH, Jarvis JUM (1990) Social suppression of ovarian cyclicity in captive and wild colonies of naked mole-rats (*Heterocephalus glaber*). J Reprod Fert 88:559–568
Faulkes CG, Arruda MF, Monteiro da Cruz AO (2003) Matrilineal genetic structure within and among populations of the cooperatively breeding common marmoset, *Callithrix jacchus*. Mol Ecol 12:1101–1118
Fedigan LM, Asquith PJ (eds) (1991) The Monkeys of Arashiyama. Thirty-five Years of Research in Japan and the West. State Univ of New York Press, Albany, New York
Feh C (1999) Alliances and reproductive success in Camargue stallions. Anim Behav 57:705–713
Fehr E (2004) Don't lose your reputation. Nature 432:449–450
Fehr E, Fischbacher U (2002) Why social preferences matter – the impact of non-selfish motives on competition, cooperation and incentives. Econ J 112:C1–33
Fehr E, Fischbacher U (2003) The nature of human altruism. Nature 425:785–791
Fehr E, Fischbacher U (2004) Third-party punishment and social norms. Evol Hum Behav 25:63–87
Fehr E, Gächter S (2000a) Fairness and retaliation: the economics of reciprocity. J Econ Perspect 14:159–181
Fehr E, Gächter S (2000b) Cooperation and punishment in public goods experiments. Am Econ Rev 90:980–994
Fehr E, Gächter S (2002) Altruistic punishment in humans. Nature 415:137–140
Fehr E, Gächter S (2003) The puzzle of human cooperation – reply. Nature 421:912
Fehr E, Henrich J (2003) Is strong reciprocity a maladaptation? In: Hammerstein P (ed) Genetic and Cultural Evolution of Cooperation. MIT Press, Cambridge/MA, pp 55–82

Fehr E, Kirchsteiger G, Riedl A (1993) Does fairness prevent market clearing? Q J Econ 108:437–459

Fehr E, Rockenbach B (2003) Detrimental effects of sanctions on human altruism. Nature 422:137–140

Fehr E, Schmidt KM (1999) A theory of fairness, competition, and cooperation. Q J Econ 114:817–868

Fehr E, Gächter S, Kirchsteiger G (1997) Reciprocity as a contract enforcement device: experimental evidence. Econometrica 65:833–860

Fehr E, Fischbacher U, Gächter S (2002) Strong reciprocity, human cooperation, and the enforcement of social norms. Hum Nat 13:1–25

Feistner ATC, McGrew WC (1989) Food-sharing in primates: a critical review. In: Seth PK, Seth S (eds) Perspectives in Primate Biology. Today, Tomorrow's Printers and Publishers, New Delhi, pp 21–36

Fessler DMT, Haley KJ (2003) The strategy of affect. Emotions in human cooperation. In: Hammerstein P (ed) Genetic and Cultural Evolution of Cooperation. MIT Press, Cambridge/MA, pp 7–36

Fischbacher U, Gächter S (2004) Heterogeneous Social Preferences and the Dynamics of Free Riding in Public Goods. Mimeo, Univ of St. Gallen

Fischbacher U, Gächter S, Fehr E (2001) Are people conditionally cooperative? Evidence from a public goods experiment. Econ Lett 71:397–404

Fischer EA (1980) The relationship between mating system and simultaneous hermaphroditism in the coral reef fish, *Hypoplectrus nigricans*. Anim Behav 28:620–633

Fischer EA (1988) Simultaneous hermaphroditism, Tit-for-Tat, and the evolutionary stability of social systems. Ethol Sociobiol 9:119–36

Fischer MK, Hoffmann KH, Völkl W (2001) Competition for mutualists in an ant-homopteran interaction mediated by hierarchies of ant attendance. Oikos 92:531–541

Fischer RC, Richter A, Wanek W, Mayer V (2002) Plants feed ants: food bodies of myrmecophytic Piper and their significance for the interaction with *Pheidole bicornis* ant. Oecologia 133:186–192

Fishman MA (2003) Indirect reciprocity among imperfect individuals. J theor Biol 225:285–292

Flack JC, de Waal FBM (2000) "Any animal whatever": Darwinian building blocks of morality in monkeys and apes. J Conscious Stud 7:1–29

Fodor J (1983) The modularity of the mind. MIT Press, Cambridge/MA

Forsythe R, Horowitz J, Savin NE, Sefton M (1994) Fairness in simple bargaining experiments. Games Econ Behav 6:347–369

Foster KR, Shaulsky G, Strassmann JE, Queller DC, Thompson CRL (2004) Pleiotropy as a mechanism to stabilize coopeeration. Nature 431:693–696

Fragaszy DM, Feuerstein JM, Mitra D (1997) Transfers of food from adults to infants in tufted capuchins (*Cebus apella*). J Comp Psychol 111:194–200

Francis CM, Anthony ELP, Brunton JA, Kunz TH (1994) Lactation in male fruit bats. Nature 367:691–692

Frank RH (1988) Passions Within Reason: The Strategic Role of the Emotions. Norton, New York

Frank RH (1994) Microeconomics and Behavior. McGraw-Hill, New York

Frean MR (1994) The prisoner's dilemma without synchrony. Proc R Soc Lond B 257:75–79

Fredrikson WT, Sackett G (1984) Kin preferences in primates, *Macaca nemestrina*: relatedness or familiarity? J Comp Psychol 98:29–34

Fredsted T, Pertoldi C, Olesen J, Eberle M, Kappeler PM (2004) Microgeographic heterogeneity in spatial distribution and mtDNA variability of gray mouse lemurs (*Microcebus murinus*, Primates: Cheirogaleidae). Behav Ecol Sociobiol 56:393–403

French JA (1994) Alloparents in the Mongolian gerbil: impact on long-term reproductive performance of breeders and opportunities for independent reproduction. Behav Ecol 5:273–279

French JA (1997) Proximate regulation of singular breeding in callitrichid primates. In: Solomon NG, French JA (eds) Cooperative Breeding in Mammal. Cambridge Univ Press, Cambridge, pp 34–75

French JA, Abbott DH, Snowdon CT (1984) The effects of social environment on oestrogen secretion, scent marking and sociosexual behavior in tamarins (*Saguinus oedipus*). Am J Primatol 6:155–167

Friedman D, Sunder S (1994) Experimental Methods. A Primer for Economists. Princeton Univ Press, Princeton/NJ
Fuchs S (1981) Consequences of premature weaning on the reproduction of mothers and offspring in laboratory mice. Z Tierpsychol 55:19–32
Fuchs S (1982) Optimality of parental investment: the influence of nursing on the reproductive success of mother and female young house mice. Behav Ecol Sociobiol 10:39–51
Fudenberg D, Maskin E (1986) The folk theorem in repeated games with discounting or with incomplete information. Econometrica 54:533–556
Furuichi T (1997) Agonistic interactions and matrifocal dominance rank of wild bonobos (*Pan paniscus*) at Wamba. Int J Primatol 18:855–875
Furuichi T, Ihobe H (1994) Variation in male relationships in bonobos and chimpanzees. Behaviour 130:212–228
Fuster J (1989) The Prefrontal Cortex. Raven Press, New York
Gächter S, Falk A (2002) Reputation and reciprocity – consequences for the labour relation. Scand J Econ 104:1–26
Gächter S, Fehr E (1999) Collective action as a social exchange. J Econ Behav Org 39:341–369
Gächter S, Renner E (2004) Leading by Example in the Presence of Free Rider Incentives. Mimeo, Univ of Nottingham
Gächter S, Thöni C (2005) Social learning and voluntary cooperation among like-minded people. J Eur Econ Assoc, in press
Gächter S, Herrmann B, Thöni C (2003) Understanding Determinants of Social Capital. Cooperation and Informal Sanctions in a Cross-Societal Perspective. Mimeo, Univ of St. Gallen
Galdikas BMF (1988) Orangutan diet, range, and activity at Tanjung Putting, Central Borneo. Int J Primatol 9:1–35
Gale D, Shapely LS (1962) College admissions and the stability of marriage. Am Math Month 69:9–15
Garber P (1997) One for all and breeding for one: cooperation and competition as a tamarin reproductive strategy. Evol Anthropol 5:187–199
Gaston AJ (1977) Social behaviour within groups of jungle babblers *Turdoides striatus*. Anim Behav 25:828–848
Ghiglieri M (1984) The Chimpanzees of the Kibale Forest. Columbia Univ Press, New York
Gibson RM, Langen TA (1996) How do animals choose their mates? Trends Ecol Evol 11:468–470
Gigerenzer G (1997) The modularity of social intelligence. In: Whiten A, Byrne R (eds) Machiavellian Intelligence. II. Extensions and Evaluations. Cambridge Univ Press, Cambridge, pp 264–288
Gigerenzer G, Todd PM, ABC Research Group (eds) (1999) Simple heuristics that make us smart. Oxford Univ Press, Oxford
Gilchrist JS (2001) Reproduction and pup care in the communal breeding banded mongoose. PhD thesis, Cambridge Univ Press, Cambridge
Gintis H (2000) Strong reciprocity and human sociality. J theor Biol 206:169–179
Gintis H (2003) The hitchhikers guide to altruism: genes, culture and the internalization of norms. J theor Biol 220:407–418
Gintis H, Smith EA, Bowles S (2001) Costly signaling and cooperation. J theor Biol 213:103–119
Gintis H, Bowles S, Boyd R, Fehr E (2003) Explaining altruistic behavior in humans. Evol Hum Behav 24:153–172
Gittleman JL (1989) Carnivore group living: comparative trends. In: Gittleman JL (ed) Carnivore Behaviour, Ecology and Evolution. Cornell Univ Press, Ithaca/NY, pp 183–207
Glick BB, Eaton GG, Johnson DF, Worlein J (1986) Social behavior of infant and mother Japanese macaques (*Macaca fuscata*): effects of kinship, partner sex and infant sex. Int J Primatol 7:139–155
Godfray HCJ (1995a) Evolutionary theory of parent-offspring conflict. Nature 376:133–138
Godfray HCJ (1995b) Signaling of need between parents and young: parent-offspring conflict and sibling rivalry. Am Nat 146:1–24
Goldberg T, Wrangham RW (1997) Genetic correlates of social behaviour in wild chimpanzees: evidence from mitochondrial DNA. Anim Behav 54:559–570
Goldizen AW (1987a) Facultative polyandry and the role of infant carrying in wild saddle-back tamarins (*Saguinus fusicollis*). Behav Ecol Sociobiol 20:99–109

Goldizen AW (1987b) Tamarins and marmosets: communal care of offspring. In: Smuts BB, Cheney DL, Seyfarth RM, Wrangham RW, Struhsaker TT (eds) Primate Societies. Univ of Chicago Press, Chicago, pp 34–43

Goldizen AW, Terborgh J (1989) Demography and dispersal patterns of a tamarin population: possible causes. Am Nat 134:208–224

Goodall J (1968) The behaviour of free-living chimpanzees in the Gombe Stream area. Anim Behav Monogr 1:161–311

Goodall J (1971) In the Shadow of Man. Houghton Mifflin, Boston

Goodall J (1986) The Chimpanzees of Gombe: Patterns of Behavior. Harvard Univ Press, Cambridge/MA

Goodall J, Bandora A, Bergmann E, Busse C, Matama H, Mpongo E, Pierce A, Riss D (1979) Intercommunity interactions in the chimpanzee population of the Gombe National Park. In: Hamburg D, McCown L (eds) The Great Apes. Benjamin-Cummings, Menlo Park, pp 13–54

Goodwin BK, Holt MT (1999) Price transmission and asymmetric adjustment in the US beef sector. Am J Agric Econ 81:630–638

Gouzoules S (1984) Primate mating systems, kin association and cooperative behavior: evidence for kin recognition? Ybk Phys Anthropol 27:99–134

Gouzoules S, Gouzoules H (1987) Kinship. In: Smuts BB, Cheney DL, Seyfarth RM, Wrangham RW, Struhsaker TT (eds) Primate Societies. Univ of Chicago Press, Chicago, pp 299–305

Grafen A (1990) Sexual selection unhandicapped by the Fisher Process. J theor Biol 144:473–516

Grafen A (1991) Modelling in behavioural ecology. In: Krebs JR, Davies NB (eds) Behavioural Ecology: An Evolutionary Approach. Blackwell, Oxford, pp 5–31

Gray SJ, Jensen SP, Hurst JL (2000) Structural complexity of territories: effects on preference, use of space and territorial defence in commensal house mice (*Mus domesticus*). Anim Behav 60:765–772

Green L, Myerson J, Holt DD, Slevin JR, Estle SJ (2004) Discounting of delayed food rewards in pigeons and rats: is there a magnitude effect? J Exp Anal Behav 81:39–50

Greene E, Lyon BE, Muehter VR, Ratcliffe L, Oliver SJ, Boag PT (2000) Disruptive sexual selection for plumage coloration in a passerine bird. Nature 407:1000–1003

Greenwood PJ (1980) Mating systems, philopatry and dispersal in birds and mammals. Anim Behav 28:1140–1162

Greig D, Travisano M (2004) The prisoner's dilemma and polymorphism in yeast and SUC genes. Proc R Soc Lond B 271:S25–26

Griffin AS, West SA (2002a) Kin discrimination and the benefit of helping in cooperatively breeding vertebrates. Science 302:634–636

Griffin AS, West SA (2002b) Kin selection: fact and fiction. Trends Ecol Evol 17:15–21

Griffin AS, West SA (2003) Kin discrimination and the benefit of helping in cooperatively breeding vertebrates. Science 302:634–636

Gürerk Ö, Irlenbusch B, Rockenbach B (2004) On the evolement of institutions in social dilemmas. Mimeo, Univ of Erfurt

Güth W, Schmittberger R, Schwartze B (1982) An experimental analysis of ultimatum bargaining. J Econ Behav Org 3:367–388

Güth W, Levati MV, van der Heijden E, Sutter M (2004) Leadership and cooperation in public goods experiments. Discussion Paper 28-2004, Papers on Strategic Interaction, Max Planck Institute for Research into Economic Systems

Gust DA, Gordon TP (1994) The absence of a matrilineally based dominance system in sooty mangabeys, *Cercocebus torquatus atys*. Anim Behav 47:589–94

Grutter AS (2004) Cleaner fish use tactile dancing behavior as a preconflcit management strategy. Curr Biol 14:1080–1083

Haig D (2002) Kinship and Genomic Imprinting. Rutgers Univ Press, New Brunswick

Haig D (2003) On intrapersonal reciprocity. Evol Hum Behav 24:418–425

Hamilton WD (1963) The evolution of altruistic behavior. Am Nat 97:354–356

Hamilton WD (1964) The genetical evolution of social behavior. I and II. J theoret Biol 7:17–52

Hamilton WD (1970) Selfish and spiteful behavior in an evolutionary model. Nature 228:1218–1220

Hamilton WD (1971) Geometry for the selfish herd. J theoret Biol 31:295–311

Hammerstein P (2003a) Genetic and Cultural Evolution of Cooperation. MIT Press, Cambridge/MA

Hammerstein P (2003b) Why is reciprocirty so rare in social animals? A protestant appeal. In: Hammerstein P (ed) Genetic and Cultural Evolution of Cooperation. MIT Press, Cambridge/MA, pp 83–93

Hammond KA, Diamond J (1992) An experimental test for a ceiling on sustained metabolic rate in lactating mice. Physiol Zool 65:952–977

Harcourt AH (1987) Dominance and fertility among female primates. J Zool Lond 213:471–487

Harcourt AH (1988) Alliances in contests and social intelligence. In: Byrne RW, Whiten A (eds) Machiavellian Intelligence. Clarendon Press, Oxford, pp 132–152

Harcourt AH (1992) Coalitions and alliances: are primates more complex than non-primates. In: Harcourt AH, de Waal FBM (eds) Coalitions and Alliances in Humans and other Animals. Oxford Univ Press, Oxford, pp 445–471

Harcourt AH, de Waal FBM (1992) Coalitions and Alliances in Human and other Animals. Oxford Univ Press, Oxford

Harcourt AH, Greenberg H (2001) Do gorilla females join males to avoid infanticide? Anim Behav 62:905–915

Harcourt AH, Stewart KJ (1987) The influence of help in contests on dominance rank in primates: hints from gorillas. Anim Behav 35:182–190

Harcourt AH, Stewart KJ (1989) Functions of alliances in contests within wild gorilla groups. Behaviour 109:176–190

Hardin G (1968) Tragedy of commons. Science 162:1243–1248

Hardin G (1998) Extensions of "the tragedy of the commons". Science 280:682–683

Hare B, Call J, Tomasello M (2001) Do chimpanzees know what conspecifics know? Anim Behav 61:139–151

Hare B, Addessi E, Call J, Tomasello M, Visalberghi E (2003) Do capuchin monkeys, *Cebus apella*, know what conspecifics do and do not see? Anim Behav 65:131–142

Harper B (2000) Beauty, stature and the labour market: a British cohort study. Oxford Bull Econ Stat 62:771–800

Hart BL, Hart LA (1992) Reciprocal allogrooming in impala, *Aepyceros melampus*. Anim Behav 44:1073–1083

Harvey PH, Bradbury JW (1991) Sexual selection. In: Krebs JR, Davies NB (eds) Behavioural Ecology: An Evolutionary Approach. Blackwell, Oxford, pp 203–233

Hashimoto C, Furuichi T, Takenaka O (1996) Matrilineal kin relationship and social behavior of wild bonobos (*Pan paniscus*): sequencing the D-loop region of mitochondrial DNA. Primates 37:305–318

Hasselmann K, Latif M, Hooss G, Azar C, Edenhofer O, Jaeger CC, Johannessen OM, Kemfert C, Welp M, Wokaun A (2003) The challenge of long-term climate change. Science 302:1923–1925

Hauert C, Doebeli M (2004) Spatial structure often inhibits the evolution of cooperation in the snowdrift game. Nature 428:643–646

Hauert C, Schuster HG (1998) Extending the iterated prisoner's dilemma without synchrony. J theoret Biol 192:155–166

Hauert C, de Monte S, Hofbauer J, Sigmund K (2002) Volunteering as red queen mechanism for cooperation in public goods games. Science 296:1129–1132

Haug M (1978) Attack by female mice on "strangers". Aggress Behav 4:133–139

Hauser MD, Carey S, Hauser LB (2000) Spontaneous number representation in semi-free-ranging rhesus monkeys. Proc R Soc Lond B 267:829–833

Hauser MD, Chen MK, Chen F, Chang E, Chuang E (2003) Give unto others: genetically unrelated cotton-top tamarin monkey preferentially give food to those who altruistically give food back. Proc R Soc Lond B 270:2363–2370

Hausfater G, Altmann J, Altmannn SA (1982) Long-term consistency of female dominance relations among female baboons (*Papio cynocephalus*). Science 217:752–754

Hawkes K (1990) Showing off: tests of an hypothesis about men's foraging goals. Ethol Sociobiol 12:29–54

Hawkes K, O'Connell J, Blurton-Jones N (2001) Hadza meat sharing. Evol Hum Behav 22:113–142

Hayaki H, Huffman M, Nishida T (1989) Dominance among male chimpanzees in the Mahale Mountains National Park, Tanzania. Primates 30:187–197
Haydock J, Koenig WD (2002) Reproductive skew in the polygynandrous acorn woodpecker. Proc Natl Acad Sci USA 99:7178–7183
Haydock J, Koenig WD (2003) Patterns of reproductive skew in the polygynandrous acorn woodpecker. Am Nat 162:277–289
Hayes LD (2000) To nest communally or not to nest communally: a review of rodent communal nesting and nursing. Anim Behav 59:677–688
Hayes LD, Solomon NG (2004) Costs and benefits of communal rearing to female prairie voles (*Microtus ochrogaster*). Behav Ecol Sociobiol 56:585–593
Heinrich J, Boyd R, Bowles S, Camerer C, Fehr E, Gintis R, McElreath R (2001) In search of *Homo economicus*: behavioural experiments in 15 small-scale societies. Am Econ Rev 91:73–78
Heinsohn RG (1991) Kidnapping and reciprocity in cooperatively breeding white-winged choughs. Anim Behav 41:1097–1100
Heinsohn RG (1992) Cooperative enhancement of reproductive success in white-winged choughs. Evol Ecol 6:97–114
Heinsohn RG (2004) Parental care, load lightening and costs. In: Koenig W, Dickinson J (eds) Ecology and Evolution of Cooperative Breeding in Birds. Cambridge Univ Press, Cambridge, pp 67–80
Heinsohn RG, Packer C (1995) Complex cooperative strategies in group-territorial African lions. Science 269:1260–1262
Heinsohn RG, Legge S (1999) The cost of helping. Trends Ecol Evol 14:53–57
Hemelrijk CK (1990a) Models of and tests for reciprocity, unidirectionality and other social interaction patterns at a group level. Anim Behav 39:1013–1029
Hemelrijk CK (1990b) A matrix partial correlation test used in investigations of reciprocity and other social interaction patterns at group level. J theoret Biol 143:405–420
Hemelrijk CK (1994) Support for being groomed in long-tailed macaques, *Macaca fascicularis*. Anim Behav 48:479–481
Hemelrijk CK, Ek A (1991) Reciprocity and interchange of grooming and "support" in captive chimpanzees. Anim Behav 41:923–935
Hemelrijk CK, van Laere GJ, van Hooff JARAM (1992) Sexual exchange relationships in captive chimpanzees? Behav Ecol Sociobiol 30:269–275
Hemelrijk CK, Meier C, Martin RD (1999) "Friendship" for fitness in chimpanzees? Anim Behav 58:1223–1229
Henrich J (2004) "Inequity aversion" in capuchins? Nature 428:139
Henrich J, Boyd R (2001) Why people punish defectors: weak conformist transmission can stabilize costly enforcement of norms in cooperative dilemmas. J theoret Biol 208:79–89
Henrich J, Boyd R, Bowles S, Camerer C, Fehr E, Gintis H, McElreath R (2001) In search of *Homo economicus* – behavioral experiments in 15 small-scale societies. Am Econ Rev 91:73–78
Henrich J, Boyd R, Bowles S, Camerer C, Fehr E, Gintis H, McElreath R, Alvard M, Barr A, Ensminger J, Henrich NS, Hill K, Gil-White F, Gurven M, Marlowe FW, Patton JQ, Tracer D (2005) "Economic man" in cross-cultural perspective: behavioral experiments in 15 small-scale societies. Behav Brain Sci, in press
Henzi SP, Barrett L (1999) The value of grooming to female primates. Primates 40:47–59
Henzi SP, Barrett L (2002) Infants as a commodity in a baboon market. Anim Behav 63:915–921
Henzi SP, Barrett L (2003) Evolutionary ecology, sexual conflict, and behavioral differentiation among baboon populations. Evol Anthropol 12:217–230
Henzi SP, Byrne RW, Whiten A (1992) Patterns of movement by baboons in the Drakensberg mountains: primary responses to the environment. Int J Primatol 13:601–629
Henzi SP, Lycett JE, Weingrill T (1997) Cohort size and the allocation of social effort by female mountain baboons. Anim Behav 54:1235–1243
Henzi SP, Weingrill T, Barrett L (1999) Male behaviour and the evolutionary ecology of chacma baboons. South Afr J Sci 95:240–242
Henzi SP, Barrett L, Gaynor D, Greeff T, Weingrill T, Hill RA (2003) Effect of resource competition on the long-term allocation of grooming by female baboons: evaluating Seyfarth's model. Anim Behav 66:931–938

Hill D (1994) Affiliative behavior among adult males of the genus *Macaca*. Behaviour 130:293–308
Hill K (2002) Altruistic cooperation during foraging by the Ache and the evolved human predisposition to cooperate. Hum Nat 13:105–128
Hirschleifer J (1987) On the emotions as guarantors of threats and promises. In: Dupre J (ed) The Latest on the Best: Essays in Evolution and Optimality. MIT Press, Cambridge/MA, pp 307–326
Hodge SJ (2003) The evolution of cooperation in the communal breeding banded mongoose. PhD thesis, Univ of Cambridge, Cambridge
Hoeksema JD, Bruna EM (2000) Pursuing the big questions about interspecific mutualism: a review of theoretical approaches. Oecologia 125:321–330
Hoeksema JD, Kummel M (2003) Ecological persistence of the plant-mycorrhizal mutualism: a hypothesis from species coexistence theory. Am Nat 162:S40–50
Hoeksema JD, Schwartz MW (2003) Expanding comparative-advantage biological market models: contingency of mutualism on partners' resource requirements and acquisition trade-offs. Proc R Soc Lond B 270:913–919
Hoekstra RF (2003) Power in the genome. Who suppresses the outlaw? In: Hammerstein P (ed) Genetic and Cultural Evolution of Cooperation. Cambridge/MA, MIT Press, pp 257–270
Hohmann G, Fruth B (2002) Dynamics in social organization of bonobos (*Pan paniscus*). In: Boesch B, Hohmann G, Marchant LF (eds) Behavioural Diversity in Chimpanzees and Bonobos. Cambridge Univ Press, Cambridge, pp 138–155
Hohmann G, Gerloff U, Tautz D, Fruth B (1999) Social bonds and genetic ties: kinship, association and affiliation in a community of bonobos (*Pan paniscus*). Behaviour 136:1219–1235
Horn AG, Leonard MC (2002) Efficacy and design of begging signals. In: Wright J, Leonard M (eds) The Evolution of Begging. Kluwer Academic, Dordrecht/NL, pp 127–142
Hooff JARAM van (2000) Relationships among non-human primate males: a deductive approach. In: Kappeler PM (ed) Primate Males. Cambridge Univ Press, Cambridge, pp 183–191
Hooff JARAM van (2001) Conflict, reconciliation and negotiation in non-human primates: the value of long-term relationships. In: Noë R, van Hooff JARAM, Hammerstein P (eds) Economics in Nature. Cambridge Univ Press, Cambridge, pp 67–90
Hooff JARAM van, van Schaik CP (1992) Cooperation in competition: the ecology of primate bonds. In: Harcourt AH, de Waal FBM (eds) Coalitions and Alliances in Humans and other Animals. Oxford Univ Press, Oxford, pp 357–389
Hooff JARAM van, van Schaik CP (1994) Male bonds: affiliative relationships among nonhuman primate males. Behaviour 130:309–337
Hoogland JL, Tamarin RH, Levy CK (1989) Communal nursing in prairie dogs. Behav Ecol Sociobiol 24:91–95
Hrdy SB (1976) The care and exploitation of nonhuman primate infants by conspecifics other than the mother. In: Rosenblatt J, Hinde RA, Shaw E, Beer C (eds) Advances in the Study of Behavior. Vol 6. Academic Press, New York, pp 101–158
Hrdy SB (1977) The Langurs of Abu: Female and Male Strategies of Reproduction. Harvard Univ Press, Harvard/MA
Hrdy SB, Hrdy DB (1976) Hierarchical relations among female Hanuman langurs (Primates: Colobinae, *Presbytis entellus*). Science 193:913–915
Huck M, Loettker P, Boehle U-R, Heymann EW (2005) Paternity and kinship patterns in polyandrous moustached tamarins (*Saguinus mystax*). Am J Phys Anthropol, in press
Humphrey NK (1976) The social function of intellect. In: Bateson PPG, Hinde RA (eds) Growing Points in Ethology. Cambridge Univ Press, Cambridge, pp 303–321
Hunte W, Horrocks JA (1986) Kin and non-kin interventions in the aggressive disputes of vervet monkeys. Behav Ecol Sociobiol 20:257–263
Hutchins M, Barash DP (1976) Grooming in primates: implications for its utilitarian function. Primates 17:145–150
Hyman J (2002) Conditional strategies in territorial defense: do Carolina wrens play tit-for-tat? Behav Ecol 13:664–669
Inglett BJ, French JA, Simmons LG, Vires KW (1989) Dynamics of intrafamily aggression and social reintegration in lion tamarins. Zoo Biol 8:67–78
Isaac G (1978) The food sharing behavior of protohuman hominids. Sci Am 238:90–108

Isaac M, Walker J (1988) Communication and free-riding behavior: the voluntary contribution mechanism. Econ Inq 26:585–608
Isbell LA (1991) Contest and scramble competition: patterns of female aggression and ranging behavior among primates. Behav Ecol 2:143–155
Izawa K (1980) Social behavior of the wild black-capped capuchin (*Cebus apella*). Primates 21:443–467
Izawa K (1994) Group division of wild black-capped capuchins. Field Stud New World Monkeys, La Macarena, Colombia 9:5–14
Izzo TJ, Vasconcelos HL (2002) Cheating the cheater: domatia loss minimizes the effects of ant castration in an Amazonian ant-plant. Oecologia 133:200–205
Jack KM, Fedigan L (2003) Male dominance and reproductive success in white-faced capuchins (*Cebus capucinus*). Am J Phys Anthropol 36:S121–122
Jack KM, Fedigan L (2004a) Male dispersal patterns in white-faced capuchins, *Cebus capucinus*. Part 1: patterns and causes of natal emigration. Anim Behav 67:761–769
Jack KM, Fedigan L (2004b) Male dispersal patterns in white-faced capuchins, *Cebus capucinus*. Part 2: patterns and causes of secondary dispersal. Anim Behav 67:771–782
Jacoby S (1983) Wild Justice: The Evolution of Revenge. Harper, Row, New York
Johnson D, Stopka P, Knights S (2003) The puzzle of human cooperation. Nature 421:911–912
James CD, Hoffman MT, Lightfoot DC, Forbes GS, Whitford WG (1994) Fruit abortion in *Yucca elata* and its implications for the mutualistic association with yucca moths. Oikos 69:207–216
Jameson EW (1998) Prepartum mammogenesis milk production and optimal litter size. Oecologia 114:282–291
Jamieson IG (1989) Behavioral heterochrony and the evolution of birds' helping at the nest: an unselected consequence of communal breeding? Am Nat 133:394–406
Jamieson IG, Craig JL (1987) Critique of helping behavior in birds: a departure from functional explanations. Vol 7. In: Bateson P, Klopfer P (eds) Perspectives in Ethology. Plenum Press, New York, pp 79–98
Janeway CA, Travers P (1997) Immunology. Current Biology. Ltd/Garland Publishing Inc, New York
Janson CH (1988) Food competition in brown capuchin monkeys (*Cebus apella*): quantitative effects of group size and tree productivity. Behaviour 105:53–76
Janson CH (1992) Evolutionary ecology of primate social structure. In: Smith EA, Winterhalder B (eds) Evolutionary Ecology and Human Behavior. Aldine de Gruyter, New York, pp 95–130
Janus M (1989) Reciprocity in play, grooming and proximity in sibling and non-sibling young rhesus monkeys. Int J Primatol 10:243–261
Jarman PJ (1974) The social organisation of antelopes in relation to their ecology. Behaviour 48:215–267
Jarvis JUM (1981) Eusociality in a mammal: cooperative breeding in naked mole-rat colonies. Science 212:571–573
Jennions MD, Macdonald DW (1994) Cooperative breeding in mammals. Trends Ecol Evol 9:89–93
Johnson JA (1987) Dominance rank in juvenile olive baboons *Papio anubis*: the influence of gender, size, maternal rank and orphaning. Anim Behav 35:1694–1708
Johnson DDP, Krüger O (2004) The good of wrath: supernatural punishment and the evolution of cooperation. Polit Theol 5:159–176
Johnson MS, Thomson SC, Speakman JR (2001) Limits to sustained energy intake. II. Inter-relationships between resting metabolic rate, life-history traits and morphology in *Mus musculus*. J Exp Biol 204:1937–1946
Johnson DDP, Stopka P, Bell J (2002) Individual variation evades the prisoner's dilemma. BMC Evol Biol 2:1–8
Johnson DDP, Stopka P, MacDonald DW (2004) Ideal flea constraints on group living: unwanted public goods and the emergence of cooperation. Behav Ecol 15:181–186
Johnstone RA (1995) Sexual selection, honest advertisement and the handicap principle: reviewing the evidence. Biol Rev 70:1–65
Johnstone RA (2000) Models of reproductive skew: a review and synthesis. Ethology 106:5–26
Johnstone RA (2004) Begging and sibling competition: how should offspring respond to their rivals? Am Nat 163:388–406

Johnstone RA, Bshary R (2002) From parasitism to mutualism: partner control in asymmetric interactions. Ecol Lett 5:634–639
Johnstone RA, Woodroffe R, Cant MA, Wright J (1999) Reproductive skew in multi-member groups. Am Nat 153:315–331
Kagel J, Roth AE (eds) (1995) Handbook of Experimental Economics. Princeton Univ Press, Princeton/NJ
Kahan JP, Rapoport A (1984) Theories of Coalition Formation. Erlbaum, Hillsdale/NJ
Kampe KKW, Frith CD, Dolan RJ, Frith U (2001) Psychology: reward value of attractiveness and gaze. Nature 413:589
Kaplan JR (1977) Patterns of fight interference in free-ranging rhesus monkeys. Am J Phys Anthropol 47:279–288
Kaplan JR (1978) Fight interference and altruism in rhesus monkeys. Am J Phys Anthropol 49:241–249
Kaplan H, Hill K, Lancaster J, Hurtado A (2000) A theory of human life history evolution: diet, intelligence and longevity. Evol Anthropol 9:156–185
Kappeler PM, van Schaik CP (1992) Methodological and evolutionary aspects of reconciliation among primates. Ethology 92:51–69
Kappeler PM, Wimmer B, Zinner D, Tautz D (2002) The hidden matrilineal structure of a solitary lemur: implications for primate social evolution. Proc R Soc Lond 269:1755–1763
Kapsalis E (2003) Matrilineal kinship and primate behavior. In: Chapais B, Berman C (eds) Kinship and Behavior in Primates. Oxford Univ Press, Oxford, pp 153–176
Kapsalis E, Berman CM (1996) Models of affiliative relationships among free-ranging rhesus monkeys (*Macaca mulatta*). I. Criteria for kinship. Behaviour 133:1209–1234
Karama S, Roch Lecours A, Leroux J-M, Bourgouin P, Beaudoin G, Joubert S, Beauregard M (2002) Areas of brain activation in males and females during viewing of erotic film excerpts. Hum Brain Mapp 16:1–13
Kareem AM, Barnard CJ (1982) The importance of kinship and familiarity in social interactions between mice. Anim Behav 30:594–601
Keeley L (1996) War before Civilization. Oxford Univ Press, New York
Keller L, Reeve HK (1994) Partitioning of reproduction in animal societies. Trends Ecol Evol 9:98–103
Keller L, Waller D (2002) Inbreeding effects in wild populations. Trend Ecol Evol 17:230–241
Keser C, van Winden F (2000) Conditional cooperation and voluntary contributions to public goods. Scand J Econ 102:23–29
Keverne EB, Martensz ND, Tuite B (1989) Beta-endorphin concentrations in cerebrospinal fluid of monkeys are influenced by grooming relationships. Psychoneuroendocrinology 14:155–161
Kiers ET, Rousseau RA, West SA, Denison RF (2003) Host sanctions and the legume-rhizobium mutualism. Nature 425:78–81
Kilner RM (2002) The evolution of complex begging displays. In: Wright J, Leonard ML (eds) The Evolution of Nestling Begging: Competition, Cooperation and Communication. Kluwer, Dordrecht/NL
King-Casas B, Tomlin D, Anen C, Camerer CF, Quartz SR, Montague PR (2005) Getting to know you: reputation and trust in a two-person economic exchange. Science 308:78–83
Knauft B (1991) Violence and sociality in human evolution. Curr Anthropol 32:391–428
Knutson B, Adams CM, Fong GW, Hommer D (2001) Anticipation of increasing monetary reward selectively recruits nucleus accumbens. J Neurosci 21:RC159:1–5
Knutson B, Fong GW, Adams CM, Varner JL, Hommer D (2001) Dissociation of reward anticipation and outcome with event-related fMRI. NeuroReport 12:3683–3687
Koenig A (2000) Competitive regimes in forest-dwelling Hanuman langur females (*Semnopithecus entellus*). Behav Ecol Sociobiol 48:93–109
König B (1989) Behavioural ecology of kin recognition in house mice. Ethol Ecol Evol 1:99–110
König B (1993) Maternal investment of communally nursing female house mice (*Mus musculus domesticus*). Behav Process 30:61–74
König B (1994a) Communal nursing in mammals. Verh Dtsch Zool Ges 87:115–127

König B (1994b) Components of lifetime reproductive success in communally and solitarily nursing house mice – a laboratory study. Behav Ecol Sociobiol 34:275–283

König B (1994c) Fitness effects of communal rearing in house mice: the role of relatedness and familiarity. Anim Behav 48:1449–1457

König B (1997) Cooperative care of young in mammals. Naturwissenschaften 84:95–104

König B, Markl H (1987) Maternal care in house mice. I. The weaning strategy as a means for parental manipulation of offspring quality. Behav Ecol Sociobiol 20:1–9

König B, Riester J, Markl H (1988) Maternal care in house mice (*Mus musculus*): II. The energy cost of lactation as a function of litter size. J Zool Lond 216:195–210

Koenig WD (1988) Reciprocal altruism in birds: a critical review. Ethol Sociobiol 9:73–84

Koenig WD, Dickinson J (2004) Ecology and Evolution of Cooperative Breeding in Birds. Cambridge Univ Press, Cambridge

Koenig WD, Mumme RL (1987) Population Ecology of the Cooperatively Breeding Acorn Woodpecker. Princeton Univ Press, Princeton/NJ

Koenig WA, Pitelka FA, Carmen WJ, Mumme RL, Stanback MT (1992) The evolution of delayed dispersal in cooperative breeders. Q Rev Biol 67:111–150

Kokko H, Johnstone RA (1999) Social queuing in animal societies: a dynamic model of reproductive skew. Proc R Soc Lond B 266:571–578

Kokko H, Johnstone RA, Clutton-Brock TH (2001) The evolution of cooperative breeding through group augmentation. Proc R Soc Lond B 268:187–196

Kokko H, Johnstone RA, Wright J (2002) The evolution of parental and alloparental effort in cooperatively breeding groups: when should helpers pay to stay? Behav Ecol 13:291–300

Komdeur J (1994) Experimental evidence for helping and hindering by previous offspring in the cooperative breeding Seychelles warbler, *Acrocephalus seychellensis*. Behav Ecol Sociobiol 34:175–186

Konner M (2004) The ties that bind. Nature 429:705

Koyama NF (2001) The long-term effects of reconciliation in Japanese macaques (*Macaca fuscata*). Ethology 107:975–987

Kreps D, Milgrom P, Roberts J, Wilson R (1982) Rational cooperation in the finitely repeated prisoners' dilemma. J Econ Theor 27:245–252

Kropotkin P (1972 [1902]) Mutual Aid: A Factor of Evolution. New York Univ Press, New York

Kuester J, Paul A, Arnemann J (1994) Kinship, familiarity and mating avoidance in Barbary macaques, *Macaca sylvanus*. Anim Behav 48:1183–1194

Kummel M, Salant SW (2004) The economics of mutualisms: optimal utilisation of mycorrhizal mutualistic partners by plants. Working paper presented during the 5th Toulouse Conference on Environment and Resource Economics "Advances in Economics and Biology" (Toulouse June 1–2, 2004)

Kummer H (1991) Evolutionary transformations of possessive behavior. In: Rudmin FW (ed) To Have Possessions: A Handbook on Ownership and Property. Special Issue of J Soc Behav Pers 6:75–83

Kurland JA (1977) Kin Selection in the Japanese Monkey. Contributions to Primatology. Vol. 12. Karger, Basel

Kutsukake N, Castles DL (2001) Reconciliation and variation in post-conflict stress in Japanese macaques (*Macaca fuscata fuscata*): testing the integrated hypothesis. Anim Cogn 4:259–268

Kutsukake N, Castles DL (2004) Reconciliation and post-conflict third-party affiliation among wild chimpanzees in the Mahale Mountains, Tanzania. Primates 45:157–165

La Cerra P, Bingham R (1998) The adaptive nature of the human neurocognitive architecture: an alternative model. Proc Natl Acad Sci USA 95:11290–11294

Lacey EA, Sherman PW (1991) Social organisation of naked mole-rat colonies: evidence for divisions of labour. In: Sherman PW, Jarvis JUM, Alexander RD (eds) The Ecology of the Naked Mole Rat. Princeton Univ Press, Princeton/NJ, pp 275–336

Lacey EA, Sherman PW (1997) Cooperative breeding in naked mole-rats. In: Solomon NG, French JA (eds) Cooperative Breeding in Mammals. Cambridge Univ Press, Cambridge, pp 267–301

Launhardt K, Borries C, Hardt C, Epplen JT, Winkler P (2001) Paternity analysis of alternative male reproductive routes among the langurs (*Semnopithecus entellus*) of Ramnagar. Anim Behav 61:53–64

Lazaro-Perea C (2001) Intergroup interactions in wild common marmosets, *Callithrix jacchus*: territorial defence and assessment of neighbours. Anim Behav 62:11–21

Lazaro-Perea C, De Fatima Arruda M, Snowdon CT (2004) Grooming as a reward? Social function of grooming between females in cooperatively breeding marmosets. Anim Behav 67:627–636

Ledyard JO (1995) Public goods: a survey of experimental research. In: Kagel JH, Roth AE (eds) Handbook of Experimental Economics. Princeton Univ Press, Princeton/NJ, pp 111–194

Lee PC, Oliver JI (1979) Competition, dominance and the acquisition of rank in juvenile yellow baboons (*Papio cynocephalus*). Anim Behav 27:576–585

Lehmann L, Perrin N (2003) Inbreeding avoidance through kin recognition: choosy females boost male dispersal. Am Nat 162:638–652

Leimar O, Axén A (1993) Strategic behaviour in an interspecific mutualism: interactions between lycaenid larvae and ants. Anim Behav 46:1177–1182

Leimar O, Hammerstein P (2001) Evolution of cooperation through indirect reciprocity. Proc R Soc Lond B 268:745–753

Leinfelder I, de Vries H, Deleu R, Nelissen M (2001) Rank and grooming reciprocity among females in a mixed-sex group of captive hamadryas baboons. Am J Primatol 55:25–42

Leonard ML, Horn AG, Eden SF (1989) Does juvenile helping enhance breeder reproductive success? A removal experiment on moorhen. Behav Ecol Sociobiol 25:357–361

Lewis RJ (2002) Beyond dominance the importance of leverage. Q Rev Biol 77:149–164

Lewis SE, Pusey AE (1997) Factors influencing the occurrence of communal care in plural breeding mammals. In: Solomon NG, French JA (eds) Cooperative Breeding in Mammals. Cambridge Univ Press, Cambridge, pp 335–363

Lidicker WZJ (1976) Social behaviour and density regulation in house mice living in large enclosures. J Anim Ecol 45:677–697

Lim MM, Wang Z, Olazábal DE, Ren X, Terwilliger EF, Young LJ (2004) Enhanced partner preference in a promiscuous species by manipulating the expression of a single gene. Nature 429:754–757

Limongelli L, Boysen S, Visalberghi E (1995) Comprehension of cause-effect relationships in a tool-using task by chimpanzees. J Comp Psychol 109:18–26

Lindström J (1999) Early development and fitness in birds and mammals. Trends Ecol Evol 14:343–347

Linklatter WL, Cameron EZ (2000) Tests for cooperative behaviour between stallions. Anim Behav 60:731–743

Ljungberg T, Westlund K, Lindqvist Forsberg AJ (1999) Conflict resolution in 5-year-old boys: does postconflict affiliative behaviour have a reconciliatory role? Anim Behav 58:1007–1016

Lomnicki A (1988) Population Ecology of Individuals. Princeton Univ Press, Princeton/NJ

Lotem A, Fishman MA, Stone L (1999) Evolution of cooperation between individuals. Nature 400:226–227

Lucas JR, Creel SR, Waser PM (1997) Dynamic optimisation and cooperative breeding: an evaluation of future fitness. In: Solomon NG, French JA (eds) Cooperative Breeding in Mammals. Cambridge Univ Press, Cambridge, pp 171–198

Maestripieri D (1993) Vigilance costs of allogrooming in macaque mothers. Am Nat 141:744–753

Maestripieri D, Schino G, Aureli F, Troisi A (1992) A modest proposal: displacement activities as indicators of emotions in primates. Anim Behav 44:967–979

Magrath RD, Whittingham LA (1997) Subordinate males are more likely to help if unrelated to the breeding female in cooperatively breeding white-browed scrubwrens. Behav Ecol Sociobiol 41:185–192

Magrath RD, Johnstone RA, Heinsohn RG (2004) Reproductive skew. In: Koenig W, Dickinson J (eds) Ecology and Evolution of Cooperative Breeding in Birds. Cambridge Univ Press, Cambridge, pp 157–176

Magurran AE, Irving PW, Henderson PA (1996) Is there a fish alarm pheromone? A study and critique. Proc R Soc Lond B 263:1551–1556

Maklakov AA (2002) Snake-directed mobbing in a cooperative breeder: anti-predator behaviour or self-advertizement for the formation of dispersal coalitions. Behav Ecol Sociobiol 52:372–378
Malcolm JR, Marten K (1982) Natural selection and the communal rearing of pups in African wild dogs, *Lycaon pictus*. Behav Ecol Sociobiol 10:1–13
Mann MA, Miele JL, Kinsley CH, Svare B (1983) Postpartum behavior in the mouse: the contribution of suckling stimulation to water intake, food intake and body weight regulation. Physiol Behav 31:633–638
Manning CJ, Wakeland EK, Potts WK (1992) Communal nesting patterns in mice implicate MHC genes in kin recognition. Nature 360:581–583
Manning CJ, Dewsbury DA, Wakeland EK, Potts WK (1995) Communal nesting and communal nursing in house mice, *Mus musculus domesticus*. Anim Behav 50:741–751
Manser MB (1998) The evolution of auditory communication in suricates *Suricata suricatta*. PhD thesis, Univ of Cambridge, Cambridge
Manson JH, Rose LM, Perry S, Gros-Louis J (1999) Dynamics of female-female relationships in wild *Cebus capucinus*: data from two Costa Rican sites. Int J Primatol 20:679–706
Manson JH, Navette CD, Silk JB, Perry S (2004) Time-matched grooming in female primates? New analyses from two species. Anim Behav 67:493–500
Marlowe F (2003) A critical period for provisioning by Hadza men: implications for pair bonding. Evol Hum Behav 24:217–229
Marr DL, Pellmyr O (2003) Effect of pollinator-inflicted ovule damage on floral abscission in the yucca-yucca moth mutualism: the role of mechanical and chemical factors. Oecologia 136:236–243
Masclet D, Noussair C, Tucker S, Villeval M-C (2003) Monetary and non-monetary punishment in the voluntary contributions mechanism. Am Econ Rev 93:366–380
Mason JH, Perry S (2000) Correlates of self-directed behaviour in wild white-faced capuchins. Ethology 106:301–307
Massey A (1977) Agonistic aids and kinship in a group of pig-tail macaques. Behav Ecol Sociobiol 2:31–40
Matheson M, Bernstein I (2000) Grooming, social bonding, and agonistic aiding in rhesus monkeys. Am J Primatol 51:177–186
Matsuzawa T (2001) Primate foundations of human intelligence: a view of tool use in nonhuman primates and fossil hominids. In: Matsuzawa T (ed) Primate Origins of Human Cognition and Behavior. Springer, Tokyo, pp 3–25
Maynard Smith J (1964) Group selection and kin selection. Nature 201:1145–1147
Maynard Smith J (1982) Evolution and the Theory of Games. Cambridge Univ Press, Cambridge
Maynard-Smith J (1983) Game theory and the evolution of cooperation. In: Bendall DS (ed) Evolution from Molecules to Men. Cambridge Univ Press, Cambridge, pp 445–456
Maynard-Smith J, Szathmary E (1995) The Major Transitions in Evolution. Freeman, Oxford
McCracken GF (1984) Communal nursing in Mexican free-tailed bat maternity colonies. Science 223:1090–1091
McDonald DB (1989a) Cooperation under sexual selection: age-graded changes in a lekking bird. Am Nat 134:709–730
McDonald DB (1989b) Correlates of male mating success in a lekking bird with male-male cooperation. Anim Behav 37:1007–1022
McElreath R, Clutton-Brock TH, Fehr E, Fessler DMT, Hagen EH, Hammerstein P, Kosfeld M, Millinski M, Silk JB, Tooby J, Wilson MI (2003) Group report: the role of cognition and emotion in cooperation. In: Hammerstein P (ed) Genetic and Cultural Evolution of Cooperation. MIT Press, Cambridge/MA, pp 125–152
McGrew WC (1975) Patterns of plant food sharing by wild chimpanzees. In: Kawai M, Kondon S, Ehara A (eds) Contemporary Primatology: Proceedings of the Fifth Congress of the International Primatological Society. Karger, Basel, pp 304–309
McGuire AM (2003) "It was nothing" – extending evolutionary models of altruism by two social cognitive biases in judgments of the costs and benefits of helping. Soc Cogn 21:363–394
McNamara JM, Gasson CE, Houston AI (1999) Incorporating rules for responding into evolutionary games. Nature 401:368–371

Mech DL (1970) The Wolf. Natural History Press, New York
Meikle DB, Vessey SH (1981) Nepotism among rhesus monkey brothers. Nature 29:160–161
Mendres KA, de Waal FBM (2000) Capuchins do cooperate: the advantage of an intuitive task. Anim Behav 60:523–529
Mesterton-Gibbons M, Dugatkin LA (1992) Cooperation among unrelated individuals: evolutionary factors. Q Rev Biol 67:267–281
Michod RE (1982) The theory of kin selection. Ann Rev Ecol Syst 13:23–55
Michod RE (2003) Cooperation and conflict mediation during the origin of multicellularity. In: Hammerstein P (ed) Genetic and Cultural Evolution of Cooperation. MIT Press, Cambridge/MA, pp 291–307
Milinski M (1987) TIT FOR TAT in sticklebacks and the evolution of cooperation. Nature 325:433–437
Milinski M, Külling D, Kettler R (1990) Tit for Tat: stickleback (*Gastereus aculeatus*) "trusting" a cooperative partner. Behav Ecol 1:7–11
Milinski M, Semmann D, Krambeck HJ (2002a) Reputation helps solve the "tragedy of the Commons". Nature 415:424–426
Milinski M, Semmann D, Krambeck HJ (2002b) Donors to charity gain in both indirect reciprocity and political reputation. Proc R Soc Lond B 269:881–883
Milinski M, Semmann D, Bakker TCM, Krambeck HJ (2001) Cooperation through indirect reciprocity: image scoring or standing strategy? Proc R Soc Lond B 268:2495–2501
Miller EK, Cohen JD (2001) An integrative theory of pre-frontal cortex function. Ann Rev Neurosci 24:167–202
Miller GF, Todd PM (1999) Mate choice turns cognitive. Trends Cogn Sci 2:190–198
Mills MGL (1990) Kalahari Hyenas: The Comparative Behavioural Ecology of Two Species. Chapman & Hall, London
Milton K (1988) Foraging behaviour and the evolution of primate intelligence. In: Byrne RW, Whiten A (eds) Machiavellian Intelligence. Clarendon Press, Oxford, pp 285–305
Mitani JC, Watts DP (1999) Demographic influences on the hunting behavior of chimpanzees. Am J Phys Anthropol 109:439–454
Mitani JC, Watts DP (2001) Why do chimpanzees hunt and share meat? Anim Behav 61:915–924
Mitani JC, Watts DP (2002) Demographic and social constraints on male chimpanzee behaviour. Anim Behav 64:727–737
Mitani JC, Merriwether D, Zhang C (2000) Male affiliation, cooperation and kinship in wild chimpanzees. Anim Behav 59:885–893
Mitani JC, Watts DP, Muller M (2002a) Recent developments in the study of wild chimpanzee behavior. Evol Anthropol 11:9–25
Mitani JC, Watts DP, Pepper JW, Merriwether AD (2002b) Demographic and social constraints on male chimpanzee behaviour. Anim Behav 64:727–737
Mitchell CL (1990) The ecological basis for female social dominance: a behavioral study of the squirrel monkey (*Saimiri sciureus sciureus*) in the wild. PhD thesis, Princeton Univ Press, Princeton/NJ
Mitchell CL (1994) Migration alliances and coalitions among adult male South American squirrel monkeys (*Saimiri sciureus*). Behaviour 130:169–190
Mithen S (1996) The Prehistory of Mind: The Cognitive Origin of Art and Science. Thames & Hudson, London
Moehlman PD (1979) Jackal helpers and pup survival. Nature 277:382–383
Mohtashemi M, Mui L (2003) Evolution of indirect reciprocity by social information: the role of trust and reputation in evolution of altruism. J theor Biol 223:523–531
Morin PA, Goldberg TL (2004) Determination of genealogical relationships from genetic data: a review of methods and applications. In: Chapais B, Berman C (eds) Kinship and Behavior in Primates. Oxford Univ Press, Oxford, pp 15–45
Morin PA, Moore JJ, Chakraborty R, Jin L, Goodall J, Woodruff DS (1994) Kin selection, social structure, gene flow and the evolution of chimpanzees. Science 265:1193–1201
Moss G, Colman AM (2001) Choices and preferences: experiments on gender differences. J. Brand Manage 9:89–98

Moxnes E, van der Heijden E (2003) The effect of leadership in a public bad experiment. J Confl Resolut 47:776–795

Müller C, König B (submitted) Energetic peak load reduction in lactating wild-bred house mice can diminish reproductive costs. Front Zool

Mueller UG, Poulin J, Adams RMM (2004) Symbiont choice in a fungus-growing ant (*Attini, Formicidae*). Behav Ecol 15:357–364

Mulder RA, Langmore NE (1993) Dominant males punish helpers for temporary defection in superb fairy-wrens. Anim Behav 45:830–833

Muller M, Wrangham RW (2001) The reproductive ecology of male hominoids. In: Ellison PT (ed) Reproductive Ecology and Human Evolution. Aldine de Gruyter, Hawthorne/NY, pp 397–427

Muramatsu R, Hanoch Y (2005) Emotions as a mechanism for boundedly rational agents: the fast and frugal way. J Econ Psychol 26:201–221

Muroyama Y (1991) Mutual reciprocity of grooming in female Japanese macaques (*Macaca fuscata*). Behaviour 119:161–170

Muroyama Y (1994) Exchange of grooming for allomothering in female pata monkeys. Behaviour 128:103–119

Myerson J, Green L, Hanson JS, Holt DD, Estle SJ (2003) Discounting delayed and probabilistic rewards: processes and traits. J Econ Psychol 24:619–635

Neill DB (2001) Optimality under noise: higher memory strategies for the alternating prisoner's dilemma. J theor Biol 211:159–180

Netto WJ, van Hooff JARAM (1986) Conflict interference and the development of dominance relationships in immature *Macaca fasicularis*. In: Else JG, Lee PV (eds) Primate Ontogeny, Cognition and Social Behaviour. Cambridge Univ Press, Cambridge, pp 291–300

Neuhäusser-Wespy F, König B (2000) Living together – feeding apart. How to measure individual food consumption in a social species. Behav Res Methods Instruments Comput 32:169–172

Newcomer DM, de Farcy DD (1985) White-faced capuchin (*Cebus capucinus*) predation on a nestling coati (*Nasua narica*). J Mammal 66:185–186

Nicolson NA (1987) Infants, mothers and other females. In: Smuts BB, Cheney DL, Seyfarth RM, Wrangham RW, Struhsaker TT (eds) Primate Societies. Univ of Chicago Press, Chicago, pp 330–342

Nieder A, Freedman DJ, Miller EK (2002) Representation of the quantity of visual Items in the primate prefrontal portex. Science 297:1708–1711

Nievergelt CM, Digby LJ, Ramakrishnan U, Woodruff D (2000) Genetic analysis of group composition and breeding system in a wild common marmoset (*Callithrix jacchus*) population. Int J Primatol 21:1–20

Nimchinsky EA, Gilissen E, Allman JM, Perl DP, Erwin JM, Hof PR (1999) A neuronal morphologic type unique to humans and great apes. Proc Natl Acad Sci USA 96:5268–5273

Nishida T (1968) The social group of wild chimpanzees in the Mahale Mountains. Primates 9:167–224

Nishida T (1979) The social structure of chimpanzees in the Mahale Mountains. In: Hamburg DA, McCown ER (eds) The Great Apes. Benjamin/Cummings, Menlo Park/CA, pp 73–121

Nishida T (1983) Alpha status and agonistic alliance in wild chimpanzees (*Pan troglodytes schweinfurthii*). Primates 24:318–336

Nishida T (1990) The Chimpanzees of the Mahale Mountains: Sexual and Life History Strategies. Univ of Tokyo Press, Tokyo

Nishida T, Hasegawa-Hiraiwa M. (1987) Chimpanzees and bonobos: cooperative relationships among males. In: Smuts BB, Cheney DL, Seyfarth RM, Wrangham RW, Struhsaker TT (eds) Primate Societies. Univ of Chicago Press, Chicago, pp 165–177

Nishida T, Hosaka K (1996) Coalition strategies among adult male chimpanzees of the Mahale Mountains, Tanzania. In: McGrew WC, Marchant LF, Nishida T (eds) Great Ape Societies. Cambridge Univ Press, Cambridge, pp 114–134

Nishida T, Kawanaka K (1972) Inter-unit-group relationships among wild chimpanzees of the Mahale mountains. Kyoto Univ Afr Stud 7:131–169

Nishida T, Uehara S, Nyondo R (1983) Predatory behavior among wild chimpanzees of the Mahale Mountains. Primates 20:1–20

Nishida T, Hasegawa T, Hayaki H, Takahata Y, Uehara S (1992) Meat-sharing as a coalition strategy by an alpha male chimpanzee? In: Nishida T, McGrew WC, Marler P, Pickford M, de Waal F (eds) Topics in Primatology. Vol 1: Human Origins. Karger, Basel, pp 159–174

Noë R (1990) A veto game played by baboons: a challenge to the use of the prisoner's dilemma as a paradigm for reciprocity and cooperation. Anim Behav 39:78–90

Noë R (1992) Alliance formation among male baboons: shopping for profitable partners. In: Harcourt AH, de Waal FBM (eds) Coalitions and Alliances in Humans and other Animals. Oxford Univ Press, Oxford, pp 285–322

Noë R (1994) A model of coalition formation among male baboons with fighting ability as the crucial parameter. Anim Behav 47:211–213

Noë R (2001) Biological markets: partner choice as the driving force behind the evolution of mutualisms. In: Noë R, Hammerstein P, van Hoof JARAM (eds) Economics in Nature: Social Dilemmas, Mate Choice and Biological Markets. Cambridge Univ Press, Cambridge, pp 93–118

Noë R (2005) Cooperation experiments: coordination through communication versus acting apart together. Anim Behav, in press

Noë R, Hammerstein P (1994) Biological markets: supply and demand determine the effect of partner choice in cooperation, mutualism and mating. Behav Ecol Sociobiol 35:1–11

Noë R, Hammerstein P (1995) Biological markets. Trends Evol Ecol 10:336–339

Noë R, Sluijter AA (1995) Which adult male savanna baboons form coalitions? Int J Primatol 16:77–105

Noë R, van Schaik CP, van Hooff JARAM (1991) The market effect: an explanation for pay-off asymmetries among collaborating animals. Ethology 87:97–118

Noordwijk MA van, van Schaik CP (1985) Male migration and rank acquisition in wild long-tailed macaques. Anim Behav 33:849–861

Noordwijk MA van, van Schaik CP (1999) The effects of dominance rank and group size on female lifetime reproductive success in wild long-tailed macaques, *Macaca fascicularis*. Primates 40:105–130

Noordwijk MA van, van Schaik CP (2001) Career moves: transfer and rank challenge decisions by male long-tailed macaques. Behaviour 138:359–395

Noordwijk MA van, van Schaik CP (2004) Sexual selection and the careers of primate males: paternity concentration, dominance acquisition tactics and transfer decisions. In: Kappeler PM, van Schaik CP (eds) Sexual Selection in Primates: New and Comparative Perspectives. Cambridge Univ Press, Cambridge, pp 208–229

Noussair C, Tucker S (2005) Combining monetary and social sanctions to promote cooperation. Econ Inq, in press

Nowak MA (1990) Stochastic strategies in the prisoner's dilemma. Theroet Popul Biol 38:93–112

Nowak MA, May RM (1992) Evolutionary games and spatial chaos. Nature 359:826–829

Nowak MA, Sigmund K (1992) Tit for tat in heterogeneous populations. Nature 355:250–253

Nowak MA, Sigmund K (1993) A strategy of win-stay, lose-shift outperforms tit-for-tat in the prisoner's dilemma game. Nature 364:56–58

Nowak MA, Sigmund K (1994) The alternating prisoner's dilemma. J theoret Biol 168:219–226

Nowak MA, Sigmund K (1998a) Evolution of indirect reciprocity by image scoring. Nature 393:573–577

Nowak MA, Sigmund K (1998b) The dynamics of indirect reciprocity. J theor Biol 194:561–574

Nowak MA, Sigmund K (2004) Evolutionary dynamics of biological games. Science 303:793–799

Nowak MA, Sigmund K (in press) Evolution of indirect reciprocity. Nature

Nowak MA, Sasaki A, Taylor C, Fudenberg D (2004) Emergence of cooperation and evolutionary stability in finite populations. Nature 428:646–649

Nunn CL (2000) Collective benefits, free-riders and male extragroup conflict. In: Kappeler PM (ed) Primate Males. Cambridge Univ Press, Cambridge

Nunn CL, Deaner R (2004) Patterns of participation and free riding in territorial conflicts among ringtailed lemurs (*Lemur catta*). Behav Ecol Sociobiol 57:50–61

Nunn CL, Lewis RJ (2001) Social dilemmas and human behaviour. In: Noë R, van Hooff JARAM, Hammerstein P (eds) Economics in Nature. Social Dilemmas, Mate Choice and Biological Markets. Cambridge Univ Press, Cambridge

O'Brien TG (1993) Allogrooming behavior among adult female wedge-capped capuchin monkeys. Anim Behav 46:499–510
O'Brien TG, Robinson JG (1993) Stability of relationships in female wedge-capped capuchin monkeys: effects of age, rank and relatedness. In: Pereira ME, Fairbanks LA (eds) Juvenile Primates: Life History, Development and Behavior. Oxford Univ Press, New York, pp 197–210
Ofek H (2001) Second Nature: Economic Origins of Human Evolution. Cambridge Univ Press, Cambridge
Oftedal OT (1984) Milk composition, milk yield and energy output at peak lactation: a comparative review. Symp Zool Soc Lond 51:33–85
Oftedal OT, Gittleman JL (1989) Patterns of energy output during reproduction in carnivores. In: Gittleman JL (ed) Carnivore Behavior, Ecology and Evolution. Chapman & Hall, London, pp 355–378
Ohtsuki H (2004) Reactive strategies in indirect reciprocity. J theor Biol 227:299–314
Oliveira-Castro JM (2003) Effects of base price upon search behavior of consumers in a supermarket: an operant analysis. J Econ Psychol 24:637–652
Oosterbeek H, Sloof R, van de Kuilen G (2004) Cultural differences in ultimatum game experiments: evidence from a meta-analysis. Exp Econ 7:171–188
O'Riain MJ, Jarvis JUM, Faulkes CG (1996) A dispersive morph in the naked mole-rat. Nature 380:619–621
O'Riain MJ, Bennett NC, Brotherton PNM, McIlrath G, Clutton-Brock TH (2000a) Reproductive suppression and inbreeding avoidance in wild populations of cooperatively breeding meerkats (*Suricata suricatta*). Behav Ecol Sociobiol 48:471–477
O'Riain MJ, Jarvis JUM, Alexander R, Buffenstein R, Peeters C (2000b) Morphological castes in a vertebrate. Proc Natl Acad Sci USA 97:13194–13197
Ostner J, Kappeler PM (2004) Male life history and the unusual sex ratios of redfronted lemur (*Eulemur fulvus rufus*) groups. Anim Behav 67:249–259
Ostrom E (1999) Governing the Commons. Cambridge Univ Press, Cambridge
Ostrom E (2001) Social dilemmas and human behaviour. In: Noë R, van Hooff JARAM, Hammerstein P (eds) Economics in Nature. Social Dilemmas, Mate Choice and Biological Markets. Cambridge Univ Press, Cambridge
Ostrom E (2003) Trust, reciprocity: interdisciplinary lessons from experimental research. In: Ostrom E, Walker J (eds) The Russel Sage Foundation Series on Trust. Vol VI. Russel Sage Foundation, New York
Ostrom E, Walker J, Gardner R (1992) Covenants with and without a sword: self-governance is possible. Am Polit Sci Rev 86:404–417
Owens DD, Owens MJ (1984) Helping behaviour in brown hyenas. Nature 308:843–845
Packer C (1977) Reciprocal altruism in *Papio anubis*. Nature 265:441–443
Packer C, Pusey A (1982) Cooperation and competition within coalitions of male lions: kin selection or game theory? Nature 296:740–742
Packer C, Scheel D, Pusey AE (1990) Why lions form groups: food is not enough. Am Nat 136:1–19
Packer C, Gilbert DA, Pusey AE, O'Brien SJ (1991) A molecular genetic analysis of kinship and cooperation in African lions. Nature 351:562–565
Packer C, Lewis S, Pusey AE (1992) A comparative analysis of non-offspring nursing. Anim Behav 43:265–282
Packer C, Collins DA, Sindimwo A, Goodall J (1995) Reproductive constraints on aggressive competition in female baboons. Nature 373:60–63
Page T, Putterman L, Unel B (2005) Voluntary association in public goods experiments: reciprocity, mimicry and efficiency. Econ J, in press
Pagnoni G, Zink CF, Montague PR, Berns GS (2002) Activity in human ventral striatum locked to errors of reward prediction. Nat Neurosci 5:97–98
Palfrey TR, Prisbrey JE (1997) Anomalous behavior in public goods experiments: how much and why? Am Econ Rev 87:829–846
Palombit RA (1999) Infanticide and the evolution of pairbonds in nonhuman primates. Evol Anthropol 7:117–128

Palombit RA, Cheney DL, Seyfarth RM (2001) Female-female competition for male "friends" in wild chacma baboons, *Papio cycocephalus ursinus*. Anim Behav 61:1159–1171
Panchanathan K, Boyd R (2003) A tale of two defectors: the importance of standing for evolution of indirect reciprocity. J theor Biol 224:115–126
Panchanathan K, Boyd R (2004) Indirect reciprocity can stabilize cooperation without the second-order free rider problem. Nature 432:499–502
Pandit SA, van Schaik CP (2003) A model for leveling coalitions among male primates: towards a theory of egalitarianism. Behav Ecol Sociobiol 55:161–168
Panksepp J (2005) Affective cosnciousness: core emotional feelings in animals and humans. Conscious Cogn, in press
Parr LA, Matheson MD, Bernstein IS, de Waal FBM (1997) Grooming down the hierarchy: allogrooming in captive brown capuchin monkeys, *Cebus apella*. Anim Behav 54:361–367
Passingham R (1993) The Frontal Lobes and Voluntary Action. Oxford Univ Press, Oxford
Paul A, Kuester J (2003) The impact of kinship on mating and reproduction. In: Chapais B, Berman CM (eds) Kinship and Behavior in Primates. Oxford Univ Press, New York, pp 271–291
Pawlowski B, Dunbar RIM (1999) Impact of market value on human mate choice decisions. Proc R Soc Lond B 266:281–285
Pawlowski B, Dunbar RIM (2001) Human mate chocie strategies. In: Noë R, van Hooff JARAM, Hammerstein P (eds) Economics in Nature. Social Dilemmas, Mate Choice and Biological Markets. Cambridge Univ Press, Cambridge
Payne HFP (in preparation) Competition, tolerance and grooming among samango monkey and chacma baboon females. PhD thesis, Univ of Natal, South Africa
Payne HFP, Lawes MJ, Henzi SP (2003) Competition and the exchange of grooming in female samango monkeys (*Cercopithecus mitis erytharcus*). Behaviour 140:453–471
Peck JR (1993) Friendship and the evolution of co-operation. J theor Biol 162:195–228
Pelikán J (1981) Patterns of reproduction in the house mouse. Symp Zool Soc Lond 47:205–229
Pellmyr O, Huth CJ (1994) Evolutionary stability of mutualism between yuccas and yucca moths. Nature 372:257–260
Penn DJ (2002) The scent of genetic compatibility: sexual selection and the major histocompatibility complex. Ethology 108:1–21
Pennycuik PR, Johnston PG, Westwood NH, Reisner AH (1986) Variation in numbers in a house mouse population housed in a large outdoor enclosure: seasonal fluctuations. J Anim Ecol 55:371–391
Pereira ME (1988) Agonistic interactions of juvenile savanna baboons. I. Fundamental features. Ethology 79:195–217
Pereira ME (1989) Agonistic interactions of juvenile savanna baboons. II. Agonistic support and rank acquisition. Ethology 80:152–171
Pereira ME (1993) Agonistic interaction, dominance relation and ontogenetic trajectories in ring-tailed lemurs. In: Pereira ME, Fairbanks LA (eds) Juvenile Primates. Oxford Univ Press, New York, pp 285–305
Perry S (1998) Male-male social relationships in wild white-faced capuchins, *Cebus capucinus*. Behaviour 135:139–172
Perry S, Rose L (1994) Begging and transfer of coati meat by white-faced capuchin monkeys, *Cebus capucinus*. Primates 35:409–415
Petit O, Abegg C, Thierry B (1997) A comparative study of aggression and conciliation in three cercopithecine monkeys (*Macaca fuscata, Macaca nigra, Papio papio*). Behaviour 134:415–432
Pfann GA, Biddle JE, Hamermesh DS, Bosman CM (2000) Business success and businesses' beauty capital. Econ Lett 67:201–207
Pfeiffer T, Rutte C, Killingback T, Taborsky M, Bonhoeffer S (2005) Evolution of cooperation through generalized reciprocity. Proc R Soc Lond B, in press
Pollock G, Dugatkin LA (1992) Reciprocity and the emergence of reputation. J theor Biol 159:25–37
Pope TR (1990) The reproductive consequences of male cooperation in the red howler monkey: paternity exclusion in multi-male and single-male troops using genetic markers. Behav Ecol Sociobiol 27:439–446

Pope TR (1998) Effects of demographic change on group kin structure and gene dynamics of populations of red howling monkeys. J Mammal 79:692–712
Pope TR (2000a) Reproductive success increases with degree of kinship in cooperative coalitions of female red howler monkeys (*Alouatta seniculus*). Behav Ecol Sociobiol 48:253–267
Pope TR (2000b) The evolution of male philopatry in neotropical monkeys. In: Kappeler PM (ed) Primate Males. Cambridge Univ Press, Cambridge, pp 219–235
Posch M, Pichler A, Sigmund K (1999) The efficiency of adapting aspiration levels. Proc R Soc Lond B 266:1427–1435
Potts R (2004) Palaoenvironmental basis of cognitive evolution in Great Apes. Am J Primatol 62:209–228
Poundstone W (1992) Prisoner's Dilemma. Anchor Books, New York
Preuschoft S, van Schaik CP (2000) Dominance and communication: conflict management in various social settings. In: Aureli F, de Waal FBM (eds) Natural Conflict Resolution. Univ of California Press, Berkeley, pp 77–105
Preuschoft S, Wang X, Aureli F, de Waal FBM (2002) Reconciliation in captive chimpanzees: a reevaluation with controlled methods. Int J Primatol 23:29–50
Price EC (1992a) Changes in the activity of captive cotton-top tamarins (*Saguinus oedipus*) over the breeding cycle. Primates 33:99–106
Price EC (1992b) The costs of infant carrying in captive cotton-top tamarins. Int J Primatol 13:125–141
Price ME (2005) Monitoring, reputation and "greenbeard" reciprocity in a Shuar work team. J Org Behav, in press
Prud'homme J, Chapais B (1996) Development of intervention behavior in Japanese macaques: testing the targeting hypothesis. Int J Primatol 17:429–443
Pusey AE (1979) Intercommunity transfer of chimpanzees in Gombe National Park. In: Hamburg D, McCown L (eds) The Great Apes. Benjamin-Cummings, Menlo Park, pp 465–479
Pusey AE, Packer C (1987) Dispersal and philopatry. In: Smuts BB, Cheney DL, Seyfarth RM, Wrangham RW, Struhsaker TT (eds) Primate Societies. Univ of Chicago Press, Chicago, pp 250–266
Pusey AE, Packer C (1994) Non-offspring nursing in social carnivores: minimizing the costs. Behav Ecol 5:362–374
Pusey AE, Packer C (1997) The ecology of relationships. In: Krebs JR, Davies NB (eds) Behavioural Ecology: An Evolutionary Approach. Blackwell, Oxford, pp 254–283
Queller DC, Strassmann JE (1998) Kin selection and social insects. BioScience 48:165–175
Quervain DJ-F de, Fischbacher U, Treyer V, Scellhammer M, Schnyder U, Buck A, Fehr E (2004) The neural basis of altruistic punishment. Science 305:1254–1258
Rabin M (1993) Incorporating fairness into game theory and economics. Am Econ Rev 83:1281–1302
Radespiel U, Sarikaya Z, Zimmermann E, Bruford MW (2001) Sociogenetic structure in a free-living nocturnal primate population: sex-specific differences in the grey mouse lemur (*Microcebus murinus*). Behav Ecol Sociobiol 50:493–502
Rajpurohit LS, Sommer V (1993) Juvenile male emigration from natal one-male trops in Hanuman langurs. In: Pereira ME, Fairbanks LA (eds) Juvenile Primates. Oxford Univ Press, New York, pp 86–103
Rahaman H, Parsatharathy MD (1978) Behavioural variants of bonnet macaque (*Macaca radiata*) inhabiting cultivated gardens. J Bombay Nat Hist Soc 75:405–425
Rajala AK, Hantula DA (2000) Towards a behavioural ecology of consumption: delay-reduction effects on foraging in a simulated internet mall. Manag Decis Econ 21:145–158
Rankin FR (2003) Communication in ultimatum games. Econ Lett 81:267–271
Rapaport L (2001) Food transfer among adult lion tamarins: mutualism, reciprocity or one-sided relationships? Int J Primatol 22:611–628
Rapaport A, Chammah AM (1965) Prisoner's Dilemma. Univ of Michigan Press, Ann Arbor
Rawlins RG, Kessler MJ (1986) The Cayo Santiago Macaques: History, Behavior and Biology. State Univ of New York Press, Albany
Rawls J (1971) A Theory of Justice. Harvard Univ Press, Cambridge/MA

Reeve HK, Keller L (1995) Partitioning of reproduction in mother-daughter versus sibling associations: a test of optimal skew theory. Am Nat 145:119–132
Reeve HK, Sherman PW (1991) Intracolonial aggression and nepotism by the breeding female naked mole-rat. In: Sherman PW; Jarvis JUM, Alexander, RD (eds) The Biology of the Naked Mole-Rat. Princeton Univ Press, Princeton/NJ, pp. 337–357
Reeve HK, Emlen S, Keller L (1998) Reproductive sharing in animal societies: reproductive incentives or incomplete control by dominant breeders? Behav Ecol 9:267–278
Rege M, Telle K (2004) The impact of social approval and framing on cooperation in public good situations. J Public Econ 88:1625–1644
Rendall D (2004) "Recognizing" kin: mechanisms, media, minds, modules and muddles. In: Chapais B, Berman C (eds) Kinship and Behavior in Primates. Oxford Univ Press, Oxford, pp 295–316
Rhijn JG van (1973) Behavioural dimorphism in male ruffs, *Philomachus pugnax* (L). Behaviour 47:153–229
Rhijn JG van (1983) On the maintenance and origin of alternative strategies in the Ruff, *Philomachus pugnax*. Ibis 125:482–498
Richerson PJ, Boyd R (2001) Institutional evolution in the Holocene: the rise of complex societies. In: Runciman WG (ed) The Origin of Human Social Institutions. Proc Br Acad 110:197–204
Richerson PJ, Boyd RT, Henrich J (2003) Cultural evolution of human cooperation. In: Hammerstein P (ed) Genetic and Cultural Evolution of Cooperation. MIT Press, Cambridge/MA, pp 357–388
Ridley AR (2003) The causes and consequences of helping behaviour in the cooperatively breeding Arabian babbler. PhD thesis, Univ of Cambridge, Cambridge
Riolo R, Cohen MD, Axelrod R (2001a) Evolution of cooperation without reciprocity. Nature 414:441–443
Riolo R, Cohen MD, Axelrod R (2001b) Does similarity breed cooperation. Nature 418:500
Riss D, Goodall J (1977) The recent rise to the alpha rank in a population of free-living chimpanzees. Fol Primatol 27:134–151
Roberts G, Sherratt TN (1998) Development of cooperative relationships through increasing investment. Nature 394:175–179
Roberts G, Sherratt TN (2002) Does similarity breed cooperation? Nature 418:499–500
Roberts RL, Jenkins KT, Lawler JT, Wegner F, Norcross JL, Bernhards DE, Newman JD (2001) Prolactin levels are elevated after infant carrying in parentally inexperienced common marmosets. Physiol Behav 72:713–720
Roberts WA (2002) Are animals stuck in time? Psychol Bull 128:473–489
Robinson JG, Janson CH (1987) Capuchins, squirrel monkeys and atelines: socioecological convergence with Old World primates. In: Smuts BB, Cheney DL, Seyfarth RM, Wrangham RW, Struhsaker TT (eds) Primate Societies. Univ of Chicago Press, Chicago, pp 69–82
Rodseth L, Wrangham RW (2003) Human kinship: a continuation of politics by other means? In: Chapais B, Berman CM (eds) Kinship and Behavior in Primates. Oxford Univ Press, New York, pp 389–419
Ron T, Henzi SP, Motro U (1994) A new model of fission in primate troops. Anim Behav 47:223–226
Ron T, Henzi SP, Motro U (1996) Do female chacma baboons compete for a safe spatial position in a woodland habitat in Zululand, South Africa? Behaviour 133:475–490
Rood JP (1986) Ecology and social evolution in the mongooses. In: Rubenstein DI, Wrangham RW (eds) Ecological Aspects of Social Evolution. Princeton Univ Press, Princeton/NJ, pp 131–152
Rood JP (1990) Group size, survival, reproduction and routes to breeding in dwarf mongooses. Anim Behav 39:566–572
Rose L (1997) Vertebrate predation and food-sharing in *Cebus* and *Pan*. Int J Primatol 18:727–765
Rosenkranz MA, Jackson DC, Dalton KM, Dolski I, Ryff CD, Singer BH, Muller D, Kalin NH, Davidson RJ (2003) Affective style and in vivo immune response: neurobehavioral mechanisms. Proc Natl Acad Sci USA 100:11148–11152
Roth AE (1995) Bargaining experiments. In: Kagel JH, Roth AE (eds) Handbook of Experimental Economics. Princeton Univ Press, Princeton/NJ
Rothstein SI, Pierotti R (1988) Distinctions among reciprocal altruism, kin selection and cooperation and a model for the initial evolution of beneficent behavior. Ethol Sociobiol 9:189–209
Roulin A, Heeb P (1999) The immunological function of allosuckling. Ecol Lett 2:319–324

Rowell T, Wilson C, Cords M (1991) Reciprocity and partner preference in grooming of female blue monkeys. Int J Primatol 12:319–336
Ruiter JR de, van Hooff JARAM, Scheffrahn W (1994) Social and genetic aspects of paternity in wild long-tailed macaques (*Macaca fascicularis*). Behaviour 129:203–224
Russell AF (2004) Mammals: comparisons and contrasts. In: Koenig W, Dickinson J (eds) Ecology and Evolution of Cooperative Breeding in Birds. Cambridge Univ Press, Cambridge, pp 210–227
Russell AF, Hatchwell BJ (2001) Experimental evidence for kin-biased helping in a cooperatively breeding vertebrate. Proc R Soc Lond B 268:2169–2174
Russell AF, Clutton-Brock TH, Brotherton PNM, Sharpe LL, McIlrath GM, Dalerum FD, Cameron EZ, Barnard JA (2002) Factors affecting pup growth and survival in cooperatively breeding meerkats *Suricata suricatta*. J Anim Ecol 71:700–709
Russell AF, Sharpe LL, Brotherton PNM, Clutton-Brock TH (2003a) Cost minimization by helpers in cooperative vertebrates. Proc Natl Acad Sci USA 100:3333–3338
Russell AF, Brotherton PNM, McIlrath GM, Sharpe LL, Clutton-Brock TH (2003b) Breeding success in cooperative meerkats: effects of helper number and maternal state. Behav Ecol 14:486–492
Russell AF, Carlson AA, McIlrath GM, Sharpe L, Clutton-Brock TH (2005) Morphological modifications of breeders in cooperative mammals. Evolution, in press
Rutte C, Taborsky M (in review) Generalized reciprocity in female rats
Sachs J, Mueller UG, Wilcox TP, Bull JJ (2004) The evolution of cooperation. Q Rev Biol 79:135–160
Sade DS (1972) Sociometrics of *Macaca mulatta*: linkages and cliques in grooming matrices. Fol Primatol 18:196–223
Sally D (1995) Conversation and cooperation in social dilemmas: a meta-analysis of experiments from 1958 to 1992. Ration Soc 7:58–92
Sanfey AG, Rilling JK, Aronson JA, Nystrom LE, Cohen JD (2003) The neural basis of economic decision-making in the ultimatum game. Science 300:1755–1758
Savage A (1990) The reproductive biology of the cotton-top tamarin (*Saguinus oedipus oedipus*) in Colombia. PhD thesis, Univ of Wisconsin, Madison
Sayler A, Salmon M (1969) Communal nursing in mice: influence of multiple mothers on the growth of the young. Science 164:1309–1310
Schaffner CM, Caine NG (2000) The peacefulness of cooperatively breeding primates. In: Aureli F, de Waal FBM (eds) Natural Conflict Resolution. Univ of California Press, Berkeley, pp 155–169
Schaffner CM, French JA (1997) Group size and aggression: "recruitment incentives" in a cooperatively breeding primate. Anim Behav 54:171–180
Schaffner CM, Aureli F, Caine NG (2005) Following the rules: why small groups of tamarins do not reconcile conflicts. Fol Primatol 76:67–76
Schaik CP van (1983) Why are diurnal primates living in groups? Behaviour 87:120–144
Schaik CP van (1989) The ecology of social relationships amongst female primates. In: Standen V, Foley R (eds) Comparative Socioecology: The Behavioural Ecology of Humans and other Mammals. Blackwell, Oxford, pp 195–218
Schaik CP van (1996) Social evolution in primates: the role of ecological factors and male behaviour. Proc Brit Acad 88:9–31
Schaik CP van (2000) Infanticide by male primates: the sexual selection hypothesis revisited. In: van Schaik CP, Janson CH (eds) Infanticide by Males and its Implications. Cambridge Univ Press, Cambridge, pp 27–60
Schaik CP van (2003) Local traditions in orangutans and chimpanzees: social learning and social tolerance. In: Fragaszy DM, Perry S (eds) The Biology of Traditions: Models and Evidence. Cambridge Univ Press, Cambridge, pp 297–328
Schaik CP van, Aureli F (2000) The natural history of valuable relationships in primates. In: Aureli F, de Waal FBM (eds) Natural Conflict Resolution. California Press, Berkeley, pp 307–333
Schaik CP van, Hörstermann M (1994) Predation risk and the number of adult males in a primate group: a comparative test. Behav Ecol Sociobiol 35:261–272
Schaik CP van, Kappeler PM (1997) Infanticide risk and the evolution of permanent male-female association in non-human primates: a new hypothesis and comparative test. Proc R Soc Lond B 264:1687–1694

Schaik CP van, van Hooff JARAM (1996) Toward an understanding of the orangutan's social system. In: McGrew WC, Marchant LF, Nishida T (eds) Great Ape Societies. Cambridge Univ Press, Cambridge, pp 3–15
Schaik CP van, van Noordwijk MA (1986) The hidden costs of sociality: intra-group variation in feeding strategies in Sumatran long-tailed macaques (*Macaca fascicularis*). Behaviour 99:296–315
Schaik CP van, van Noordwijk MA, Warsono B, Sutriono E (1983) Party size and early detection of predators in Sumatran forest primates. Primates 24:211–221
Schaik CP van, Pandit SA, Vogel ER (2005) A model for within-group coalitionary aggression among males. Behav Ecol Sociobiol, in press
Schaller GB (1972) The Serengeti Lion. Univ of Chicago Press, Chicago
Schaub H (1996) Testing kin altruism in long-tailed macaques (*Macaca fascicularis*) in a food-sharing experiment. Int J Primatol 17:445–467
Schino G (1998) Reconciliation in domestic goats. Behaviour 135:343–356
Schino G (2000) Beyond the primates: expanding the reconciliation horizon. In: Aureli F, de Waal FBM (eds) Natural Conflict Resolution. Univ of California Press, Berkeley, pp 225–242
Schino G (2001) Grooming, competition and social rank among female primates: a meta analysis. Anim Behav 62:265–271
Schino G, Scucchi S, Maestripieri D, Turillazzi PG (1988) Allogrooming as a tension- reduction mechanism: a behavioral approach. Am J Primatol 16:43–50
Schino G, Perretta G, Taglioni AM, Monaco V, Troisi A (1996) Primate displacement activities as an ethopharmacological model of anxiety. Anxiety 2:186–191
Schoech SJ, Mumme RL, Wingfield JC (1996) Prolactin and helping behaviour in the cooperatively breeding Florida scrub-jay, *Aphelocoma c. coerulescens*. Anim Behav 52:445–456
Schoech SJ, Reynolds SJ, Boughton RK (2004) Endocrinology. In: Koenig W, Dickinson J (eds) Ecology and Evolution of Cooperative Breeding in Birds. Cambridge Univ Press, Cambridge, pp 128–141
Schwartz MW, Hoeksema JD (1998) Specialization and resource trade: biological markets as a model of mutualisms. Ecology 79:1029–1038
Seabright P (2004) The Company of Strangers. A Natural History of Economic Life. Princeton Univ Press, Princeton/NJ
Sefton M, Shupp R, Walker J (2002) The effect of rewards and sanctions in provision of public goods. CeDEx Working Paper 2002-2, Univ of Nottingham
Seinen I, Schram A (2001) Social status and group norms: indirect reciprocity in a helping experiment. Discussion Paper TI2001-003/1, Tinbergen Institute, Amsterdam
Seinen I, Schram A (2005) Social status and group norms: indirect reciprocity in a helping experiment. Eur Econ Rev, in press
Sekulik R (1983) Male relationships and infant deaths in red howler monkeys (*Alouatta seniculus*). Primates 24:475–485
Sella G, Premoli MC, Turri F (1997) Egg trading in the simultaneously hermaphroditic polychaete worm *Ophryotrocha gracilis* (Huth). Behav Ecol 8:83–86
Selten R, Stoecker R (1986) End behavior in sequences of finite prisoner's dilemma supergames. A learning theory approach. J Econ Behav Org 7:47–70
Semendeferi K, Damasio H, Frank R, van Hoesen GW (1997) The evolution of the frontal lobes: a volumetric analysis based on three-dimensional reconstructions of magnetic resonance scans of human and ape brains. J Hum Evol 32:375–388
Semmann D, Krambeck HJ, Milinski M (2003) Volunteering leads to rock-paper-scissors dynamics in a public goods game. Nature 425:390–393
Semmann D, Krambeck HJ, Milinski M (2004) Strategic investment in reputation. Behav Ecol Sociobiol 56:248–252
Semmann D, Krambeck HJ, Milinski M (2005) Reputation is valuable within and outside one's own social group. Behav Ecol Sociobiol 57:611–616
Setchell JM, Lee PC, Wickings EJ, Dixson AF (2002) Reproductive parameters and maternal investment in mandrills. Int J Primatol 23:51–68
Sethi R, Somanathan E (2003) Understanding reciprocity. J Econ Behav Org 50:1–27

Seyfarth RM (1977) A model of social grooming among adult female monkeys. J theoret Biol 65:671–698

Seyfarth RM (1980) The distribution of grooming and related behaviours among adult female vervet monkeys. Anim Behav 28:798–813

Seyfarth RM (1983) Groming and social competition in primates. In: Hinde RA (ed) Primate Social Relationships: An Integrated Approach. Sinauer Associates, Sunderland/MA, pp 182–190

Seyfarth RM, Cheney DL (1984) Grooming, alliances and reciprocal altruism in vervet monkeys. Nature 308:541–543

Seyfarth RM, Cheney DL (1988) Empirical tests of reciprocity theory: problems in assessment. Ethol Sociobiol 9:181–187

Sherman PW (1977) Nepotism and the evolution of alarm calls. Science 197:1246–1253

Sherman PW, Lacey EA, Reeve HW, Keller L (1995) The eusociality continuum. Behav Ecol 6:102–108

Sherman PW, Reeve HK, Pfennig DW (1997) Recognition systems. In: Krebs JR, Davies NB (eds) Behavioural Ecology: An Evolutionary Approach. Blackwell, Oxford, pp 69–96

Shettleworth SJ (1998) Cognition, Evolution and Behavior. Oxford Univ Press, New York

Sigg H, Stolba A, Abegglen JJ, Dasser V (1982) Life history of hamadryas baboons: physical development, infant mortality, reproductive parameters and family relationships. Primates 23:473–487

Sigmund K, Hauert C, Nowak MA (2001) Reward and punishment. Proc Natl Acad Sci USA 98:10757–10762

Silk JB (1982) Altruism among female *Macaca radiata*: explanations and analysis of patterns of grooming and coalition formation. Behaviour 79:162–188

Silk JB (1984) Local resource competition and the evolution of male-biased sex ratios. J theoret Biol 108:203–213

Silk JB (1987) Social behaviour in evolutionary perspective. In: Smuts BB, Cheney DL, Seyfarth RM, Wrangham RW, Struhsaker TT (eds) Primate Societies. Chicago, Chicago Univ Press, pp 318–329

Silk JB (1992a) Patterns of intervention in agonistic contests among male bonnet macaques. In: Harcourt AH, de Waal FBM (eds) Coalitions and Alliances in Humans and other Animals. Oxford Univ Press, Oxford, pp 215–232

Silk JB (1992b) The patterning of intervention among male bonnet macaques: reciprocity, revenge and loyalty. Curr Anthropol 33:318–325

Silk JB (1993) Does participation in coalitions influence dominance relationships among male bonnet macaques? Behaviour 126:171–189

Silk JB (1994) Social relationships of male bonnet macaques: male bonding in a matrilineal society. Behaviour 130:271–291

Silk JB (1996) Why do primates reconcile. Evol Anthropol 5:39–42

Silk JB (1997) The function of peaceful post-conflict contacts among primates. Primates 38:265–279

Silk JB (2001) Ties that bond: the role of kinship in primate societies. In: Stone L (ed) New Directions in Anthropological Kinship. Rowman and Littlefield, Boulder/CO, pp 71–92

Silk JB (2002a) Kin selection in primate groups. Int J Primatol 23:849–875

Silk JB (2002b) Grunts, girneys and good intentions: the origins of strategic commitment in nonhuman primates. In: Nesse R (ed) Commitment: Evolutionary Perspectives. Russell Sage Press, New York, pp 138–157

Silk JB (2002c) Practice random acts of aggression and senseless acts of intimidation: the logic of status contests in social groups. Evol Anthropol 11:221–225

Silk JB, Cheney DL, Seyfarth RM (1996) The form and function of post-conflict interactions between female baboons. Anim Behav 52:259–268

Silk JB, Cheney DL, Seyfarth RM (1999) The structure of social relationships among female savannah baboons in Moremi Reserve, Botswana. Behaviour 136:679–703

Silk JB, Alberts SC, Altmann J (2003) Social bonds of female baboons enhance infant survival. Science 302:1231–1234

Silk JB, Alberts SC, Altmann J (2004) Patterns of coalition formation by adult female baboons in Amboseli, Kenya. Anim Behav 67:573–582

Simms EL, Taylor DL (2002) Partner choice in nitrogen-fixation mutualisms of legumes and rhizobia. Integr Comp Biol 42:369–380
Simonds PE (1974) Sex differences in bonnet macaque networks and social structure. Arch Sex Behav 3:151–166
Simpson M (1973) Social grooming of male chimpanzees. In: Crook J, Michael R (eds) Comparative Ecology and Behaviour of Primates. Academic Press, London, pp 411–505
Sinervo B, Clobert J (2003) Morphs, dispersal behavior, genetic similarity and the evolution of cooperation. Science 300:1949–1951
Singleton GR (1983) The social and genetic structure of a natural colony of house mice, *Mus musculus*, at Healesville Wildlife Sanctuary. Aust J Zool 31:155–166
Singleton I, van Schaik CP (2002) The social organisation of a population of Sumatran orang-utans. Fol Primatol 73:1–20
Skoyles JR, Sagan D (2002) Up from Dragons: The Evolution of Human Intelligence. McGraw-Hill, New York
Smiseth PT, Moore AJ (2004) Behavioral dynamics between caring males and females in a beetle with facultative biparental care. Behav Ecol 15:621–628
Smith V (2003) Constructivist and ecological rationality in economics. Am Econ Rev 93:465–508
Smith K, Alberts SC, Altmann J (2003) Wild female baboons bias their social behaviour towards paternal half-sisters. Proc R Soc Lond B 270:503–510
Smucny DA, Price CS, Byrne EA (1997) Post-conflict affiliation and stress reduction in captive rhesus macaques. Adv Ethol 32:157
Smuts BB, Smuts RW (1993) Male aggression and sexual coercion of females in nonhuman primates and other animals: evidence and theoretical implications. Adv Stud Behav 22:1–63
Smuts BB, Watanabe JM (1990) Social relationships and ritualized greetings in adult male baboons (*Papio cynocephalus anubis*). Int J Primatol 11:147–172
Sober E, Wilson DS (1998) Unto Others: The Evolution and Psychology of Unselfish Behavior. Harvard Univ Press, Cambridge/MA
Solomon NG (1991) Current indirect fitness benefits associated with philopatry in juvenile prairie voles. Behav Ecol Sociobiol 29:277–282
Solomon NG, French JA (ed.) (1997) Cooperative Breeding in Mammals. Cambridge Univ Press, Cambridge
Solomon NG, Getz LL (1997) Examination of alternative hypotheses for cooperative breeding in rodents. In: Solomon NG, French NG (eds) Cooperative Breeding in Mammals. Cambridge Univ Press, Cambridge
Soltis J, Boyd R, Richerson PJ (1995) Can group-functional behaviors evolve by cultural group selection? An empirical test. Curr Anthropol 36:473–494
Southwick CH (1955) Regulatory mechanisms of house mouse populations: social behavior affecting litter survival. Ecology 36:627–634
Spence M (1973) Job market signalling. Q J Econ 87:355–374
Sprague DS, Suzuki S, Takahashi H, Sato S (1998) Male life history in natural populations of Japanese macaques: migration, dominance rank and troop participation of males in two habitats. Primates 39:351–363
Stacey PB, Koenig WD (1990) Cooperative Breeding in Birds. Cambridge Univ Press, Cambridge
Stammbach E (1987) Desert, forest and montane baboons. In: Smuts BB, Cheney DL, Seyfarth RM, Wrangham RW, Struhsaker TT (eds) Primate Societies. Univ of Chicago Press, Chicago, pp 112–120
Stamps JA (1991) Why evolutionary issues are reviving an interest in proximate behavioral mechanisms. Am Zool 31:338–348
Stander PE (1992) Cooperative hunting in lions: the role of the individual. Behav Ecol Sociobiol 29:445–454
Stanford CB (1995) The influence of chimpanzee predation on group size and anti-predator behaviour in red colobus monkeys. Anim Behav 49:577–587
Stanford CB (1998) Chimpanzee and Red Colobus: The Ecology of Predator and Prey. Harvard Univ Press, Cambridge/MA
Stanford CB (1999) The Hunting Apes. Princeton Univ Press, Princeton/NJ

Stanford CB, Wallis J, Matama H, Goodall J (1994a) Patterns of predation by chimpanzees on red colobus monkeys in Gombe National Park, Tanzania, 1982–1991. Am J Phys Anthropol 94:213–229
Stanford CB, Wallis J, Mpongo E, Goodall J (1994b) Hunting decisions in wild chimpanzees. Behaviour 131:1–20
Starin ED (1994) Philopatry and affiliation among red colobus. Behaviour 130:253–270
Steenbeek R, Sterck EHM, de Vries H, van Hooff JARAM (2000) Costs and benefits of the one-male, age-graded and all-male phases in wild Thomas's langur groups. In: Kappeler PM (ed) Primate Males. Cambridge Univ Press, Cambridge, pp 130–145
Stephens DW (2000) Cumulative benefit games: achieving cooperation when players discount the future. J theor Biol 205 (1):1–16
Stephens DW, Krebs JR (1986) Foraging Theory. Princeton Univ Press, Princeton/NJ
Stephens DW, McLinn CM, Stevens JR (2002) Discounting and reciprocity in an iterated prisoner's dilemma. Science 298:2216–2218
Sterck EHM, van Hooff JARAM (2000) The number of males in langur groups: monopolizability of females or demographic processes? In: Kappeler PM (ed) Primate Males. Cambridge Univ Press, Cambridge, pp 120–129
Sterck EHM, Watts DP, van Schaik CP (1997) The evolution of female social relationships in nonhuman primates. Behav Ecol Sociobiol 41:291–309
Stevens JR (2004) The selfish nature of generosity: harassment and food sharing in primates. Proc R Soc Lond B 271:451–456
Stevens JR, Gilby IC (2004) A conceptual framework for nonkin food sharing: timing and currency of benefits. Anim Behav 67:603–614
Stevens JR, Hauser M (2004) Why be nice? Psychological constraints on the evolution of cooperation. Trends Cogn Sci 8:60–65
Stewart KJ, Harcourt AH (1987) Gorillas: variation in female relationships. In: Smuts BB, Cheney DL, Seyfarth RM, Wrangham RW, Struhsaker TT (eds) Primate Societies. Univ of Chicago Press, Chicago, pp 155–164
Strassmann JE, Solis CR, Hughes CR, Goodnight KF, Queller DC (1997) Colony life-history and demography of a swarm-founding social wasp. Behav Ecol Sociobiol 40:71–77
Strier KB (2000) From binding brotherhoods to short-term sovereignty: the dilemma of male Cebidae. In: Kappeler PM (ed) Primate Males. Cambridge Univ Press, Cambridge, pp 72–83
Strier KB (2004) Patrilineal kinship and primate behavior. In: Chapais B, Berman C (eds) Kinship and Behavior in Primates. Oxford Univ Press, New York, pp 177–199
Strier KB, Carvalho DS, Bejar NO (2000) Prescription for peacefulness. In: Aureli F, de Waal FBM (eds) Natural Conflict Resolution. Univ of California Press, Berkeley, pp 315–317
Strier KB, Dib LT, Figueira JEC (2002) Social dynamics of male muriquis (*Brachyteles arachnoides hypoxanthus*). Behaviour 139:315–342
Struhsaker TT (1997) Ecology of an African Rain Forest. Univ Press of Florida, Gainesville
Struhsaker TT (2000) Variation in adult sex ratios of red colobus monkey groups: implications for interspecific comparisons. In: Kappeler PM (ed) Primate Males. Cambridge Univ Press, Cambridge, pp 108–119
Struhsaker TT, Leland L (1987) Colobine infanticide. In: Smuts BB, Cheney DL, Seyfarth RM, Wrangham RW, Struhsaker TT (eds) Primate Societies. Univ of Chicago Press, Chicago, pp 83–97
Suddendorf TT, Corballis MC (1997) Mental time travel and the evolution of the human mind. Genet Soc Gen Psychol Monogr 123:133–167
Sugden R (1986) The Economics of Rights, Cooperation and Welfare. Basil Blackwell, Oxford
Sugiyama Y (1971) Characteristics of the social life of bonnet macaques (*Macaca radiata*). Primates 12:247–266
Swedell L (2002) Affiliation among females in wild hamadryas baboons (*Papio hamadryas hamadryas*). Int J Primatol 23:1205–1226
Takahashi H (1997) Huddling relationships in night sleeping groups among wild Japanese macaques in Kinkayan island during winter. Primates 38:57–68
Tanaka I, Takefushi H (1993) Elimination of external parasites (lices) is the primary function of grooming in free-ranging Japanese macaques. Anthropol Sci 101:187–193

Tardif SD (1997) The bioenergetics of parental behaviour and the evolution of alloparental care in marmosets and tamarins. In: Solomon NG, French JA (eds) Cooperative Breeding in Mammals. Cambridge Univ Press, Cambridge, pp 11–33
Taylor CE, McGuire MT (1988) Reciprocal altruism: 15 years later. Ethol Sociobiol 9:67–72
Tebbich S, Taborsky M, Winkler H (1996) Social manipulation causes cooperation in keas. Anim Behav 52:1–10
Teleki G (1973) The Predatory Behavior of Wild Chimpanzees. Bucknell Univ Press, Brunswick
Thakar JD, Kunte K, Chauhan AK, Watve AV, Watve MG (2003) Nectarless flowers: ecological correlates and evolutionary stability. Oecologia 136:565–570
Thierry B (1986) A comparative study of aggression and response to aggression in three species of macaque. In: Else JG, Lee PC (eds) Primate Ontogeny, Cognition and Social Behavior. Cambridge Univ Press, Cambridge, pp 307–313
Thierry B (1990) Feedback loop between kinship and dominance: the macaque model. J theoret Biol 145:511–521
Thierry B (2000) Covariation of conflict management patterns across macaque species. In: Aureli F, de Waal FBM (eds) Natural Conflict Resolution. Univ of California Press, Berkeley, pp 106–128
Thierry B, Wunderlich D, Gueth C (1989) Possession and transfer of objects in a group of brown capuchins (*Cebus apella*). Behaviour 110:294–305
Thompson RKR, Oden DL (2000) Categorical perception and conceptual judgments by non-human primates: the paleological monkey and the analogical ape. Cogn Sci 24:363–396
Tinbergen N (1959) Comparative studies on the behaviour of gulls (Laridae): a progress report. Behaviour 15:1–70
Todd PM, Gigerenzer G (2000) Précis of simple heuristics that make us smart. Behav Brain Sci 23:727–780
Todd PM, Gigerenzer G (2003) Bounding rationality to the world. J Econ Psychol 24:143–165
Tomasello M, Call J (1997) Primate Social Cognition. Oxford Univ press, Oxford
Tremblay L, Schultz W (1999) Relative reward preference in primate orbitofrontal cortex. Nature 398:704–708
Trivers RL (1971) The evolution of reciprocal altruism. Q Rev Biol 46:25–57
Trivers RL (1972) Parental investment and sexual selection. In: Campbell B (ed) Sexual Selection and the Descent of Man. Aldine, Chicago, pp 136–179
Trivers RL (1974) Parent-offspring conflict. Am Zool 14:249–264
Trivers RL (1981) Sociobiology and politics. In: White E (ed) Sociobiology and Human Politics. Lexington Books, Lexington, pp 1–43
Trivers RL (1985) Social Evolution. Benjamin/Cummings, Menlo Park
Trivers R (2000) The elements of a scientific theory of self-deception. Ann NY Acad Sci 907:114–131
Trivers R (2002) Natural Selection and Social Theory: Selected Papers of Robert Trivers. Oxford Univ Press, New York
Trivers R (2004) Mutual benefits at all levels of life. Science 304:965
Trivers RL, Hare H (1976) Haplodiploidy and the evolution of the social insects. Science 30:253–269
Troisi A (2002) Displacement activities as a behavioral measure of stress in nonhuman primates and human subjects. Stress 5:47–54
Tulving E (2002) Episodic memory: from mind to brain. Ann Rev Psychol 53:1–25
Turner PE, Chao L (1999) Prisoner's dilemma in an RNA virus. Nature 398:441–443
Uehara S, Nishida T, Hamai M, Hasegawa T, Hayaki H, Huffman M, Kawanaka K, Kobayashi S, Mitani J, Takahata Y, Takasaki H, Tsukahara T (1992) Characteristics of predation by the chimpanzees in the Mahale Mountains National Park, Tanzania. In: Nishida T, McGrew WC, Marler P, Pickford M, de Waal F (eds) Topics in Primatology. Vol 1: Human Origins. Univ of Tokyo Press, Tokyo, pp 143–158
Veblen T (1899) The Theory of the Leisure Class: An Economic Study of Institutions. The Macmillan Company, New York
Vehrencamp SL (1983a) A model for the evolution of despotic versus egalitarian societies. Anim Behav 31:667–682
Vehrencamp SL (1983b) Optimal degree of skew in cooperative societies. Am Zool 23:327–335

Velicer GC, Yu YN (2003) Evolution of novel cooperative swarming in the bacterium *Myxococcus xanthus*. Nature 425:75–78

Vervaecke H, de Vries H, van Elsacker L (2000) The pivotal role of rank in grooming and support behavior in a captive group of bonobos (*Pan paniscus*). Behaviour 137:1463–1485

Vigilant L, Hofreiter M, Siedel H, Boesch C (2001) Paternity and relatedness in wild chimpanzee communities. Proc Natl Acad Sci USA 98:12890–12895

Visalberghi E, Limongelli (1994) Lack of comprehension of cause-effect relations in tool using capuchin monkeys. J Comp Psychol 108:15–22

Visalberghi E, Quarantotti BP, Tranchida F (2000) Solving a cooperation task without taking into account the partner's behavior: the case of capuchin monkeys (*Cebus apella*). J Comp Psychol 114:297–301

Vleck CM, Mays NA, Dawson JW, Goldsmith A (1991) Hormonal correlates of parental and helping behavior in cooperatively breeding Harris' hawks, *Parabuteo unicinctus*. Auk 108:638–648

Voland E (2000) Contributions of family reconstruction studies to evolutionary reproductive ecology. Evol Antropol 9:134–146

Voland E, Dunbar RIM (1995) Resource competition and reproduction. The relationship between economic and parental strategies in the Krummhörn population (1720–1874). Hum Nat 6:33–49

Voland E, Engel C (1990) Female choice in humans: a conditional mate selection strategy of the Krummhoern women (Germany, 1720–1874). Ethology 84:144–154

Vries H de, Netto W, Hanegraaf P (1993) MatMan: a program for the analysis of sociometric matrices and behavioural transition matrices. Behaviour 125:157–175

Waal FBM de (1982 [1998]) Chimpanzee Politics: Power and Sex among Apes. Johns Hopkins Univ Press, Baltimore

Waal FBM de (1986a) Class structure in a rhesus monkey group; the interplay between dominance and tolerance. Anim Behav 34:1033–1040

Waal FBM de (1986b) The integration of dominance and social bonding in primates. Q Rev Biol 61:459–479

Waal FBM de (1989) Food sharing and reciprocal obligations among chimpanzees. J Hum Evol 18:433–459

Waal FBM de (1991a) Complementary methods and convergent evidence in the study of primate social cognition. Behaviour 118:297–320

Waal FBM de (1991b) Rank distance as a central feature of rhesus monkey social organization: a sociometric analysis. Anim Behav 41:383–398

Waal FBM de (1992a) Coalitions as part of reciprocal relations in the Arnhem chimpanzee colony. In: Harcourt AH, de Waal FBM (eds) Coalitions and Alliances in Humans and other Animals. Oxford Univ Press, Oxford, pp 233–257

Waal FBM de (1992b) The chimpanzees sense of social regularity and its relation to the human sense of justice. In: Masters RD, Gruter M (eds) The Sense of Justice: Biological Foundations of Law. Sage, Newbury Park, pp 241–255

Waal FBM de (1993) Reconciliation among primates: a review of empirical evidence and unresolved issues. In: Mason WA, Mendoza SP (eds) Primate Social Conflict. SUNY Press, Albany, pp 111–144

Waal FBM de (1996a) Conflict as negotiation. In: McGrew WC, Marchant LF, Nishida T (eds) Great Ape Societies. Cambridge Univ Press, Cambridge, pp 159–172

Waal FBM de (1996b) Good Natured: The Origins of Right and Wrong in Humans and other Animals. Harvard Univ Press, Cambridge/MA

Waal FBM de (1997a) The chimpanzee's service economy: food for grooming. Evol Hum Behav 18:375–386

Waal FBM de (1997b) Food transfers through mesh in brown capuchins. J Comp Psychol 111:370–378

Waal FBM de (2000a) The first kiss: foundations of conflict resolution research in animals. In: Aureli F, de Waal FBM (eds) Natural Conflict Resolution. Univ of California Press, Berkeley, pp 15–33

Waal FBM de (2000b) Primates – a natural heritage of conflict resolution. Science 289:586–590

Waal FBM de (2000c) Attitudinal reciprocity in food sharing among brown capuchin monkeys. Anim Behav 60:253–361
Waal FBM de (2003) Social syntax: the if-then structure of social problem solving. In: de Waal FBM, Tyack PL (eds) Animal Social Complexity. Harvard Univ Press, Cambridge/MA, pp 230–248
Waal FBM de, Berger ML (2000) Payment for labour in monkeys. Nature 404:563
Waal FBM de, Davis JM (2003) Capuchin cognitive ecology: cooperation based on projected returns. Neuropsychologia 41:221–228
Waal FBM de, Harcourt AH (1992) Coalitions and alliances: a history of ethological research. In: Harcourt AH, deWaal FBM (eds) Coalitions and Alliances in Humans and other Animals. Oxford Univ Press, Oxford, pp 1–19
Waal FBM de, Luttrell LM (1986) The similarity principle underlying social bonding among female rhesus monkeys. Fol Primatol 46:215–34
Waal FBM de, Luttrell LM (1988) Mechanisms of social reciprocity in three primate species: symmetrical relationship characteristics or cognition? Ethol Sociobiol 9:101–118
Waal FBM de, Ren R (1988) Comparison of the reconciliation behavior of stumptail and rhesus macaques. Ethology 78:129–142
Waal FBM de, van Hooff JARAM (1981) Side-directed communication and agnoistic interactions in chimpanzees. Behaviour 77:164–198
Waal FBM de, van Roosmalen A (1979) Reconciliation and consolation among chimpanzees. Behav Ecol Sociobiol 5:55–66
Waal FBM de, Yoshihara D (1983) Reconciliation and redirected affection in rhesus monkeys. Behaviour 85:224–241
Waal FBM de, van Hooff JARAM, Netto WJ (1976) An ethological analysis of types of agnostic interaction in a captive group of java-monkeys (*Macaca fascicularis*). Primates 17:257–290
Waal FBM de, Luttrell LM, Canfield ME (1993) Preliminary data on voluntary food sharing in brown capuchin monkeys. Am J Primatol 29:73–78
Wahaj SA, Guse KR, Holekamp KE (2001) Reconciliation in spotted hyena (*Crocuta crocuta*). Ethology 107:1057–1074
Walters JR (1980) Interventions and the development of dominance relationships in female baboons. Fol Primatol 34:61–89
Walters JR (1987) Kin recognition in nonhuman primates. In: Fletcher DF, Michener CD (eds) Kin Recognition in Animals. John Wiley, New York, pp 359–393
Wamboldt MZ, Gelhard RE, Insel TR (1988) Gender differences in caring for infant *Cebuella pygmaea*: the role of infant age and relatedness. Dev Psychobiol 21:187–202
Wang ZX, Novak MA (1992) Influence of the social environment on parental behaviour and pup development of meadow voles (*Microtus pennsylvanicus*) and prairie voles (*Microtus ochrogaster*). J Comp Psychol 106:163–171
Waser PM (1988) Resources, philopatry and social interactions among mammals. In: Slobodchikoff CN (ed) The Ecology of Social Behavior. Academic Press, San Diego, pp 109–130
Watanabe K (1979) Alliance formation in a free-ranging troop of Japanese macaques. Primates 20:459–474
Watanabe M (1999) Attraction is relative not absolute. Nature 398:661–662
Watanabe K (2001) A review of 50 years of research on the Japanese monkeys of Koshima: status and dominance. In: Matsuzawa T (ed) Primate Origins of Human Cognition and Behavior. Springer, Tokyo, pp 405–417
Watts DP (1991) Harassment of immigrnt female mountain gorillas by resident females. Ethology 89:135–153
Watts DP (1994) Social relationships of immigrant and resident female mountain gorillas. II. Relatedness, residence and relationships between females. Am J Primatol 32:13–30
Watts DP (1995) Post-conflict social events in wild mountain gorillas (Mammalia, Hominoidea) I. Social interactions between opponents. Ethology 100:139–157
Watts DP (1996) Comparative socio-ecology of gorillas. In: McGrew WC, Marchant LF, Nishida T (eds) Great Ape Societies. Cambridge Univ Press, Cambridge, pp 16–28
Watts DP (1997) Agonistic interventions in wild mountain gorilla groups. Behaviour 134:23–57

Watts DP (1998) Coalitionary mate-guarding by male chimpanzees at Ngogo, Kibale National Park, Uganda. Behav Ecol Sociobiol 44:43–55

Watts DP (2000a) Causes and consequences of variation in male mountain gorilla life histories. In: Kappeler PM (ed) Primate Males. Cambridge Univ Press, Cambridge, pp 169–179

Watts DP (2000b) Grooming between male chimpanzees at Ngogo, Kibale National Park. I. Partner number and diversity and grooming reciprocity. Int J Primatol 21:189–210

Watts DP (2002) Reciprocity and interchange in the social relationships of wild male chimpanzees. Behaviour 139:343–370

Watts DP, Mitani J (2000) Hunting behavior of chimpanzees at Ngogo, Kibale National Park, Uganda. Int J Primatol 23:1–28

Watts DP, Mitani J (2001) Boundary patrols and intergroup encounters in wild chimpanzees. Behaviour 138:299–327

Watts DP, Pusey AE (1993) Behavior of juvenile and adolescent great apes. In: Pereira ME, Fairbanks LA (eds) Juvenile Primates. Oxford Univ Press, New York, pp 148–171

Wedekind C, Braithwaite VA (2002) The long-term benefits of human generosity in indirect reciprocity. Curr Biol 12:1012–1015

Wedekind C, Milinski M (2000) Cooperation through image scoring in humans. Science 288:850–852

Weingrill T, Lycett JE, Henzi SP (2000) Consortship and mating success in chacma baboons (*Papio cynocephalus ursinus*). Ethology 106:1033–1044

West SA, Murray MG, Machado CA, Griffin AS, Herre EA (2001) Testing Hamilton's Rule with competition between relatives. Nature 409:510–513

West SA, Pen I, Griffin A (2002) Cooperation and competition between relatives. Science 296:72–75

Westlund K, Ljungberg T, Borefelt U, Abrahamsson C (2000) Post-conflict affiliation in common marmosets (*Callithrix jacchus jacchus*). Am J Primatol 52:31–46

White SM (2001) Juvenile development and conflicts of interest in meerkats. PhD thesis, Cambridge Univ Press, Cambridge

Whiten A, Byrne RW (1997) Machiavellian intelligence. II: Extensions and Evaluations. Cambridge Univ Press, Cambridge

Wickler W (1963) Zum Problem der Signalbildung am Beispiel der Verhaltens-Mimikry zwischen *Aspidontus* und *Labroides* (Pisces, Acanthopterygii). Z Tierpsychol 20:657–679

Widdig A (2000) Coalition formation among male Barbary macaques (*Macaca fascicularis*). Am J Primatol 50:37–51

Widdig A, Streich WJ, Tembrock G (2000) Coalition formation among male barbary macaques (*Macaca sylvanus*). Am J Primatol 50:37–51

Widdig A, Nürnberg P, Krawczak M, Streich WJ, Bercovitch FB (2001) Paternal relatedness and age proximity regulate social relationships among adult female rhesus macaques. Proc Natl Acad Sci USA 98:13769–13773

Widdig A, Nürnberg P, Krawczak M, Streich WJ, Bercovitch FB (2002) Affiliation and aggression among adulgt female rhesus macaques: a genetic analysis of paternal cohorts. Behaviour 139:371–391

Widowski TM, Ziegler TW, Elonson AM, Snowdon CT (1990) The role of males in stimulating reproductive function in female cotton-top tamarins (*Saguinus oedipus*). Anim Behav 40:731–741

Wiggins DA, Morris RD (1986) Criteria for female choice of mates: courtship feeding and parental care in the common tern. Am Nat 128:126–129

Wilkinson GS (1984) Reciprocal food sharing in the vampire bat. Nature 308:181–184

Wilkinson GS (1987) Altruism and cooperation in bats. In: Fenton NB, Racey P, Rayner JMV (eds) Recent Advances in the Study of Bats. Cambridge Univ Press, Cambridge, pp 299–323

Wilkinson GS (1988) Reciprocal altruism in bats and other mammals. Ethol Sociobiol 9:85–100

Wilkinson GS (1992) Communal nursing in the evening bat *Nycticeius humeralis*. Behav Ecol Sociobiol 31:225–235

Wilkinson GS, Baker AEM (1988) Communal nesting among genetically similar house mice. Ethology 77:103–114

Wilkinson DM, Sherratt TN (2001) Horizontally acquired mutualisms, an unsolved problem in ecology? Oikos 92:377–384

Williams GC (1966) Adaptation and Natural Selection: A Critique of Some Current Evolutionary Thought. Princeton Univ Press, Princeton/NJ
Williams JM, Liu H-Y, Pusey AE (2002) Costs and benefits of grouping for female chimpanzees at Gombe. In: Boesch B, Hohmann G, Marchant LF (eds) Behavioural Diversity in Chimpanzees and Bonobos. Cambridge Univ Press, Cambridge, pp 192–203
Wilson DS (1983) The group selection controversy: history and current status. Ann Rev Ecol Syst 14:159–187
Wilson EO (1975) Sociobiology. Harvard Univ Press, Cambridge/MA
Wilson ML, Wrangham RW (2003) Intergroup relations in chimpanzees. Ann Rev Anthropol 32:363–392
Wimmer B, Kappeler PM (2002) The effects of sexual selection and life history on the genetic structure of redfronted lemur, *Eulemur fulvus rufus*, groups. Anim Behav 64:557–568
Wimmer B, Tautz D, Kappeler PM (2002) The genetic population structure of the gray mouse lemur (*Microcebus murinus*), a basal primate from Madagascar. Behav Ecol Sociobiol 52:166–175
Winston JS, Strange BA, O'Doherty J, Dolan RJ (2002) Automatic and intentional brain responses during evaluation of trustworthiness of faces. Nat Neurosci 5:277–283
Wittig RA, Boesch C (2003) The choice of post-conflict interactions in wild chimpanzees (*Pan troglodytes*). Behaviour 140:1527–1559
Wittig RM, Boesch C (2005) How to repair relationships – reconciliation in wild chimpanzees (*Pan troglodytes*). Ethology 111:736–763
Woodroffe R, Macdonald DW (1995) Female/female competition in European badgers, *Meles meles*: effects on breeding success. J Anim Ecol 64:12–20
Woodruff DS (2004) Noninvasive genotyping and field studies of free-ranging nonhuman primates. In: Chapais B, Berman C (eds) Kinship and Behavior in Primates. Oxford Univ Press, Oxford, pp 46–68
Woolley-Barker T (1998) Genetic structure reflects social organization in hybrid hamadryas and anubis baboons. Am J Phys Anthropol 26:S235
Woolley-Barker T (1999) Social organization and genetic structure in a baboon hybrid zone. PhD thesis, New York Univ, New York
Wormell D (1994) Relationships in male-female pairs of pied tamarins *Saguinus bicolor bicolor*. Dodo 30:133–144
Wrangham RW (1975) The behavioural ecology of chimpanzees in Gombe National Park, Tanzania. PhD thesis, Cambridge Univ, Cambridge
Wrangham RW (1980) An ecological model of female-bonded primate groups. Behaviour 75:262–300
Wrangham RW (1982) Mutualism, kinship and social evolution. In: Group KsCS (ed) Current Problems in Sociobiology. Cambridge Univ Press, Cambridge, pp 269–289
Wrangham RW (1999) Evolution of coalitionary killing. Ybk Phys Anthropol 42:1–30
Wrangham RW (2000) Why are male chimpanzees more gregarious than mothers? A scramble competition hypothesis. In: Kappeler PM (ed) Primate Males. Cambridge Univ Press, Cambridge, pp 248–258
Wright S (1932) The roles of mutation, inbreeding, crossbreeding and selection in evolution. Proc Sixth Int Congr Genet 1:356–366.
Wynne CDL (2004) Fair refusal by capuchin monkeys. Nature 428:140
Wynne-Edwards V (1962) Animal Dispersion in Relation to Social Behaviour. Oliver, Boyd, Edinburgh
Yamada MA (1963) study of blood-relationship in the natural society of the Japanese macaque – an analysis of co-feeding, grooming and playmate relationships in Minoo-B-troop. Primates 4:43–65
Yamagishi T (1986) The provision of a sanctioning system as a public good. J Pers Soc Psychol 51:110–116
York AD, Rowell TE (1988) Reconciliation following aggression in patas monkeys, *Erythrocebus patas*. Anim Behav 36:502–509
Young AJ (2003) Subordinate tactics in cooperative meerkats: breeding, helping and dispersal. PhD thesis, Univ of Cambridge, Cambridge

Young PH (2003) The power of norms. In: Hammerstein P (ed) Genetic and Cultural Evolution of Cooperation. MIT Press, Cambridge/MA, pp 389–399
Zahavi A (1975) Mate selection – a selection for a handicap. J theor Biol 53:205–214
Zahavi A (1977a) The cost of honesty (further remarks on the handicap principle). J theor Biol 67:603–605
Zahavi A (1977b) The testing of a bond. Anim Behav 25:246–247
Zahavi A (1991) Arabian babblers: the quest for social status in a cooperative breeder. In: Stacey PBK, Koenig WD (eds) Cooperative Breeding of Birds: Long-Term Studies of Ecology and Behavior. Cambridge Univ Press, Cambridge, pp 105–130
Zahavi A (1997) The Handicap Principle: A Missing Piece of Darwin's Puzzle. Oxford Univ Press, New York
Zak PJ (2004) Neuroeconomics. Phil Trans R Soc Lond B 359:1737–1748
Zak PJ, Kurzban J, Matzner WT (2004) The neurobiology of trust. Ann NY Acad Sci 1032:224–227
Zak PJ, Kurzban J, Matzner WT (submitted) Oxytocin is associated with interpersonal trust in humans. Proc Natl Acad Sci

Subject Index